I. N. Melnikova
A. V. Vasilyev
Short-Wave Solar Radiation in the Earth's Atmosphere
Calculation, Observation, Interpretation

Irina N. Melnikova
Alexander V. Vasilyev

Short-Wave Solar Radiation in the Earth's Atmosphere

Calculation, Observation, Interpretation

with 60 Figures, 3 in color, and 19 Tables

 Springer

Professor Dr. Irina N. Melnikova
Russian Academy of Sciences
Research Center of Ecological Safety
Korpusnaya ul. 18
197110 St. Petersburg
Russian Federation

Dr. Alexander V. Vasilyev
St. Petersburg State University
Institute of Physics
Ulyanovskaya 1
198504 St. Petersburg
Russian Federation

Library of Congress Control Number: 2004103071

ISBN 3-540-21452-6 Springer Berlin Heidelberg New York

Springer is a part of Springer Science+Business Media
springeronline.com
© Springer-Verlag Berlin Heidelberg 2005
Printed in Germany

Cover design: E. Kirchner, Heidelberg
Production: Almas Schimmel
Typesetting: LE-TeX Jelonek, Schmidt & Vöckler GbR, Leipzig
Printing: Mercedes-Druck, Berlin
Binding: Stein + Lehmann, Berlin

Printed on acid-free paper   32/3141/as 5 4 3 2 1 0

Preface

Solar radiation has a decisive influence on climate and weather formation when passing through the atmosphere and interacting with the atmospheric components (gases, atmospheric aerosols, and clouds). The part of solar radiation that reaches the surface is a source of the existence and development of the biosphere because it regulates all biological processes. It should be mentioned that the part of solar radiation energy corresponding to the spectral region 0.35–1.0 μm is about 66% and to the spectral region 0.25–2.5 μm is more than 96% according to (Makarova et al. 1991). Thus, the study of the interaction between the atmosphere and the clouds and solar radiation in the short-wave range is especially interesting.

Numerous spectral solar radiation measurements have been made by the Atmospheric Physics Department, the Physics Faculty of Leningrad (now St. Petersburg) State University and in the Voeykov Main Geophysical Observatory under the guidance of academician Kirill Kondratyev for about 30 years from 1960. The majority of radiation observations were made during airborne experiments under clear sky condition (Kondratyev et al. 1974; Vasilyev O et al. 1987a; Kondratyev et al. 1975; Kondratyev et al. 1973; Vasilyev O et al. 1987b; Kondratyev and Ter-Markaryants 1976) and only 10 experiments were accomplished with an overcast sky (Kondratyev, Ter-Markaryants 1976); Vasilyev 1994 et al.; Kondratyev, Binenko 1984; Kondratyev, Binenko (1981). The results obtained have received international acknowledgment and currently this research direction is of special interest all over the world (King 1987; King et al. 1990; Asano 1994; Hayasaka et al. 1994; Kostadinov et al. 2000).

The airborne radiative observations were made over desert and water surfaces using the improved spectral instrument in the 1980s. As a result of 10-years of observations volume of the data set became very large. However, computer resources were not adequate for the instantaneous processing at that time. All the data were finally processed only at the end of the 1990s and now we have a rich database of the spectral values of the radiative characteristics (semispherical fluxes, intensity and spectral brightness coefficients) obtained under different atmospheric conditions. The database contains about 30,000 spectra including 2203 spectra of the upward and downward semispherical fluxes obtained during the airborne atmospheric sounding.

The inverse problem of atmospheric optics has been solved using the numerical method in the case of the interpretation of the observational results of the clear sky measurements and using the analytical method of the theory of radiation transfer in the case of overcast skies.

The interpretation of the radiative experiments under clear and overcast sky conditions is discussed in different sections because the mathematical methods of the description differ extensively. In addition, the extended (hundreds of kilometers) and stable (up to several days) cloudiness is worthy of special consideration because of its strong influence on the energy budget of the atmosphere and on climate formation.

It is necessary to set adequate optical parameters of the atmosphere for the practical problems of climatology, for distinguishing backgrounds and contrasts in the atmosphere and on the surface, and for the problems of the radiative regime of artificial and natural surfaces. The values obtained from the observational data are highly suitable in these cases. Unfortunately, to the present, theoretical values of the initial parameters are mostly used in the numerical simulations which leads to an incorrect estimation of the absorption of solar radiation in the atmosphere (especially when cloudy). The influence of the interaction of the atmospheric aerosols and cloudiness with solar radiation is taken into account in the numerical simulations of the global changes of the surface temperature only as the rest term for the coincidence between the calculated and observed values. The analysis of the database convinces us that solar radiation absorption in the dust and cloudy atmosphere is more significant than has been considered. Many authors have classified the experimental excess values of solar shortwave radiation absorption in clouds they obtained as an effect of "anomalous" absorption. This terminology indicates an underestimation of this absorption. Thus, the correct interpretation of the observational data, based on radiation transfer theory and the construction of the optical and radiative atmospheric models is of great importance.

Our results provide the spectral data of the solar irradiance measurements in the energetic units, the spectral values of the atmospheric optical parameters obtained from these experimental data and the spectral brightness coefficients of the surfaces of different types in figures and tables.

Let us point out the main results indicating the chapters where they are presented:

Chapter 1 reviews the definition of the characteristics of solar radiation and optical parameters describing the atmosphere and surface. The basic information about the interaction between solar radiation and atmospheric components (gases, aerosols and clouds) is cited as well.

In Chap. 2, the details of the radiative characteristic calculations in the atmosphere are considered. For the radiance and irradiance calculation, the Monte-Carlo method is chosen in the clear sky cases and the analytical method of the asymptotic formulas of the theory of radiation transfer is used for the overcast sky cases. Special attention is paid to the error analysis and applicability ranges of the methods. Different initial conditions of the cloudy atmosphere (the one-layer cloudiness, vertically homogeneous and heterogeneous, multilayer, the conservative scattering, accounting for the true absorption of radiation) are discussed as well.

In Chap. 3, the results of solar shortwave radiance and irradiance observation in the atmosphere are shown in detail. The authors have described both the instruments were used, as well as the special features of the measurements. Observational error analysis with the ways to minimizing the errors have been scrutinized. The methods of the data processing for obtaining the characteristics of solar radiation in the energetic units are elucidated. The examples of the vertical profiles of the spectral semispherical (upward and downward) fluxes observed under different atmospheric conditions are presented in figures in the text and in tables in Appendix 1. The results of the airborne, ground and satellite observations for the overcast skies are considered together with the contemporary views on the effect of the anomalous absorption of shortwave radiation in clouds.

In Chap. 4, the basic methods of procuring atmospheric optical parameters from the observational data of solar radiation are summarized. The application of the least-square technique for solving the atmospheric optics inverse problem is fully discussed. The influence of the observational errors on the accuracy of the solution is described and the methodology for its regularization is proposed. It is also shown how to choose the atmospheric parameters which are possible to retrieve from the radiative observations.

Chapter 5 is concerned with the methods and conditions of the inverse problem solving for clear sky conditions considered together with the results obtained. The vertical profiles and the spectral dependencies of the relevant parameters of the atmosphere and surface are shown in figures in the text and in tables in Appendix 1.

In Chap. 6, the analytical method for the retrieval of the stratus cloud optical parameters from the data of the ground, airborne and satellite radiance and irradiance observations including the full set of necessary formulas is elaborated. The example of the relevant formulas derivation for the case of using the data of the irradiance at the cloud top and bottom is demonstrated in Appendix 2. The analysis of the correctness of the inverse problem, existence, uniqueness and stability of the solution is performed and the uncertainties of the method are studied.

Chapter 7 provides the actual conditions of the cloud optical parameter retrieval from the data of the ground, airborne and satellite (ADEOS-1) observations. The spectral and vertical dependencies of the optical parameters are presented in figures in the text and in tables in Appendix 1. The analysis of the numerical values is accomplished, and the empirical hypothesis, which explains both the features revealed by the results and the anomalous absorption in clouds, is proposed. The book concludes with a summary of the results obtained.

The authors have wrote Chaps. 1 and 3 together, Sect. 2.1 and Chaps. 4 and 5 was written by Alexander Vasilyev, Chaps. 2 (excluding Sect. 2.1), 6 and 7 – by Irina Melnikova. The authors' intention was to present the material clearly for this book so that it would be useful for a large range of readers, including students, involved in the fields of atmospheric optics, the physics of the atmosphere, meteorology, climatology, the remote sounding of the atmosphere and surface and the distinguishing of backgrounds and contrasts of the natural and artificial objects in the atmosphere and on the surface.

It should be emphasized that the majority of the observations were made by the team headed by Vladimir Grishechkin (the Laboratory of Shortwave Solar Radiation of the Atmospheric Department of the Faculty of Physics, St. Petersburg State University). The authors would like to express their profound gratitude to Anatoly Kovalenko, Natalya Maltseva, Victor Ovcharenko, Lyudmila Poberovskaya, Igor Tovstenko and others who took part in the preparation of the instruments, the carrying out of the observations and the data processing. Unfortunately, our colleagues Pavel Baldin, Vladimir Grishechkin, Alexei Nikiforov and Oleg Vasilyev prematurely passed away. We dedicate the book to the memory of our friends and colleagues.

The authors very grateful to academician Kirill Kondratyev, Professors Vladislav Donchenko and Lev Ivlev, Victor Binenko and Vladimir Mikhailov for the fruitful discussions and valuable recommendations.

References

Asano S (1994) Cloud and radiation studies in Japan. Cloud radiation interactions and their parameterisation in climate models. In: WCRP-86 (WMO/TD no 648). WMO, Geneva, pp 72–73

Hayasaka T, Kikuchi N, Tanaka M (1994) Absorption of solar radiation by stratocumulus clouds: aircraft measurements and theoretical calculations. J Appl Meteor 33:1047–1055

King MD (1987) Determination of the scaled optical thickness of cloud from reflected solar radiation measurements. J Atmos Sci 44:1734–1751

King MD, Radke L, Hobbs PV (1990) Determination of the spectral absorption of solar radiation by marine stratocumulus clouds from airborne measurements within clouds. J Atmos Sci 47:894–907

Kondratyev KYa, Ter-Markaryants NE (eds) (1976) Complex radiation experiments (in Russian). Gydrometeoizdat, Leningrad

Kondratyev KYa, Vasilyev OB, Grishechkin VS et al (1973) Spectral shortwave radiation inflow in the troposphere within spectral ranges 0.4–2.4 μm. I. Observational and processing methodology (in Russian). In Main geophysical observatory studies 322:12–23

Kondratyev KYa, Vasilyev OB, Grishechkin VS et al (1974) Spectral shortwave radiation inflow in the troposphere and their variability (in Russian). Izv. RAS, Atmospheric and Ocean Physics 10:453–503

Kondratyev KYa, Vasilyev OB, Ivlev LS et al (1975) Complex observational studies above the Caspian Sea (CAENEX-73) (in Russian). Meteorology and Hydrology, pp 3–10

Kondratyev KYa, Binenko VI (eds) (1981) The first Global Experiment PIGAP. vol 2. Polar aerosol, extended cloudiness and radiation. Gidrometeoizdat, Leningrad

Kondratyev KYa, Binenko VI (1984) Impact of Clouds on Radiation and Climate (in Russian). Gidrometeoizdat, Leningrad

Kostadinov I, Giovanelli G, Ravegnani F, Bortoli D, Petritoli A, Bonafè U, Rastello ML, Pisoni P (2000) Upward and downward irradiation measurements on board "Geophysica" aircraft during the APE-THESEO and APE-GAIA field campaigns. In: IRS'2000. Current problems in Atmospheric Radiation. Proceedings of the International Radiation Symposium, St.Petersburg, Russia, pp 1185–1188

Makarova EA, Kharitonov AV, Kazachevskaya TV (1991) Solar irradiance (in Russian). Nauka, Moscow

Vasilyev AV, Melnikova IN, Mikhailov VV (1994) Vertical profile of spectral fluxes of scattered solar radiation within stratus clouds from airborne measurements (Bilingual). Izv RAS, Atmosphere and Ocean. Physics 30:630–635

Vasilyev OB, Grishechkin VS, Kondratyev KYa (1987a) Spectral radiation characteristics of the free atmosphere above Lake Ladoga (in Russian). In: Complex remote lakes monitoring. Nauka, Leningrad, pp 187–207

Vasilyev OB, Grishechkin VS, Kovalenko AP et al (1987b) Spectral informatics – measuring system for airborne and ground study of shortwave radiation field in the atmosphere (in Russian). In Complex remote lakes monitoring. Nauka, Leningrad, pp 225–228

Contents

About the Authors

Irina N. Melnikova, Doctor of Science in Physics, Head of the Laboratory for Global Climate Change of the Research Center for Ecological Safety of the Russian Academy of Science. For about twenty years has been working in the Department for Atmospheric Physics of the Research Institute for Physics of St. Petersburg State University. She has taken part in the process of getting the results from the solar radiation measurements. It helps to understand better all specifics of the data interpretation. The work after the authority of Professor Igor N. Minin has allowed her to master the methods of the radiation transfer theory and to interpret the experimental results basing on the strict theory. Currently known as a leading specialist in the problem of the interaction of the solar radiation, cloudy atmosphere and atmospheric aerosols. The collaboration with Academician Kirill Ya. Kondratyev has allowed an understanding of the significance of the problems in question for the global climate change. Studies during 1998–1999 as a visiting Professor in the Center for Climate Systems Research of the University of Tokyo and collaboration with Professor T. Nakajima was extremely useful for the assimilation of the experience of satellite data processing.

Contacts: e-mail: **Irina.Melnikova@pobox.spbu.ru**

Alexander V. Vasiljev, Candidate of Science in Physics, Associated Professor of the Physical Faculty, St. Petersburg State University. He has taken part in the airborne observations and in carrying out the ground and ship radiation measurements, has elaborated algorithms and computer codes of the radiation data processing. Knows in detail the procedures of the instruments preparation and accomplishing radiation characteristic observations and data processing. Currently works in the Laboratory for Aerosols under the guidance of Professor Lev S. Ivlev and known as a good specialist in atmospheric aerosols optics. The collaboration with the Laboratory for atmospheric heat radiation headed by Professor Yuri M. Timofeev helped him to master new numerical methods of the inverse problems solution of the atmospheric optics. This extensive experience gives him the ability to understand all features of getting quality results and of their interpretation.

Contacts: e-mail : **vsa@lich.phys.spbu.ru**

About the Authors

CHAPTER 1

Solar Radiation in the Atmosphere

1.1
Characteristics of the Radiation Field in the Atmosphere

In accordance with the contemporary conceptions, light (radiation) is an electromagnetic wave showing quantum properties. Thus, strictly speaking, the processes of light propagation in the atmosphere should be described within the ranges of electrodynamics and quantum mechanics. Nevertheless, it is suitable to abstract from the electromagnetic nature of light to solve a number of problems (including the problems described in this book) and to consider radiation as an energy flux. Light characteristics governed by energy are called *the radiative characteristics*. This approach is usual for optics because the frequency of the electromagnetic waves within the optical ranges is huge and the receiver registers only energy, received during many wave periods (not a simultaneous value of the electro-magnetic intensity). The electromagnetic nature of solar radiation including the property of the electromagnetic waves to be transverse is bound up with the phenomenon of *polarization*, which is revealing in the relationship of the process of the interaction between radiation and substance (refraction, scattering and reflection) and configuration of the electric vector oscillations on a plane, which is normal to the wave propagation direction. Further, we are using *the approximation of unpolarized radiation*. The evaluation of the accuracy of this approximation will be discussed further concerning the specific problems considered in this book.

The following main types of radiation (and their energy) are distinguished in radiation transferring throughout the atmosphere: *direct* radiation (radiation coming to the point immediately from the Sun); *diffused solar* radiation (solar radiation scattered in the atmosphere); *reflected solar* radiation from surface; *self-atmospheric* radiation (*heat atmospheric* radiation) and *self-surface* radiation (*heat* radiation). The total combination of these radiations creates the *radiation field* in the Earth atmosphere, which is characterized with energy of radiation coming from different directions within different spectral ranges. As is seen from above, it is possible to divide all radiation into solar and self (heat) radiation. In this book, we are considering only solar radiation in the spectral ranges 0.3–1.0 µm, where it is possible to neglect the energy of heat radiation of the atmosphere and surface, comparing with solar energy. Further with this spectral range we will be specifying *the short-wave spectral range*. Solar radiation integrated with respect to the wavelength over the considered

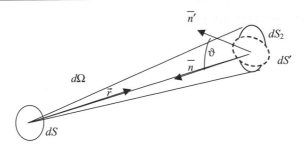

Fig. 1.1. To the definition of the intensity and to the flux of radiation (radiance and irradiance)

spectral region will be called *total radiation*. Meanwhile, it should be noted that further definitions of the radiative characteristics are not linked within this limitation and could be used either for heat or for microwave ranges.

The notion of a monochromatic parallel beam (the plane electromagnetic wave of one concrete wavelength and one strict direction) is widely used in optics for the theoretical description of different processes (Sivukhin 1980). Usually solar radiation is set just in that form to describe its interactions with different objects. The principle of an independency of the monochromatic beams under their superposition is postulated, i.e. the interaction of the radiation beams coming from different directions with the object is considered as a sum of independent interactions along all directions. The physical base of the independency principle is an incoherence of the natural radiation sources[1] (Sivukhin 1980).

This standard operation is naturally used for the radiation field, i.e. the consideration of it as a sum of non-interacted parallel monochromatic beams. Furthermore, radiation energy can't be attributed to a single beam, because if energy were finite in the wavelength and direction intervals, it would be infinitesimal for the single wavelength and for the single direction. For characterizing radiation, it is necessary to pass from energy to its distribution over spectrum and directions.

Consider an emitting object (Fig. 1.1) implying not only the radiation source but also an object reflecting or scattering external radiation. Pick out a surface element dS, encircle the solid angle $d\Omega$ around the normal r to the surface. Then radiation energy would be proportional to the area dS, the solid angle $d\Omega$, as well as to the wavelength ranges $[\lambda, \lambda + d\lambda]$ and the time interval $[t, t + dt]$. The factor of the proportionality of radiation energy to the values dS, $d\Omega$, $d\lambda$ and dt would be specified *an intensity of the radiation* or *radiance* $I_\lambda(r, t)$ at the wavelength λ to the direction r at the moment t according to (Sobolev 1972;

---

[1] It should be noted that monochromatic radiation is impossible in principle. It follows from the mathematical properties of the Fourier transformation: a spectrum consisting of one frequency is possible only with the time-infinite signal. Furthermore, the principle of the independency is not valid for the monochromatic beams because they always interfere. It is possible to remove both these contradictions if we consider monochromatic radiation not as a physical but as a mathematical object, i.e. as a real radiation expansion into a sum (integral Fourier) of the harmonic terms. The separate item of this expansion is interpreted as monochromatic radiation.

Hulst 1980; Minin 1988), namely:

$$I_\lambda(r, t) = \frac{dE}{dS d\Omega d\lambda dt} \tag{1.1}$$

In many cases, we are interested not in energy emitted by the object but in energy of the radiation field that is coming to the object (for example to the instrument input). Then it would be easy to convert the above specification of radiance. Consider the emitting object and set the second surface element of the equal area $dS_2 = dS$ at an arbitrary distance (Fig. 1.1). Let the system to be situated in a vacuum, i.e. radiation is not interacting during the path from dS to dS_2. Let the element dS_2 to be perpendicular to the direction r, then the solid angle at which the element dS_2 is seen from dS at the direction r is equal to the solid angle at which the element dS is seen from dS_2 at the opposite direction $(-r)$. The energies incoming to the surface elements dS and dS_2 are equal too thus; we are getting the consequence from the above definition of the intensity. The factor of the proportionality of emitted energy dE to the values dS, $d\Omega$, $d\lambda$ and dt is called an intensity (radiance) $I_\lambda(r, t)$ incoming from the direction r to the surface element dS perpendicular to r at the wavelength λ at the time t, i.e. (1.1). Point out the important demand of the perpendicularity of the element dS to the direction r in the definition of both the emitting and incoming intensity.

The definition of the intensity as a factor of the proportionality tends to have some formal character. Thus, the "physical" definition is often given: the intensity (radiance) is energy that incomes per unit time, per unit solid angle, per unit wavelength, per unit area perpendicular to the direction of incoming radiation, which has the units of watts per square meter per micron per steradian. This definition is correct if we specify energy to correspond not to the real unit scale (sec, sterad, μm, cm^2) but to the differential scale dt, $d\Omega$, $d\lambda$, dS, which is reduced then to the unit scale. Equation (1.1) is reflecting this obstacle.

Let the surface element dS', which radiation incomes to, not be perpendicular to the direction r but form the angle ϑ with it (Fig. 1.1). Specify *the incident angle* (the angle between the inverse direction $-r$ and the normal to the surface) as $\vartheta = \angle(n, -r)$. In that case defining the intensity as a factor of the proportionality we have to use the projection of the element dS' on a plane perpendicular to the direction of the radiation propagation in the capacity of the surface element dS. This projection is equal to $dS = dS' \cos\vartheta$. Then the following could be obtained from (1.1):

$$dE = I_\lambda(r, t) dt\, d\lambda\, d\Omega\, dS' \cos\vartheta . \tag{1.2}$$

It is suitable to attribute the sign to energy defined above. Actually, if we fix one concrete side of the surface dS' and assume the normal just to this side as a normal n then the angle ϑ varies from 0 to π, and the cosine from +1 to −1. Thus, incoming energy is positive and emitted energy is negative. It has transparent physical sense of the positive source and the negative sink of energy for the surface dS'. Now specify *the irradiance (the radiation flux of energy)* $F_\lambda(t)$ according to (Sobolev 1972; Hulst 1980; Minin 1988) (often it is specified as net spectral energy flux) as a factor of the proportionality of radiation energy dE' incoming within a particular infinitesimal interval of wavelength $[\lambda, \lambda + d\lambda]$ and time $[t, t + dt]$ to the surface dS' from *the all directions* to values dt, $d\lambda$, dS's i.e.:

$$F_\lambda(t) = \frac{dE'}{dt\,d\lambda\,dS'} \ . \tag{1.3}$$

Adduce the "physical" definition of the irradiance that is often used instead of the "formal" one expressed by (1.3). Radiation energy incoming per unit area per unit time, per unit wavelength is called a radiation flux or irradiance. This definition corresponds correctly to (1.3) provided the meaning that energy is equivalent to the difference of incoming and emitted energy and uses the differential scale of area, time and wavelength. Proceeding from this interpretation, we will further use the term *energy* as a synonym of the *flux* implying the value of energy incoming per unit area, time and wavelength.

To characterize the direction of incoming radiation to the element dS' in addition to the angle ϑ, introduce the azimuth angle φ, which is counted off as an angle between the projection of the vector r to the plane dS and any direction on this plane ($0 \le \varphi \le 2\pi$). That is to say in fact that we are using the spherical coordinates system. Energy dE' incoming to the surface dS' from all directions is expressed in terms of energy from a concrete direction $dE(\vartheta, \varphi)$ as:

$$dE' = \int_{\Omega=4\pi} dE(\vartheta, \varphi)d\Omega \ ,$$

where the integration is accomplished over the whole sphere. Using the well-known expression for an element of the solid angle in the spherical coordinates $d\Omega = d\varphi \sin\vartheta d\vartheta$ we will get:

$$dE' = \int_0^{2\pi} d\varphi \int_0^{\pi} dE(\vartheta, \varphi) \sin\vartheta d\vartheta \ .$$

After the substituting of this expression to (1.3) with accounting (1.2) we will get the formula to express the irradiance:

$$F_\lambda(t) = \int_0^{2\pi} d\varphi \int_0^{\pi} I_\lambda(\vartheta, \varphi, t) \cos\vartheta \sin\vartheta d\vartheta \ . \tag{1.4}$$

In addition to direction (ϑ, φ), wavelength λ and time t the solar radiance in the atmosphere depends on placement of the element dS. Owing to the sphericity of the Earth and its atmosphere, it is convenient to put the position of this element in the spherical coordinate system with its beginning in the Earth's center.

Nevertheless, taking into account that the thickness of the atmosphere is much less than the Earth's radius is, in a number of problems the atmosphere could be considered by convention as a plane limited with two infinite boundaries: the bottom – a ground surface and the top – a level, above which the interaction between radiation and atmosphere could be neglected. Further, we are considering only *the plane-parallel atmosphere approximation*. The grounds of the approximation for the specific problems are given in Sect. 1.3. Then the position of the element dS could be characterized with Cartesian coordinates (x, y, z) choosing the altitude as axis z (to put the z axis perpendicular to the top and bottom planes from the bottom to the top). Thus, in a general case the radiance in the atmosphere could be written as $I_\lambda(x, y, z, \vartheta, \varphi, t)$. Under the natural radiation sources (in particular – the solar one) we could neglect the behavior of the radiance in the time domain comparing with the time scales considered in the concrete problems (e. g. comparing with the instrument registration time). The radiation field under such conditions is called a *stationary* one. Further, it is possible to ignore the influence of the horizontal heterogeneity of the atmosphere on the radiation field comparing with the vertical one, i. e. don't consider the dependence of the radiance upon axes x and y. This radiation field is called a *horizontally homogeneous* one. Further, we are considering only stationary and horizontally homogeneous radiation fields. Besides, following the traditions (Sobolev 1972; Hulst 1980; Minin 1988) the subscript λ is omitted at the monochromatic values if the obvious wavelength dependence is not mentioned. Taking into account the above-mentioned assumptions, the formula linking the radiance and irradiance (1.4) is written as:

$$F(z) = \int_0^{2\pi} d\varphi \int_0^{\pi} I(z, \vartheta, \varphi) \cos\vartheta \sin\vartheta d\vartheta \, . \qquad (1.5)$$

It is natural to count off the angle ϑ from the selected direction z in the atmosphere. This angle is called the zenith incident angle (it characterizes the inclination of incident radiation from the zenith). The angle ϑ is equal to zero if radiation comes from the zenith, and it is equal to π if the radiation comes from nadir. As before we are counting off the azimuth angle from an arbitrary direction on the plane, parallel to the boundaries of the atmosphere. Then the integral (1.5) could be written as a sum of two integrals: over upper and lower hemisphere:

$$F(z) = F^{\downarrow}(z) + F^{\uparrow}(z) \, ,$$

$$F^{\downarrow}(z) = \int_0^{2\pi} d\varphi \int_0^{\pi/2} I(z, \vartheta, \varphi) \cos\vartheta \sin\vartheta d\vartheta \, , \qquad (1.6)$$

$$F^{\uparrow}(z) = \int_0^{2\pi} d\varphi \int_{\pi/2}^{\pi} I(z, \vartheta, \varphi) \cos\vartheta \sin\vartheta d\vartheta \, .$$

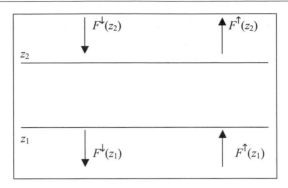

Fig. 1.2. Definition of net radiant flux

The value $F^\downarrow(z)$ is called the *downward flux* (*downwelling irradiance*), the value $F^\uparrow(z)$ – an *upward flux* (*upwelling irradiance*), both are also called *semispherical fluxes* expressed in watts per square meter (per micron). The physical sense of these definitions is evident. The downward flux is radiation energy passing through the level z down to the ground surface and the upward flux is energy passing up from the ground surface. The downward flux is always positive ($\cos \vartheta > 0$), upward is always negative ($\cos \vartheta < 0$). In practice (for example during measurements) it is advisable to consider both fluxes as positive ones. We will follow this tradition. Then for the upward flux in (1.6) the value of $\cos \vartheta$ is to be taken in magnitude, and the total flux will be equal to the difference of the semispherical fluxes $F(z) = F^\downarrow(z) - F^\uparrow(z)$. This value is often called a (*spectral*) net *radiant flux* expressed in watts per square meter (per micron).

Consider two levels in the atmosphere, defined by the altitudes z_1 and z_2 (Fig. 1.2). Obtain solar radiation energy $B(z_1, z_2)$ (per unit area, time and wavelength) absorbed by the atmosphere between these levels. Manifestly, it is necessary to subtract outcoming energy from the incoming:

$$B(z_1, z_2) = F^\downarrow(z_2) + F^\uparrow(z_1) - F^\downarrow(z_1) - F^\uparrow(z_2) = F(z_2) - F(z_1) \,. \tag{1.7}$$

The value $B(z_1, z_2)$ is called a *radiative flux divergence in the layer between levels z_1 and z_2*. It is extremely important value for studying atmospheric energetics because it determines the warming of the atmosphere, and it is also important for studying the atmospheric composition because the spectral dependence of $B(z_1, z_2)$ allows us to estimate the type and the content of specific absorbing materials (atmospheric gases and aerosols) within the layer in question. Hence, the values of the semispherical fluxes determining the radiative flux divergence are also of greatest importance for the mentioned class of problems.

To provide the possibility of comparing the radiative flux divergences in different atmospheric layers we need to normalize the value $B(z_1, z_2)$ to the thickness of the layer:

$$b(z_1, z_2) = B(z_1, z_2)/(z_2 - z_1) \,. \tag{1.8}$$

We would like to point out that the definition of the normalized radiative flux divergences (1.8) with taking into account (1.7) gives the possibility of its theoretical consideration as a continuous function of the altitude after its writing as a derivation of the net flux $b(z) = \partial F(z)/\partial z$.

When we have defined the intensity and the flux above, we scrutinized the radiation field, i.e. the situation when radiation spreads on different directions. Actually, it is possible to amount to nothing more than this definition because no strictly parallel beam exists owing to the wave properties of light (Sivukhin 1980). Nevertheless, radiation emitted by some objects could be often approximated as one directional beam without losses of the accuracy. Incident solar radiation incoming to the top of the atmosphere is practically always considered as one-directional radiation in the problems in question. Actually, it is possible to neglect the angular spread of the solar beam because of the infinitesimal radiuses of the Earth and the Sun compared with the distance between them. Thus, we are considering the case of the plane parallel horizontally homogeneous atmosphere illuminated by a parallel solar beam. Some difficulties are appearing during the application of the above definitions to this case because we must attribute certain energy to the one-directional beam.

The radiance definition corresponding to (1.1) is not applicable in this case because it does not show the dependence of energy dE upon solid angle $d\Omega$ [formally following (1.1) we would get the zero intensity]. As for the irradiance definition (1.3), it is applicable. Thus, it makes sense to examine the irradiance of the strictly one-directional beams. Then the dependence of energy dE' upon the area of the surfaces dS' projection in (1.3) appears for differently oriented surfaces dS', which gives the follows:

$$F(\vartheta) = F_0 \cos \vartheta , \qquad (1.9)$$

where F_0 is the irradiance for the perpendicular incident beam, $F(\vartheta)$ is the irradiance for the incident angle ϑ.

The incident flux F_0 is of fundamental importance for atmospheric optics and energetics. This flux is radiation energy incoming to the top of the atmosphere per unit area, per unit intervals of the wavelength and time in the case of the average distance between the Sun and the Earth, and it is called a *spectral solar constant*. Figure 1.3 illustrates the *solar constant* F_0 as a function of wavelength. Concerning the radiance of the parallel incident beam, we can define it formally using (1.5). Actually, for accomplishing (1.5) and (1.9), it is necessary to assume the following:

$$I(\vartheta, \varphi) = F_0 \delta(\vartheta - \vartheta_0)\delta(\varphi - \varphi_0) , \qquad (1.10)$$

where $\delta()$ is the delta function (Kolmogorov and Fomin 1999), ϑ_0, φ_0 are the solar zenith angle and the azimuth angle which are determining the direction of the incident parallel beam. Remember that the delta function is defined as:

$$\int_a^b f(x)\delta(x - x_0)dx = f(x_0) .$$

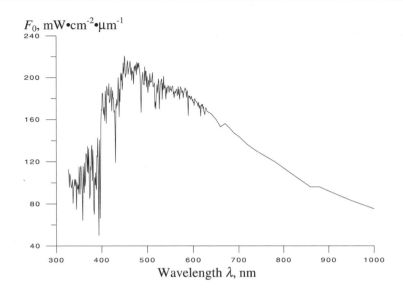

Fig. 1.3. Spectral extraterrestrial solar flux according to Makarova et al. (1991)

No real function can have such a property, thus the delta function is just a symbolic record. Roughly speaking it does not exist without the integrals.

Basing on (1.10) in the case of the parallel beam it could be said that the irradiance incoming to the perpendicular surface is numerically equal to the radiance, however this equality is truly formal because the radiance and the irradiance have different dimensions [that's all right with dimensions in (1.10)].

In conclusion consider the theoretical aspects of the procedures of radiance and irradiance measurements. It is radiation energy that influences the register element of an instrument. It could be written as:

$$E = \int_{t_1}^{t_2} dt \int_{\lambda_1}^{\lambda_2} d\lambda \int_S dxdy$$

$$\times \int_\Omega \sin\vartheta d\vartheta d\varphi I_\lambda(x,y,\vartheta,\varphi,t) f_i^*(t) f_\lambda^*(\lambda) f_S^*(x,y) f_\Omega^*(\vartheta,\varphi) \, ,$$

where $I_\lambda(x,y,\vartheta,\varphi,t)$ is the radiance incoming to the point of the input element (input slit) of an instrument with coordinates (x,y); $[t_1,t_2]$ is the time interval of the input signal registration; $[\lambda_1,\lambda_2]$ is the registration wavelength interval; $f_t^*(t)$, $f_\lambda^*(\lambda)$, $f_S^*(x,y)$, $f_\Omega^*(\vartheta,\varphi)$ are the *instrumental functions*, which characterize a signal transformation by the instrument and they depend on time t, wavelength λ, input element point (x,y), and direction of incoming radiation (ϑ,φ) correspondingly. The integration over the area S is accomplished over the instrument input element surface, and the integration over the solid angle Ω is accomplished over the instrument-viewing angle. The instruments are calibrated so that the measured value of the radiance would be outputting instantaneously. From the theoretical point it means the normalization of the instrumental functions.

$$f_t(t) = f_t^*(t) \Big/ \int_{t_1}^{t_2} f_t^*(t)dt , \quad f_\lambda(\lambda) = f_\lambda^*(\lambda) \Big/ \int_{\lambda_1}^{\lambda_2} f_\lambda^*(\lambda)d\lambda ,$$

$$f_S(x,y) = f_S^*(x,y) \Big/ \int_S f_S^*(x,y)dxdy ,$$

$$f_\Omega(\vartheta,\varphi) = f_\Omega^*(\vartheta,\varphi) \Big/ \int_\Omega f_\Omega^*(\vartheta,\varphi) \sin\vartheta d\vartheta d\varphi$$

Then the measured value of radiance I is expressed through the real radiance $I_\lambda(x,y,\vartheta,\varphi,t)$ by the following:

$$I = \int_{t_1}^{t_2} dt \int_{\lambda_1}^{\lambda_2} d\lambda \int_S dxdy$$

$$\times \int_\Omega \sin\vartheta d\vartheta d\varphi I_\lambda(x,y,\vartheta,\varphi,t)f_t(t)f_\lambda(\lambda)f_S(x,y)f_\Omega(\vartheta,\varphi) .$$

(1.11)

Actually, the equality $I = I_0$ is valid according to (1.11) for normalized instrumental functions if $I_\lambda(x,y,\vartheta,\varphi,t) = I_0 = \text{const}$.

For the radiance measurements, the instrument viewing angle is chosen as small as possible. In this case, all the factors except the wavelength are neglected. Then the following is correct:

$$I = \int_{\lambda_1}^{\lambda_2} I_\lambda f_\lambda(\lambda)d\lambda$$

and the main instrument characteristic would be a *spectral instrumental function* $f_\lambda(\lambda)$, that will be simply called the *instrumental function*. If the radiance is slightly variable in the wavelength interval $[\lambda_1,\lambda_2]$ the influence of the specific features of the instrument on the observational process are possible not to take into account.

The function $f_\lambda(\lambda)$ plays an important role in the observation of the semi-spherical fluxes because the radiance at the instrument input changes evidently along the direction (ϑ,φ). However, comparing (1.4) and (1.11) it is easy to see that condition $f_\Omega^*(\vartheta,\varphi) = \cos\vartheta$ must be implemented specifically during the measurement of the irradiance. This demand to the instruments, which are measuring the solar irradiance, is called a Lambert's cosine law.

1.2
Interaction of the Radiation and the Atmosphere

Consider a symbolic particle (a gas molecule, an aerosol particle) that is illuminated by the parallel beam F_0 (Fig. 1.4). The process of the interaction of radiation and this particle is assembled from the *radiation scattering* on the particle and the *radiation absorption* by the particle. Together these processes constitute *the radiation extinction* (the irradiance after interaction with the particle is attenuated by the processes of scattering and absorption along the incident beam direction r_0). Let the absorbed energy be equal to E_a, scattered in all directions energy be equal to E_s, and the total attenuated energy be equal to $E_e = E_a + E_s$. If the particle interacted with radiation according to geometric optics laws and was a non-transparent one (i.e. attenuated all incoming radiation), attenuated energy would correspond to energy incoming to the projection of the particle on the plane perpendicular to the direction of incoming radiation r_0. Otherwise, this projection is called the *cross-section of the particle by plane* and its area is simply called a *cross-section*. Measuring attenuated energy E_a per wavelength and time intervals $[\lambda, \lambda + d\lambda]$, $[t, t + dt]$ according to the irradiance definition (1.3) we could find the extinction cross-section as $dE_e/(F_0 d\lambda dt)$.

However, owing to the wave quantum nature of light its interaction with the substance does not submit to the laws of geometric optics. Nevertheless, it is very convenient to introduce the relation $dE_e/(F_0 d\lambda dt)$ that has the dimension and the meaning of the area, implying the equivalence of the energy of the real interaction and the energy of the interaction with a nontransparent particle possessing the cross-section equal to $dE_e/(F_0 d\lambda dt)$ in accordance with the laws of geometric optics. Besides, it is also convenient to consider such a cross-section separately for the different interaction processes. Thus, according to the definition, the ratio of absorption energy dE_a, measured within the intervals $[\lambda, \lambda + d\lambda]$, $[t, t + dt]$, to the incident radiation flux F_0 is called an *absorption cross-section C_a*. The ratio of scattering energy dE_s to the incident radiation flux is called *a scattering cross-section C_s* and the ratio of total attenuated energy dE_s to the incident radiation flux is called *an extinction cross-section C_e*:

$$C_a = \frac{dE_a}{F_0 d\lambda dt} \ , \quad C_s = \frac{dE_s}{F_0 d\lambda dt} \ , \quad C_e = \frac{dE_e}{F_0 d\lambda dt} = C_a + C_s \ . \tag{1.12}$$

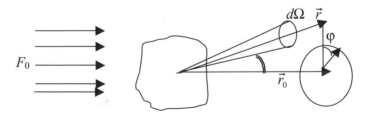

Fig. 1.4. Definition of the cross-section of the interaction

In addition to the above-mentioned, the cross-sections are defined as mono-chromatic ones at wavelength λ (for the non-stationary case – at time t as well).

Consider the process of the light scattering along direction r (Fig. 1.4). Here the value $dE_d(r)$ is the energy of scattered radiation (per intervals $[\lambda, \lambda + d\lambda]$, $[t, t+dt]$) per solid angle $d\Omega$ encircled around direction r. Define *the directional scattering cross-section* analogously to the scattering cross-section expressed by (1.12).

$$C_d(r) = \frac{dE_d(r)}{F_0 d\lambda dt d\Omega} \ .$$

(1.13)

Wavelength λ and time t are corresponding to the cross-section $C_d(r)$.

Total scattering energy is equal to the integral from $dE_d(r)$ over all directions $dE_s = \int_{4\pi} dE_d d\Omega$. Obtain the link between the cross-sections of scattering and directed scattering after substituting of $dE_d(r)$ to this integral:

$$C_s = \int_{4\pi} C_d d\Omega \ .$$

(1.14)

Passing to a spherical coordinate system as in Sect. 1.1, introduce two parameters: *the scattering angle γ* defined as an angle between directions of the incident and scattered radiation ($\gamma = \angle(r_0, r)$) and *the scattering azimuth φ* counted off an angle between the projection of vector r to the plane perpendicular to r_0 and an arbitrary direction on this plane. Then rewrite (1.14) as follows:[2]

$$C_s = \int_0^{2\pi} d\varphi \int_0^{\pi} C_d(\gamma, \varphi) \sin \gamma d\gamma \ .$$

(1.15)

The directional scattering cross-section $C_d(\gamma, \varphi)$ according to its definition could be treated as follows: as the value $C_d(\gamma, \varphi)$ is higher, then light scatters stronger to the very direction (γ, φ) comparing to other directions. It is necessary to pass to a dimensionless value for comparison of the different particles using the directional scattering cross-section. For that the value $C_d(\gamma, \varphi)$ has to be normalized to the integral C_s expressed by (1.15) and the result has to be multiplied by a solid angle. The resulting characteristic is called *a phase function* and specified with the following relation:

$$x(\gamma, \varphi) = 4\pi \frac{C_d(\gamma, \varphi)}{C_s} \ .$$

(1.16)

---

[2]It is called also "differential scattering cross-section" in another terminology and the scattering cross-section is called "integral scattering cross-section". The sense of these names is evident from (1.12)–(1.15).

The substitution of the value $C_d(\gamma, \varphi)$ from (1.15) to (1.16) gives a *normalization condition of the phase function:*

$$\frac{1}{4\pi} \int\limits_{0}^{2\pi} d\varphi \int\limits_{0}^{\pi} x(\gamma, \varphi) \sin \gamma d\gamma = 1 \,. \tag{1.17}$$

If the scattering is equal over all directions, i.e. $C_d(\gamma, \varphi) = const$, it is called *isotropic* and the relation $x(\gamma, \varphi) \equiv 1$ follows from the normalization (1.17). Thus, the multiplier 4π is used in (1.16) for convenience. In many cases, (for example the molecular scattering, the scattering on spherical aerosol particles) the phase function does not depend on the scattering azimuth. Further, we are considering only such phase functions. Then the normalization condition converts to:

$$\frac{1}{2} \int\limits_{0}^{\pi} x(\gamma) \sin \gamma d\gamma = 1 \,. \tag{1.18}$$

The integral from the phase function in limits between zero and scattering angle γ $\frac{1}{2} \int_0^\gamma x(\gamma) \sin \gamma d\gamma$ could be interpreted as a *probability of scattering to the angle interval* $[0, \gamma]$. It is easy to test this integral for satisfying all demands of the notion of the "probability". Hence the phase function $x(\gamma)$ is the *probability density of radiation scattering to the angle* γ. Often this assertion is accepted as a definition of the phase function.[3]

The real atmosphere contains different particles interacting with solar radiation: gas molecules, aerosol particles of different size, shape and chemical composition, and cloud droplets. Therefore, we are interested in the interaction not with the separate particles but with a total combination of them. In the theory of radiative transfer and in atmospheric optics it is usual to abstract from the interaction with a separate particle and to consider the atmosphere as a continuous medium for simplifying the description of the interaction between solar radiation and all atmospheric components. It is possible to attribute the special characteristics of the interaction between the atmosphere and radiation to an elementary volume (formally infinitesimal) of this continuous medium.

Scrutinize *the elementary volume* of this continuous medium $dV = dSdl$ (Fig. 1.5), on which the parallel flux of solar radiation F_0 incomes normally to the side dS. The interaction of radiation and elementary volume is reduced to the processes of scattering, absorption and radiation extenuation after radiation transfers through the elementary volume. Specify the radiation flux

---

[3] Point out that the phase function determines scattering only in the case of unpolarized incident radiation. After the scattering (both molecular and aerosol), light becomes the polarized one and the consequent scattering orders (secondary and so on) can't be described only by the phase function notion. Thus the theory of scattering, which doesn't take into account the polarization, is an approximation. In a general case, the accuracy of this approximation is estimated within 5% according to Hulst (1980). In special cases, it is necessary to test the accuracy that will be done in the following sections.

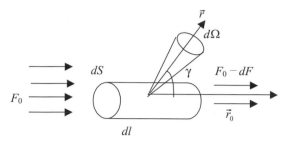

Fig. 1.5. Interaction between radiation and elementary volume of the scattering medium

as $F = F_0 - dF$ after its penetrating the elementary volume (along the incident direction r_0). Take the relative change of incident energy as an extinction characteristic:

$$\frac{dE_e}{E_0} = \frac{(F_0 - F)dSd\lambda dt}{F_0 dSd\lambda dt} = \frac{dF}{F_0}.$$

As it is manifestly proportional to the length dl in the extenuating medium, then it is possible to take the *volume extinction coefficient* α as a characteristic of radiation, attenuated by the elementary volume. This coefficient is equal to a relative change of incident energy (measured in intervals $[\lambda, \lambda + d\lambda]$, $[t, t + dt]$) normalized to the length dl (i. e. reduced to the unit length) according to the definition:

$$\alpha = \frac{dE_e}{E_0 dl} = \frac{dF}{F_0 dl}. \tag{1.19}$$

The analogous definitions of *the volume scattering σ and absorption κ coefficients* follow from the equality of extinction energy and the sum of the scattering and absorption energies.[4]

$$\sigma = \frac{dE_s}{E_0 dl}, \quad \kappa = \frac{dE_a}{E_0 dl}, \quad \alpha = \sigma + \kappa. \tag{1.20}$$

It would be possible to introduce *a volume coefficient of the directional scattering $s(r)$* considering energy $dE_d(r)$ scattered along direction r in solid angle $d\Omega$ analogously to (1.20): $s(r) = dE_d(r)/(E_0 d\Omega dl)$. However, it is not done to use this characteristic. Actually, after accounting (1.20) we are obtaining $dE_d(r) = \frac{1}{\sigma}s(r)dE_s d\Omega$ and substituting it to the relation $dE_s = \int_{4\pi} dE_d d\Omega$ that leads to the expression $\frac{1}{\sigma}\int_{4\pi} sd\Omega = 1$. It exactly corresponds to the normalizing relation (1.17) for the phase function in the spherical coordinates (Figs. 1.4 and 1.5) after the setting $s(\gamma, \varphi) = \frac{1}{4\pi}\sigma x(\gamma, \varphi)$, where $x(\gamma, \varphi)$ is the phase function of the elementary volume. As has been mentioned above, we are considering

---

[4]Notice, that the introduced volume coefficients have the dimension of the inverse length (m^{-1}, km^{-1}) and such values are usually called "linear" not "volume". Further, we will substantiate this terminological contradiction.

further the phase functions depending only upon the scattering angle γ with the normalization relation (1.18). Thus, we obtain the following relation for energy scattered along direction γ

$$dE_d(\gamma) = \frac{\sigma}{4\pi}x(\gamma)E_0 d\Omega dl \, . \tag{1.21}$$

This relation may be accepted as a definition of phase function $x(\gamma)$ of the elementary volume of the medium (however, owing to the definition formality it is often used the definition of the phase function as a probability density of radiation scattering to angle γ).

Let us link the characteristics of the interaction between radiation and a separate particle with the elementary volume. Let every particle interact with radiation independently of others. Then extinction energy of the elementary volume is equal to a sum of extinction energies of all particles in the volume. Suppose firstly that all particles are similar; they have an extinction cross-section C_e and their number concentration (number of particle in the unit volume) is equal to n. The particle number in the elementary volume is ndV. Substituting the sum of extinction energies to the extinction coefficient definition (1.19) in accordance with (1.12) and accounting the definition of the irradiance (1.3) we obtain the following:

$$\alpha = \frac{ndVC_e F_0 d\lambda dt}{F_0 dS d\lambda dt dl} = nC_e \, .$$

Thus, the volume extinction coefficient is equal to the product of particle number concentration by the extinction cross-section of one particle.[5]

If there are extenuating particles of M kinds with concentrations n_i and cross-sections $C_{e,i}$ in the elementary volume of the medium then it is valid: $dE_e = \sum_{i=1}^{M} n_i dV C_{e,i} F_0 d\lambda dt$. Analogously considering the energies of scattering, absorption and directional scattering, we are obtaining the formulas, which link the volume coefficients and cross-sections of the interaction:

$$\alpha = \sum_{i=1}^{M} n_i C_{e,i} \, , \quad \sigma = \sum_{i=1}^{M} n_i C_{s,i} \, ,$$

$$\kappa = \sum_{i=1}^{M} n_i C_{a,i} \, , \quad \sigma x(\gamma) = \sum_{i=1}^{M} n_i C_{s,i} x_i(\gamma) \, . \tag{1.22}$$

We would like to point out that the separate items in (1.22) make sense of the volume coefficients of the interaction for the separate kinds of particles. Therefore, highly important for practical problems are the "rules of summarizing" following from (1.22). These rules allow us to derive separately the coefficients

---

[5] Just by this reason, the term "volume" and not "linear" is used for the coefficient. It is defined by numerical concentration in the unit volume of the air.

of the interaction and the phase function for each of M components and then to calculate the total characteristics of the elementary volume with the formulas:

$$\alpha = \sum_{i=1}^{M} \alpha_i , \quad \sigma = \sum_{i=1}^{M} \sigma_i ,$$

$$\kappa = \sum_{i=1}^{M} \kappa_i , \quad x(\gamma) = \sum_{i=1}^{M} \sigma_i x_i(\gamma) \Big/ \sum_{i=1}^{M} \sigma_i . \tag{1.23}$$

These rules also allow calculating characteristics of the molecular and aerosol scattering and absorption of radiation in the atmosphere separately. Then (1.23) is transformed to the following:

$$\alpha = \sigma_m + \sigma_a + \kappa_m + \kappa_a ,$$

$$\sigma = \sigma_m + \sigma_a ,$$

$$\kappa = \kappa_m + \kappa_a , \tag{1.24}$$

$$x(\gamma) = \frac{\sigma_m x_m(\gamma) + \sigma_a x_a(\gamma)}{\sigma_m + \sigma_a} ,$$

where σ_m, κ_m, $x_m(\gamma)$ are the volume coefficients of the molecular scattering, absorption and molecular phase function for the atmospheric gases correspondingly and σ_a, κ_a, $x_a(\gamma)$ are the analogous aerosol characteristics.

The rules of summarizing expressed by (1.22)–(1.24) have been derived with the assumption that the particles are interacting with radiation independently. Here the following question is pertinent: is this assumption correct? From the view of geometrical optics, which we have appealed to, when introducing the cross-sections of the interaction, their areas (sections) mustn't intersect within the elementary volume, i. e. the total area of its projection to the side dS must be equal to the sum of the areas of all particles. It would be accomplished if the distances between particles were much larger than the linear sizes of the cross-sections of the interaction or, roughly speaking, much larger than the particle sizes. Dividing the elementary volume to small cubes with side d, where d is the distinctive size of the particle we are concluding that for this condition the particle number in the volume dV has to be much less than the number of cubes – $ndV \ll dV/d^3$, i. e. $n \ll 1/d^3$, where n is the particle number concentration. The second condition – the independency of the interaction between the particles and radiation – follows from the points of wave optics, according to which the independency of the interaction occurs if the distances between the particles are much larger than radiation wavelength λ and that leads to the inequality $n \ll 1/\lambda^3$. Using the values of the real molecules and aerosol particle concentrations in the atmosphere it is easy to test that the condition $n \ll 1/d^3$ is always correct, the condition $n \ll 1/\lambda^3$ is correct in the short-wave range for aerosol particles and is broken for molecules of the atmospheric gases. Nevertheless, it is assumed that light scatters not on molecules but on the air density fluctuations (thus, the air is considered as a continuous medium) and it is possible to ignore this violation (Sivukhin 1980). For the calculation of the radiation field the elementary volume is chosen so that only one interaction act may happen within the elementary volume. Such volume is different for particles of different sizes (cloud droplets size is close to 10–20 µm, for atmospheric gases molecules (more exactly – density fluctuations) – the size is about 0.5×10^{-3} µm). Thus, the diffusive medium is turned out non continuous. The violation of both conditions could occur when there are big particles in the air (for example cloud droplets). Actually taking into account the large size of the droplets (tens and hundreds of microns), there are a lot of gas molecules and small aerosol particles around these droplets and the both conditions are violated for them. Therefore, the question about the applicability of the summarizing rules in the cases mentioned above needs a special discussion.

The volume coefficients of the interaction between radiation and atmosphere are expressed through the scattering and absorption cross-sections according to relations (1.22). Thus, the most important problem will be the theoretical calculation of these cross-sections. The methods of their calculation are based on the description of the physical processes of the interaction between radiation and substance (Zuev et al. 1997). However, as we are not considering them here, the resulting formulas are adduced only, referring the reader to the cited literature.

The volume coefficient and the phase function of the molecular scattering are expressed as follows:

$$\sigma_m = \frac{8}{3}\pi^3 \frac{(m^2 - 1)^2}{n\lambda^4} \frac{6 + 3\delta}{6 - 7\delta},$$

$$x_m(\gamma) = \frac{3}{4 + 2\delta}[1 + \delta + (1 - \delta)\cos^2 \gamma],$$

$$(1.25)$$

where m is the refractive index of the air, n is the number concentration of the air molecules, λ is the radiation wavelength, δ is the depolarization factor (for the air it is equal to $\delta = 0.035$). The derivation of (1.25) is presented for example in the books by Kondratyev (1965) and Goody and Yung (1996) (the theory of the molecular scattering that is traditional for atmospheric optics) and in the book by Sivukhin (1980) (the scattering theory on the fluctuations of the air density). Using the known thermodynamic relation it is easy to obtain the number concentration:

$$n = \frac{P}{kT},$$

$$(1.26)$$

where P is the air pressure, T is the air temperature, k is the Boltzmann constant. For assuming the dependence of the air refractive index upon wavelength, pressure, temperature, and moisture, we are using the semi-empiric relation from the book by Goody and Yung (1996):

$$m - 1 = 10^{-6}\left(b(\lambda)\frac{P[+10^{-6}P(139.855 - 2.093T)]}{5.407(1 + 0.003661T)}\right.$$

$$\left. - P_w\frac{8.319 - 0.0907\lambda^{-2}}{1 + 0.003661T}\right),$$ (1.27)

$$b(\lambda) = 64.328 + \frac{29498.1}{146 - \lambda^{-2}} + \frac{255.4}{41 - \lambda^{-2}},$$

where P_w is the partial pressure of the water vapor. It should be noted that in (1.27) wavelength is measured in microns (μm), pressure – in Pascals (Pa), temperature – in degrees Celsius (°C).

Two kinds of the input data for calculating cross-section of the molecular absorption are available in the short wavelength range.

The data of the first kind are tabulated as a dependence of the experimental cross-sections upon wavelength and in some cases upon temperature, i.e. $C_{a,i}(\lambda, T)$. Regretfully, the databases of mentioned cross-sections are not freely accessible nowadays. Therefore, during the concrete calculation we have been using the database collected from Sedunov et al. (1991) and Bass and Paur (1984) together with the data taken from the base of GOMETRAN computer code (Pozanov et al. 1995; Vasilyev et al. 1998) with the kind permission of its authors Vladimir Rozanov and Yuri Timofeyev. The cross-section of the molecular absorption of the specific gas (subscript "i" is omitted) is calculated for the data of the first kind as a simple linear interpolation over the look-up table:

$$C_a(\lambda, T) = \Delta_1(\lambda, j)\Delta_1(T, k)C_a(\lambda_j, T_k) + \Delta_1(\lambda, j)\Delta_2(T, k)C_a(\lambda_j, T_{k+1})$$ (1.28)
$$+ \Delta_2(\lambda, j)\Delta_1(T, k)C_a(\lambda_{j+1}, T_k) + \Delta_2(\lambda, j)\Delta_2(T, k)C_a(\lambda_{j+1}, T_{k+1}),$$

where

$$\Delta_1(y, l) = \frac{y_{l+1} - y}{y_{l+1} - y_l}, \quad \Delta_2(y, l) = \frac{y - y_l}{y_{l+1} - y_l},$$

and numbers j and k are chosen over nodes of the table grid under the conditions $\lambda_j \leq \lambda \leq \lambda_{j+1}$, $T_k \leq T \leq T_{k+1}$. In the absence of the temperature dependence it is enough to set formally $\Delta_1(T, k) = 1$ and $\Delta_2(T, k) = 0$ in (1.28).

The data of the second kind describe the separate absorption lines of the gases (parameters of the fine structure). The theoretical aspects of the calculations using these data have been interpreted in detail, e.g. in the book by Penner (1959). For the concrete calculations, we have been using the database HITRAN-92 (Rothman et al. 1992). The volume coefficient of the molecular absorption according to the data of the second kind depends on the temperature T and air pressure P, and is calculated as:

$$\kappa_m = \sum_{i=1}^{M} n_i \left(\frac{T^*}{T}\right)^{l(i)} \sum_{j=1}^{K(i)} S_{ij} \frac{W_{ij}(T)}{W_{ij}(T^*)} f_{ij}(P, T, v, v_{ij}) ,$$

$$W_{ij}(T) = \exp\left(-\frac{c_2 E_{ij}}{T}\right)\left[1 - \exp\left(-\frac{c_2 v_{ij}}{T}\right)\right] ,$$

(1.29)

where the summarizing is accomplished over the subscript i over all gases, and it is accomplished over subscript j over all absorption lines of the specific gas; T^* is the temperature which the spectroscopic information is presented for ($T^* = 296$ K); $l(i) = 1$ for linear molecules and $l(i) = 1.5$ for other molecules, f_{ij} is the function of spectral line contour, v is the wave number, corresponds to wavelength λ ($v = 1/\lambda$), c_2 is the second radiation constant, S_{ij}, E_{ij}, v_{ij} are the spectral line parameters from the HITRAN-92 database: the intensity, transition energy in the units of the wave number and the wave number in the units of the spectral line correspondingly. There is no obvious analytical expression for the function of spectral line contour f_{ij} in the general case. Therefore, in our calculations the approximation proposed in Matveev (1972) is applied:

$$f_{ij}(P, T, v, v_{ij}) = \frac{1}{\delta_1}\left[\sqrt{\frac{\ln 2}{\pi}}(1 - x)\exp(-y^2 \ln 2) + \frac{x}{\pi(1 + y^2)}\right.$$

$$- x(1 - x)\frac{1}{\pi}\left(\frac{3}{2\ln 2} + 1 + x\right)$$

$$\left. \times \left(0.066\exp(-0.4y^2) - \frac{1}{40 - 5.5y^2 + y^4}\right)\right] ,$$

$$x = \frac{\delta_1}{\delta_2} , \quad y = \frac{v - v_{ij}}{\delta_1} ,$$

(1.30)

$$\delta_1 = \frac{1}{2}\left[\delta_2 + \sqrt{\delta_2^2 + 4\delta_3^2} + 0.05\delta_2\left(1 - \frac{2\delta_2}{\delta_2 + \sqrt{\delta_2^2 + 4\delta_3^2}}\right)\right] ,$$

$$\delta_2 = d_{ij}\frac{P}{P^*}\left(\frac{T^*}{T}\right)^{m_{ij}} ,$$

$$\delta_3 = \frac{v_{ij}}{c}\sqrt{\frac{2RT \ln 2}{\mu_i}} ,$$

where P^* is the pressure, which the spectral information is presented for ($P^* = 1013$ mbar), c is the velocity of light in a vacuum, R is the universal gas constant, μ_i is the molecular mass of gas, d_{ij}, m_{ij} are the line parameters from the HITRAN-92 database: the semi-intensity breadth of the spectral line caused by the collisions with the air molecules and the coefficient of the temperature dependence correspondingly[6]

The calculations of the aerosol scattering and absorption cross-sections so as an aerosol phase function are based on the simulations. The aerosol particles are approximated with the certain geometrical solids of the known chemical composition. Usually there are considered the homogeneous spherical particles. The calculation of the optical characteristics for such particles is accomplished according to the formulas of Mie theory, which we are not adducing here referring the reader to corresponding books. The basis of the theory and the formula derivations are presented in the books by Hulst (1957) and Bohren and Huffman (1983), the transformation to the characteristics of the elementary volume is presented in the book by Deirmendjian (1969), the applied algorithms of the calculations are presented in Bohren and Huffman (1983) and Vasilyev (1996, 1997). An important process influencing the optical characteristics of the aerosol particles especially in the troposphere is their rehydration – the absorption of the water molecules on the particle surface. It leads to the essential variations of the aerosol optical properties depending on the air humidity. The two-layer particle – "sphere in shell" – is the model for the rehydrate particle. The methods of its calculation are presented in Bohren and Huffman (1983) and Vasilyev and Ivlev (1996, 1997). The calculations are usually accomplished in advance due to their large volume and the resulting aerosol volume coefficients of the scattering and absorption together with the phase function are used in problems of radiation transfer theory as look-up tables. These data together with the incident data for the above-mentioned calculations form the base of *the aerosol models*. Nowadays there are many studies concerning the aerosol models. Here we are only mentioning that the choice or the creation of the aerosol model is definite with the features of a concrete problem. We will do this in Chapter 5 referring to the corresponding models.

The phase function of the aerosol scattering is presented in the above-mentioned calculations as a look-up table with the grid over the scattering angle. It is not convenient for some problems where the phase function needs an analytical approximation. One of the widely used approximations is a Henyey-Greenstein function (Henyey and Greenstein 1941):

$$x(\gamma) = \frac{1 - g^2}{(1 + g^2 - 2g \cos \gamma)^{3/2}} , \qquad (1.31)$$

where g is the approximation parameter ($0 \leq g < 1$). Parameter g is often called *the asymmetry factor* because it governs the degree of the phase function forward extension. The function describes the main property of the aerosol phase functions – the forward peak – (the prevalence of the scattering to the forward hemisphere $0 \leq \gamma \leq \pi/2$ over the scattering to the back hemisphere

---

[6]It should be marked that the spectroscopic data of both the first and the second kind have been obtained from the observations, so they contain errors. Moreover, the formulas for the spectral line contour either of empirical (1.30) or the theoretical (1.29) are approximations. Therefore, the calculation using (1.29) and (1.30) gives an uncertainty. Nevertheless, the molecular absorption within the short-wave range is weak enough, and we are not taking into account these uncertainties.

$\pi/2 \leq \gamma \leq \pi$) and it is very suitable for the theoretical consideration, as will be shown further. However, it describes the real phase functions with a large uncertainty (Vasilyev O and Vasilyev V 1994). Therefore, the using of this function needs a careful evaluation of the errors. The detailed consideration of this problem will be presented in Chap. 5.

1.3
Radiative Transfer in the Atmosphere

Within the elementary volume, the enhancing of energy along the length dl could occur in addition to the extinction of the radiation considered above. Heat radiation of the atmosphere within the infrared range is an evident example of this process, though as will be shown the accounting of energy enhancing is really important in the short-wave range. Value dE_r – the enhancing of energy – is proportional to the spectral $d\lambda$ and time dt intervals, to the arc of solid angle $d\Omega$ encircled around the incident direction and to the value of emitting volume $dV = dSdl$. Specify *the volume emission coefficient ε* as a coefficient of this proportionality:

$$\varepsilon = \frac{dE_r}{dVd\Omega d\lambda dt} \, . \tag{1.32}$$

Consider now the elementary volume of medium within the radiation field. In general case both the extinction and the enhancing of energy of radiation passing through this volume are taking place (Fig. 1.6). Let I be the radiance incoming to the volume perpendicular to the side dS and $I + dI$ be the radiance after passing the volume along the same direction. According to energy definition in (1.1) incoming energy is equal to $E_0 = IdSd\Omega d\lambda dt$ then the change of energy after passing the volume is equal to $dE = dIdSd\Omega d\lambda dt$. According to the law of the conservation of energy, this change is equal to the difference between enhancing dE_r and extincting dE_e energies. Then, taking into account the definitions of the volume emitting coefficient (1.32) and the volume extinction coefficient, we can define *the radiative transfer equation*:

$$\frac{dI}{dl} = -\alpha I + \varepsilon \, . \tag{1.33}$$

In spite of the simple form, (1.33) is the general transfer equation with accepting the coefficients α and ε as variable values. This derivation of the radiative transfer equation is phenomenological. The rigorous derivation must be done using the Maxwell equations.

We will move to a consideration of particular cases of transfer (1.33) in conformity with *shortwave solar radiation in the Earth atmosphere*. Within the shortwave spectral range we omit the heat atmospheric radiation against the solar one and seem to have the relation $\varepsilon = 0$. However, we are taking into account that the enhancing of emitted energy within the elementary volume could occur also owing to the scattering of external radiation coming to the

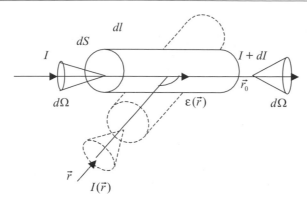

Fig. 1.6. To the derivation of the radiative transfer equation

volume along the direction of the transfer in (1.33) (i.e. along the direction normal to the side dS). Specify this direction r_0 and scrutinize radiation scattering from direction r with scattering angle γ (Fig. 1.6). Encircling the similar volume around direction r (it is denoted as a dashed line), we are obtaining energy scattered to direction r_0. Then employing precedent value of energy E_0 and definition (1.32), we are obtaining the yield to the emission coefficient corresponded to direction r:

$$d\varepsilon(r) = \frac{\frac{\sigma}{4\pi}x(\gamma)I(r)dSd\Omega d\lambda dt d\Omega dl}{dVd\Omega d\lambda dt} = \frac{\sigma}{4\pi}x(\gamma)I(r)d\Omega \ .$$

Then it is necessary to integrate value $d\varepsilon(r)$ over all directions and it leads to *the integro-differential transfer equation with taking into account the scattering:*

$$\frac{dI(r_0)}{dl} = -\alpha I(r_0) + \frac{\sigma}{4\pi}\int_{4\pi} x(\gamma)I(r)d\Omega \ . \tag{1.34}$$

Consider the geometry of solar radiation spreading throughout the atmosphere for concretization (1.34) as Fig. 1.7 illustrates. As described above in Sect. 1.1 we are presenting the atmosphere as a model of the plane-parallel and horizontally homogeneous layer. The direction of the radiation spreading is characterized with the zenith angle ϑ and with the azimuth φ counted off an arbitrary direction at a horizontal plane. Set all coefficients in (1.34) depending on the altitude (it completely corresponds to reality).

Length element dl in the plane-parallel atmosphere is $dl = -dz/\cos\vartheta$. The ground surface at the bottom of the atmosphere is neglected for the present (i.e. it is accounted that the radiation incoming to the bottom of the atmosphere is not reflected back to the atmosphere and it is equivalent to the almost absorbing surface). Within this horizontally homogeneous medium, the radiation field is also the horizontally homogeneous owing to the shift symmetry (the invariance of all conditions of the problem relatively to any horizontal displacement).

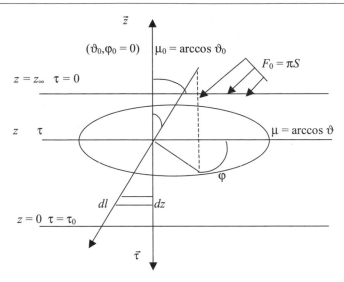

Fig. 1.7. Geometry of propagation of solar radiation in the plane parallel atmosphere

Thus, the radiance is a function of only three coordinates: altitude z and two angles, defining direction (ϑ, φ). Hence, (1.34) could be written as:

$$\frac{dI(z, \vartheta, \varphi)}{dz} \cos \vartheta = \alpha(z)I(z, \vartheta, \varphi)$$

$$- \frac{\sigma(z)}{4\pi} \int\limits_0^{2\pi} d\varphi' \int\limits_0^{\pi} x(z, \gamma)I(z, \vartheta', \varphi') \sin \vartheta' d\vartheta' \tag{1.35}$$

where scattering angle γ is an angle between directions (ϑ, φ) and $(\vartheta'\varphi')$. It is easy to express the scattering angle through ϑ, φ: to consider the scalar product of the orts in the Cartesian coordinate system and then pass to the spherical coordinates. This procedure yields the following relation known as the Cosine law for the spheroid triangles[7]:

$$\cos \gamma = \cos \vartheta \cos \vartheta' + \sin \vartheta \sin \vartheta' \cos(\varphi - \varphi') . \tag{1.36}$$

To begin with, consider the simplest particular case of transfer (1.35). Neglect the radiation scattering, i. e. the term with the integral. For atmospheric optics,

[7] Use in (1.35) of the plane atmosphere model in spite of the real spherical one is an approximation. It has been shown, that it is possible to neglect the sphericity of the atmosphere with a rather good accuracy if the angle of solar elevation is more than 10°. Then the refraction (the distortion) of the solar beams, which has been neglected during the deriving of the transfer equation is not essential. Mark that the horizontal homogeneity is not evident. This property is usually substantiated with the great extension of the horizontal heterogeneities compared with the vertical ones. However, this condition could be invalid for the atmospheric aerosols. It is more correct to interpret the model of the horizontally homogeneous atmosphere as a result of the averaging of the real atmospheric parameters over the horizontal coordinate.

it conforms to the direction of the direct radiation spreading (ϑ_0, φ_0). Actually in the cloudless atmosphere, the intensity of solar direct radiation is essentially greater than the intensity of scattered radiation. In this case, the direction of solar radiation is only one, the intensity depends only on the altitude, and the transfer equation (1.35) transforms to the following:

$$\frac{dI(z)}{dz} \cos \vartheta_0 = \alpha(z)I(z) . \tag{1.37}$$

Mark that it is always $\cos \vartheta_0 > 0$ in (1.37). Differential equation (1.37) together with boundary condition $I = I(z_\infty)$, where z_∞ is the altitude of the top of the atmosphere (the level above which it is possible to neglect the interaction between solar radiation and atmosphere) is elementary solved that leads to:

$$I(z) = I(z_\infty) \exp \left(\frac{1}{\cos \vartheta_0} \int_{z_\infty}^{z} \alpha(z')dz' \right) .$$

It is convenient to rewrite this solution as:

$$I(z) = I(z_\infty) \exp \left(-\frac{1}{\cos \vartheta_0} \int_{z}^{z_\infty} \alpha(z')dz' \right) . \tag{1.38}$$

This relation illustrates the exponential decrease of the intensity in the extinct medium and it is called Beer's Law.

Introduce the dimensionless value:

$$\tau(z) = \int_{z}^{z_\infty} \alpha(z')dz' , \tag{1.39}$$

that is called *the optical depth* of the atmosphere at altitude z. Its important particular case is *the optical thickness* of the whole atmosphere:

$$\tau_0 = \int_{0}^{z_\infty} \alpha(z')dz' . \tag{1.40}$$

Then Beer's Law is written as:

$$I(z) = I(z_\infty) \exp(-\tau(z)/\cos \vartheta_0) . \tag{1.41}$$

As it follows from definitions (1.39) and (1.40) and from "summarizing rules" (1.23), the analogous rules are correct for the optical deepness and for the optical thickness:

$$\tau(z) = \sum_{i=1}^{M} \tau_i(z) , \quad \tau_0 = \sum_{i=1}^{M} \tau_{0,i} .$$

Therefore, it is possible to specify the optical thickness of the molecular scattering, the optical thickness of the aerosol absorption etc.

According to the condition accepted in Sect. 1.1 we are considering solar radiation incoming to the plane atmosphere top as an incident solar parallel flux F_0 from direction (ϑ_0, φ_0). Then, deducing the intensity through delta-function (1.10) and substituting it to the formula of the link between the flux and intensity (1.5) it is possible to obtain Beer's Law for the solar irradiance incoming to the horizontal surface at the level z:

$$F_d(z) = F_0 \cos \vartheta_0 \exp(-\tau(z)/\cos \vartheta_0) . \tag{1.42}$$

In particular, it is accomplished for the solar direct irradiance at the bottom of the atmosphere[8]:

$$F_d(0) = F_0 \cos \vartheta_0 \exp(-\tau_0/\cos \vartheta_0) . \tag{1.43}$$

Return to the general case of the transfer equation with taking into account scattering (1.35). Accomplish the transformation to the dimensionless parameters in the transfer equation for convenience of further analysis. In accordance with the optical thickness definition (1.39) the function $\tau(z)$ is monotonically decreasing with altitude that follows from condition $\alpha(z') > 0$. In this case there is an inverse function $z(\tau)$ that is also decreasing monotonically. Using the formal substitution of function $z(\tau)$ rewrite the transfer equation and pass from vertical coordinate τ to coordinate z, moreover, the boundary condition is at the top of the atmosphere $\tau = 0$ and at the bottom $\tau = \tau_0$, and the direction of axis τ is opposite to axis z. It follows from the definition (1.39): $d\tau = -\alpha(z)dz$. Specify $\mu = \cos \vartheta$ and pass from the zenith angle to its cosine (the formal substitution $\vartheta = \arccos \mu$ with taking into account $\sin \vartheta d\vartheta = -d\mu$). Finally, divide both parts of the equation to value $\alpha(\tau)$, and instead (1.35) obtain the following equation:

$$\mu \frac{dI(\tau, \mu, \varphi)}{d\tau} = -I(\tau, \mu, \varphi) + \frac{\omega_0(\tau)}{4\pi} \int\limits_0^{2\pi} d\varphi' \int\limits_{-1}^{1} x(\tau, \chi)I(\tau, \mu', \varphi')d\mu' , \tag{1.44}$$

where

$$\omega_0(\tau) = \frac{\sigma(\tau)}{\alpha(\tau)} = \frac{\sigma(\tau)}{\sigma(\tau) + \kappa(\tau)} , \tag{1.45}$$

---

[8]Point out that according to Beer's Law the radiance in vacuum ($\alpha = 0$) does not change (the same conclusion follows immediately from the radiance definition). It contradicts to the everyday identification of radiance as a brightness of the luminous object. Actually, it is well known that the viewing brightness of stars decreases with the increasing of distance. It is evident that as the star is further, then the solid angle, in which the radiation incomes to a receiver (an eye, a telescope objective), is smaller, hence energy perceived by the instrument is smaller too. Just this energy is often identified with the brightness (and it is called radiance sometimes), although in accordance to definition (1.1) it is necessary to normalize it to the solid angle. Thus, the essence of the contradiction is incorrect using of the term "radiance". In astronomy, the notion equivalent to radiance (1.1) is the absolute star quantity (magnitude).

and the scattering angle cosine according to (1.36):

$$\chi = \mu\mu' + \sqrt{1 - \mu^2}\sqrt{1 - \mu'^2} \cos(\varphi - \varphi'). \tag{1.46}$$

For the phase function it is also suitable to pass from scattering angle γ to its cosine χ with formal substitution $\gamma = \arccos\chi$.

Dimensionless value ω_0 defined by (1.45) is called *the single scattering albedo* or otherwise *the probability of the quantum surviving* per the single scattering event. If there is no absorption ($\kappa = 0$) then the case is called *conservative scattering*, $\omega_0 = 1$. If the scattering is absent then the extinction is caused only by absorption, $\sigma = 0$, $\omega_0 = 0$ and the solution of the transfer equation is reduced to Beer's Law – (1.41)–(1.43). From consideration of these cases, the sense of value ω_0 is following: it defines the part of scattered radiation relatively to the total extinction, and corresponds to the probability of the quantum to survive and accepts the quantum absorption as its "death".

It is necessary to specify the boundary conditions at the top and bottom of the atmosphere. As it has been mentioned above, solar radiation is characterizing with values F_0, ϑ_0, φ_0 incomes to the top. Usually it is assumed $\varphi_0 = 0$, i.e. all azimuths are counted off the solar azimuth. Additionally specify $\mu_0 = \cos\vartheta_0$ and $F_0 = \pi S$.[9]

As has been mentioned above, solar radiation in the Earth's atmosphere consists of direct and scattered radiation. It is accepted not to include the direct radiation to the transfer equation and to write the equation only for the scattered radiation. The calculation of the direct radiation is accomplished using Beer's Law (1.41). Therefore, present the radiance as a sum of direct and scattered radiance $I(\tau, \mu, \varphi) = I'(\tau, \mu, \varphi) + I''(\tau, \mu, \varphi)$. From expression for the direct radiance of the parallel beam (1.10) the following is correct $I'(0, \mu, \varphi) = \pi S\delta(\mu - \mu_0)\delta(\varphi - 0)$, and it leads to $I'(\tau, \mu, \varphi) = \pi S\delta(mu - \mu_0)\delta(\varphi)\exp(-\tau/\mu_0)$ for Beer's Law. Substitute the above sum to (1.44), with taking into account the validity of (1.37) for direct radiation and properties of the delta function (Kolmogorov and Fomin 1989). Then introducing the dependence upon value μ_0 and omitting primes $I''(\tau, \mu, \mu_0, \varphi)$, we are obtaining *the transfer equation for scattered radiation.*

$$\mu\frac{I(\tau, \mu, \mu_0, \varphi)}{d\tau} = -I(\tau, \mu, \mu_0, \varphi) + \frac{\omega_0(\tau)}{4\pi}\int_0^{2\pi} d\varphi' \int_{-1}^{1} x(\tau, \chi)I(\tau, \mu', \mu_0, \varphi')d\mu'$$

$$+ \frac{\omega_0(\tau)}{4}Sx(\tau, \chi_0)\exp(-\tau/\mu_0) \tag{1.47}$$

[9] Specifying πS has the following sense. Suppose that radiation equal to radiance S from all directions incomes to the top of the atmosphere, and this radiation is called isotropic. Then, according to (1.6) linking the irradiance and radiance, the incoming to the top irradiance is equal to πS. Thus, value S is an isotropic radiance that corresponds to the same irradiance as a parallel solar beam normally incoming to the top of the atmosphere is.

where value χ is defined by (1.46) and for χ_0 the following expression is correct according to the same equation:

$$\chi_0 = \mu\mu_0 + \sqrt{1 - \mu^2}\sqrt{1 - \mu_0^2}\cos(\varphi) \tag{1.48}$$

Point out that (1.47) is written only for *the diffuse radiation*. The boundary conditions are taking into account by the third term in the right part of (1.47). The sense of this term is the yield of *the first order of the scattering* to the radiance and the integral term describes the yield of *the multiple scattering*.

The ground surface at the bottom of the atmosphere is usually called *the underlying surface* or *the surface*. Solar radiation interacts with the surface reflecting from it. Hence, the laws of the reflection as a boundary condition at the bottom of the atmosphere should be taken into account. However, it is done otherwise in the radiative transfer theory. As will be shown in the following section, there are comparatively simple methods of calculating the reflection by the surface if the solution of the transfer equation for the atmosphere without the interaction between radiation and surface is obtained. Thus, neither direct nor reflected radiation is included in (1.47). As there is no diffused radiation at the atmospheric top and bottom, the boundary conditions are following

$$\begin{aligned} I(0, \mu, \mu_0, \varphi) = 0 \quad \mu > 0\,, \\ I(\tau_0, \mu, \mu_0, \varphi) = 0 \quad \mu < 0\,. \end{aligned} \tag{1.49}$$

Transfer equation (1.47) together with (1.46), (1.48) and boundary conditions (1.49) defines the problem of the solar diffused radiance in the plane parallel atmosphere. Nowadays different methods both analytical (Sobolev 1972; Hulst 1980; Minin 1988; Yanovitskij 1997) and numerical (Lenoble 1985; Marchuk 1988) are elaborated. Our interest to the transfer equation is concerning the processing and interpretation of the observational data of the semispherical solar irradiance in the clear and overcast sky conditions. The specific numerical methods used for these cases will be exposed in Chap. 2. Now continue the analysis of the transfer equation to introduce some notions and relations, which will be used further.

The diffused radiation within the elementary volume could be interpreted as a source of radiation. It follows from the derivation of the volume emission coefficient through the diffused radiance in (1.34) if the increasing of the radiance is linked with the existence of the radiation sources. Then introduce *the source function*:

$$\begin{aligned} B(\tau, \mu, \mu_0, \varphi) = {} & \frac{\omega_0(\tau)}{4\pi} \int_0^{2\pi} d\varphi' \int_{-1}^{1} x(\tau, \chi)I(\tau, \mu', \mu_0, \varphi')d\mu' \\ & + \frac{\omega_0(\tau)}{4} Sx(\tau, \chi_0) \exp(-\tau/\mu_0)\,, \end{aligned} \tag{1.50}$$

and the transfer equation is rewritten as follows:

$$\mu \frac{dI(\tau, \mu, \mu_0, \varphi)}{d\tau} = -I(\tau, \mu, \mu_0, \varphi) + B(\tau, \mu, \mu_0, \varphi) . \tag{1.51}$$

Equation (1.51) is the linear inhomogeneous differential equation of type $dy(x)/dx = ay(x) + b(x)$. Its solution is well known:

$$y(x) = y(x_0) \exp(a(x - x_0)) + \int_{x_0}^{x} b(x') \exp(a(x - x'))dx' .$$

Applying it to (1.51) under boundary conditions (1.49), it is obtained:

$$I(\tau, \mu, \mu_0, \varphi) = \frac{1}{\mu} \int_{0}^{\tau} B(\tau', \mu, \mu_0, \varphi) \exp\left(-\frac{\tau - \tau'}{\mu}\right) d\tau' \quad \mu > 0 ,$$

$$\tag{1.52}$$

$$I(\tau, \mu, \mu_0, \varphi) = -\frac{1}{\mu} \int_{t}^{\tau_0} B(\tau', \mu, \mu_0, \varphi) \exp\left(-\frac{\tau - \tau'}{\mu}\right) d\tau' \quad \mu < 0 .$$

Certainly (1.52) are not the problem's solution because source function $B(\tau, \mu, \mu_0, \varphi)$ itself is expressed through the desired radiance. However, (1.52) allows the calculation of the radiance if the source function is known, for example in the case of the first order scattering approximation when only the second term exists in the definition of function $B(\tau, \mu, \mu_0, \varphi)$ (1.50). The expressions for the reflected and transmitted scattered radiance of the first order scattering in the homogeneous atmosphere (where the single scattering albedo does not depend on altitude) have been obtained (Minin 1988):

$$I_1(\tau, \mu, \mu_0, \varphi) = \frac{S\mu_0\omega_0}{4} x(\chi_0) \frac{1 - \exp\left[-\tau\left(\frac{1}{\mu} + \frac{1}{\mu_0}\right)\right]}{\mu + \mu_0} \quad \mu < 0 ,$$

$$I_1(\tau, \mu, \mu_0, \varphi) = \frac{S\mu_0\omega_0}{4} x(\chi_0) \frac{\exp\left(\frac{-\tau}{\mu}\right) - \exp\left(\frac{-\tau}{\mu_0}\right)}{\mu - \mu_0} \quad \mu > 0 .$$

Return to general expressions for the radiance (1.21), substitute them to source function definition (1.19), and deduce the following:

$$B(\tau, \mu, \mu_0, \varphi) = \frac{\omega_0(\tau)}{4\pi} \int_{0}^{2\pi} d\varphi' \left[\int_{0}^{1} x(\tau, \chi) \frac{d\mu'}{\mu'} \int_{0}^{\tau} B(\tau', \mu', \mu_0, \varphi') \right.$$

$$\times \exp\left(-\frac{\tau - \tau'}{\mu'}\right) d\tau' - \int_{-1}^{0} x(\tau, \chi) \frac{d\mu'}{\mu'} \int_{\tau}^{\tau_0} B(\tau', \mu', \mu_0, \varphi') \exp\left(-\frac{\tau - \tau'}{\mu'}\right) d\tau' \right]$$

$$+ \frac{\omega_0(\tau)}{4} Sx(\tau, \chi_0) \exp(-\tau/\zeta) .$$

$$\tag{1.53}$$

Equation (1.53) is the integral equation for the source function. Usually just this equation is analyzed in the radiative transfer theory but not (1.47). The desired radiance is linked with the solution of (1.53) with the simple expressions. It is possible otherwise to substitute definition (1.50) to expressions (1.52) and to obtain the integral equations for the radiance used in the numerical methods of the radiative transfer theory.

It is possible to write the integral equation for the source function (1.53) through the operator form (Hulst 1980; Lenoble 1985; Marchuk et al. 1980)

$$\mathbf{B} = \mathbf{KB} + \mathbf{q} , \tag{1.54}$$

where $\mathbf{B} = B(\tau, \mu, \mu_0, \varphi)$ is the source function, $\mathbf{q}$ is the absolute term, $\mathbf{K}$ is the integral operator. *The operator kernel* and *the absolute term* are expressed according to (1.53) as:

$$\mathbf{K} = K(\tau, \mu, \mu_0, \varphi, \tau', \mu', \varphi') = \frac{\omega_0(\tau)}{4\pi\eta'} x(\tau, \chi) \exp\left(-\frac{\tau - \tau'}{\mu'}\right)$$

$$\text{for} \quad 0 \le \tau' \le \tau 0 \le \mu' \le 1 ,$$

$$\mathbf{K} = K(\tau\mu, \mu_0, \varphi, \tau', \mu', \varphi') = -\frac{\omega_0(\tau)}{4\pi\eta'} x(\tau, \chi) \exp\left(-\frac{\tau - \tau'}{\mu'}\right) \tag{1.55}$$

$$\text{for} \quad \tau \le \tau' \le \tau_0 - 1 \le \mu \le 0 ,$$

$$K = 0 \quad \text{out of the pointed ranges,}$$

$$\mathbf{q} = q(\tau, \mu, \mu_0, \varphi) = \frac{\omega_0(\tau)}{4} Sx(\tau, \chi_0) \exp(-\tau/\mu_0) .$$

Remember that according to Kolmogorov and Fomin (1989) the operator recording is:

$$\mathbf{Ky} \equiv \int_a^b K(x, x') y(x') dx' .$$

Equation (1.54) is the Fredholm equation of the second kind. The mathematical theory of these equations is perfectly developed, e. g. Kolmogorov and Fomin (1989). The formal solution of the Fredholm equation of the second kind is presented with the Neumann series:

$$\mathbf{B} = \mathbf{q} + \mathbf{Kq} + \mathbf{K}^2\mathbf{q} + \mathbf{K}^3\mathbf{q} + \dots \tag{1.56}$$

Expression (1.56) concerning the transfer theory is an expansion of the solution (the source function) over powers of the scattering order. Actually, the item $\mathbf{q}$ is a yield of the first order scattering to the source function, the item $\mathbf{Kq}$ is the second order, $\mathbf{K}^2q = \mathbf{K(Kq)}$ is the third order etc. As kernel $\mathbf{K}$ is proportional to the single scattering albedo, the velocity of the series convergence is linked with this parameter: the higher ω_0 (the scattering is greater) the higher order uf the scattering is necessary to account in the series. Mark that, according to (1.56), source function **B** linearly depends on **q**. Hence, source function **B** (and the desired radiance) is directly proportional to value S, i.e. to the extraterrestrial solar flux. So it is often assumed $S = 1$ and finally the obtained radiance multiplied by the real value $S = F_0/\pi$.

As per (1.55) $\mathbf{q} = \mu_0\mathbf{B}\mathbf{I}_0$, where $\mathbf{I}_0 = I(0, \mu, \mu_0, \varphi) = \pi\delta(\mu - \mu_0)\delta(\varphi)$ is the extraterrestrial radiance. Consequently the desired radiance $\mathbf{I} = I(\tau, \mu, \mu_0, \varphi)$ also linearly depends on $\mathbf{I}_0$ and it is possible to formally write the following:

$$\mathbf{I} = \mathbf{T}\mathbf{I}_0 , \tag{1.57}$$

where **T** is the linear operator and the problem of calculating the radiance is reduced to the finding of the operator. As function $\mathbf{I}_0$ is the delta-function of direction (μ_0, φ_0) (where the azimuth of extraterrestrial radiation is assumed arbitrary) the radiance could be calculated for no matter how complicated an incident radiation field $I_0(\mu_0, \varphi_0)$ after obtaining the operator **T** as a function of all possible directions $\mathbf{T}(\mu_0, \varphi_0)$ due to the linearity of (1.57). The following relation is used for that:

$$\mathbf{I} = \int\limits_0^{2\pi} d\varphi_0 \int\limits_0^1 \mathbf{T}(\mu_0, \varphi_0)\mathbf{I}_0(\mu_0, \varphi_0)d\mu_0 . \tag{1.58}$$

The linearity of (1.57) is widely used in the modern radiative transfer theory including the applied calculations. It is especially convenient for describing the reflection from the surface that will be considered in the following section.

The presentation of the solution of the differential and integral equation as a series expansion over the orthogonal functions is the standard mathematical method. Certain simplification is succeeded after expanding the phase function over the series of Legendre Polynomials in the case of the radiative transfer equation. Legendre Polynomials are defined, e.g. (Kolmogorov and Fomin 1999) as,

$$P_n(z) = \frac{1}{2^n n!}\frac{d(z^2 - 1)^n}{dz} .$$

However, during the practical calculation the following recurrent formula is more appropriated:

$$P_n(z) = \frac{2n-1}{n}zP_{n-1}(z) - \frac{n-1}{n}P_{n-2}(z) \tag{1.59}$$

where $P_0(z) = 1, P_1(z) = z$.

With (1.59) the relations $P_1(z) = z, P_2(z) = 1/2(3z^2 - 1)$ etc. are obtained. Legendre Polynomials constitute the orthogonal function system within the interval $[-1, 1]$:

$$\int\limits_{-1}^1 P_n(z)P_m(z)dz = 0 , \quad \text{for} \quad n \neq m \text{ and } \int\limits_{-1}^1 P_n^2(z)dz = \frac{2}{2n+1}$$

because any function within the interval could be expanded to the series over Legendre Polynomials. The following is deduced for the phase function:

$$x(\chi) = \sum_{i=0}^{\infty} x_i P_i(\chi)$$

$$x_i = \frac{2i+1}{2} \int_{-1}^{1} x(\chi')P_i(\chi')d\chi' .$$

(1.60)

From the normalizing condition of the phase function (1.18) and from equality $P_0 = 1$ it always follows $x_0 = 1$. The first coefficient of the expansion x_1 is of an important physical sense:

$$x_1 = \frac{3}{2} \int_{-1}^{1} x(\chi)\chi d\chi = 3g .$$

(1.61)

From the phase function interpretation as a probability density of the scattering to the certain angle it follows that value $g = x_1/3$ is *the mean cosine of scattering angle*. It determines the elongation of the phase function, namely, as g is closer to unity then the phase function is more extended to the forward direction and weaker extended to the backscatter direction. In the context of parameter g the Henyey-Greenstein approximation (1.31) is appropriate. It is easy testing that its mean cosine is just equal to the parameter of the approximation and it is specified with the same sign g (but it is not otherwise, the using of sign g for the mean cosine does not imply the Henyey-Greenstein approximation is obligatory). Other expansion items of the Henyey-Greenstein function over Legendre Polynomials are also simply expressed through its parameter: $x_i = (2i + 1)g^i$. This very reason determines the wide application of the Henyey-Greenstein function but not an accuracy of the real phase function approximation.

Practically the series is to break at the certain item with number N. The value N was shown in the study by Dave (1970) to reach hundreds and even thousands to approximate the phase function with the necessary accuracy. It is not appropriate for expansion (1.60) using for the radiance calculation even with modern computers. It is the essential problem of the application of the described methodology. We would like to point out that for the molecular scattering determined by (1.25) the phase function is much more simple ($N = 2$):

$$x_m(\chi) = P_0(\chi) + \frac{1-\delta}{2+\delta}P_2(\chi) .$$

The phase function cosine χ in transfer equation (1.47) (and in all consequences from it) is a function of directions of incident and scattered radiation (1.46).

For such a function the theorem of Legendre Polynomials addition (Smirnov 1974; Korn and Korn 2000) is known. According to it the following is correct:

$$P_i\left(\mu\mu' + \sqrt{(1-\mu^2)}\sqrt{(1-\mu'^2)}\cos(\varphi - \varphi')\right)$$
$$+ P_i(\mu)P_i(\mu') + 2\sum_{m=1}^{i}\frac{(i-m)!}{(i+m)!}P_i^m(\mu)P_i^m(\mu')\cos m(\varphi - \varphi') \qquad (1.62)$$

where $P_i^m(z)$ are associated Legendre Polynomials defined as:

$$P_i^m(z) = (1-z)^{m/2}\frac{d^m P_i(z)}{dz^m} \quad \text{and} \quad P_i^0(z) = P_i(z).$$

(Letter m specifies the superscript and not a power here and further in the analogous relations). There are known recurrence relations for the practical calculation of function $P_i^m(z)$ (Korn G and Korn T 2000). Applying relation (1.31) to expansion of the phase function (1.60) it is inferred:

$$x(\chi) = \sum_{i=0}^{N}x_i P_i(\mu)P_i(\mu')$$
$$+ 2\sum_{i=1}^{N}x_i\sum_{m=1}^{i}\frac{(i-m)!}{(i+m)!}P_i^m(\mu)P_i^m(\mu')\cos m(\varphi - \varphi'). \qquad (1.63)$$

After changing the summation order in the second item of (1.63) and accounting that for $m=1$ it is valid $i = 1,\ldots,N$, and $m = 2-$ is $i = 2,\ldots,N$ etc., we finally obtain the following:

$$x(\chi) = p^0(\mu,\mu') + 2\sum_{i=1}^{N}p^m(\mu,\mu')\cos m(\varphi - \varphi'),$$
$$p^m(\mu,\mu') = \sum_{i=m}^{N}x_i\frac{(i-m)!}{(i+m)!}P_i^m(\mu)P_i^m(\mu'). \qquad (1.64)$$

Write the relations analogous to (1.64) for the radiance and source function:

$$I(\tau,\mu,\mu_0,\varphi) = I^0(\tau,\mu,\mu_0) + 2\sum_{i=1}^{N}I^m(\tau,\mu,\mu_0)\cos m\varphi,$$
$$B(\tau,\mu,\mu_0,\varphi) = B^0(\tau,\mu,\mu_0) + 2\sum_{i=1}^{N}B^m(\tau,\mu,\mu_0)\cos m\varphi, \qquad (1.65)$$

where $I^m(\tau,\mu,\mu_0)$ and $B^m(\tau,\mu,\mu_0)$ for $m = 0,\ldots,N$ are certain unknown functions. Substitute expansions (1.64) and (1.65) to expression for the source function (1.50), compute the integral over the azimuth and level items with the equal numbers m in the left-hand and right-hand parts of the equation. Only the items with equal numbers m will be nonzero in the product of the series (the phase function by the radiance) due to the orthogonality of the trigonometric functions:

$$\int_0^{2\pi} \cos m_1\varphi' \cos m_2\varphi' d\varphi' = 0 \quad \text{for} \quad m_1 \neq m_2 \,,$$

$$\pi \quad\quad\quad\quad\quad\quad\quad\quad\quad\quad\quad\quad \text{for} \quad m_1 = m_2 \,,$$

$$\int_0^{2\pi} \sin m_1\varphi' \cos m_2\varphi' d\varphi' = 0 \,.$$

Finally, obtain:

$$B^m(\tau, \mu, \mu_0) = \frac{\omega_0(\tau)}{2} \int_{-1}^{1} p^m(\tau, \mu, \mu')I^m(\tau, \mu, \mu')d\mu'$$

$$+ \frac{\omega_0(\tau)}{4} Sp^m(\tau, \mu, \mu_0) \exp(-\tau/\mu_0) \,. \tag{1.66}$$

Further from (1.51) the following equation is derived:

$$\mu \frac{dI^m(\tau, \mu, \mu_0)}{d\tau} = -I^m(\tau, \mu, \mu_0) + B^m(\tau, \mu, \mu_0) \,, \tag{1.67}$$

with boundary conditions:

$$I^m(0, \mu, \mu_0) = 0 \,, \quad \text{for} \quad \mu_0 > 0 \quad \text{and}$$

$$I^m(\tau_0, \mu, \mu_0) = 0 \quad \text{for} \quad \mu_0 < 0 \,. \tag{1.68}$$

The following relations are correct for it:

$$I^m(\tau, \mu, \mu_0) = \frac{1}{\mu} \int_0^{\tau} B^m(\tau', \mu, \mu_0) \exp\left(-\frac{\tau - \tau'}{\mu}\right) d\tau' \quad \mu > 0 \,,$$

$$I^m(\tau, \mu, \mu_0) = -\frac{1}{\mu} \int_{\tau}^{\tau_0} B^m(\tau', \mu, \mu_0) \exp\left(-\frac{\tau - \tau'}{\mu}\right) d\tau' \quad \mu < 0 \,. \tag{1.69}$$

The substitution of (1.69) to (1.66) yields the integral equation again for the source function:

$$B^m(\tau, \mu, \mu_0) = \frac{\omega_0(\tau)}{2} \int\limits_0^1 p^m(\tau, \mu, \mu') \frac{d\mu'}{\mu'} \int\limits_0^\tau B^m(\tau', \mu', \mu_0) \exp\left(-\frac{\tau - \tau'}{\mu'}\right) d\tau'$$

$$- \frac{\omega_0(\tau)}{2} \int\limits_{-1}^0 p^m(\tau, \mu, \mu') \frac{d\mu'}{\mu'} \int\limits_\tau^{\tau_0} B^m(\tau', \mu', \mu_0) \exp\left(-\frac{\tau - \tau'}{\mu'}\right) d\tau'$$

$$+ \frac{\omega_0(\tau)}{4} Sp^m(\tau, \mu, \mu_0) \exp(-\tau/\mu_0) .$$

$$(1.70)$$

Thus, passing to the phase function expansion over Legendre Polynomials (1.60) and (1.64) allows obtaining (1.66)–(1.70), where the azimuthal dependence of the functions is absent, that certainly simplifies the analysis and solution. Besides expansions of the radiance and the source function (1.65) are called expansions over *the azimuthal harmonics* and the method is called *a method of the azimuthal harmonics*.

1.4
Reflection of the Radiation from the Underlying Surface

The ratio of the irradiances reflected from the surface to the irradiances incoming to the surface is called *an albedo of the surface* and it is one of the most important characteristics of the underlying surface:

$$A = \frac{F^\uparrow(\tau_0)}{F^\downarrow(\tau_0)} .$$

$$(1.71)$$

This characteristic has a clear physical meaning – it corresponds to the part of the incoming radiation energy reflected back to the atmosphere. Actually, if value $A = 0$ then the surface absorbs all radiation (the absolutely black surface), if value $A = 1$ then, otherwise, the surface absorbs nothing and reflects all radiation (the absolutely white surface). Generalizing the notion of the albedo, we are introducing *the albedo of the system of atmosphere plus surface*, specifying it at arbitrary level τ:

$$A(\tau) = \frac{F^\uparrow(\tau)}{F^\downarrow(\tau)} .$$

$$(1.72)$$

Remember that here and below we are considering values defining the single wavelength, i. e. the spectral characteristics of the radiation field and surface. The integral albedo that is called just "albedo" for briefness (do not confuse it with *the spectral albedo*) is of great importance in atmospheric energetics.[10]

---

[10] It is necessary to point out that the albedo (like other reflection characteristics) is formally defined only for the surface without the atmosphere. In transfer theory, they are often called "true". Taking

The albedo of the surface characterizes the reflection process of radiation only as a description of the energy transformation, but doesn't tell us about the dependence of the radiance upon the reflection angle and azimuth. If the surface were an ideal plane, such dependence would be defined with the well-known laws of reflection and refraction (Sivukhin 1980). However, all natural surfaces are rough, i.e. they have different scales of the roughness and even the water surface is practically always not smooth. Therefore, considering the incoming parallel beam is more complicated in reality, notwithstanding the reflection from every micro-roughness is ordered to the classical laws of geometric optics. In particular, reflected radiation extends to all possible directions and not only to the direction according to the law: "the reflection angle is equal to the incident angle". This light reflection from natural surfaces is usually called *the diffused*.

It is possible to select three main types of diffused reflection. *The orthotropic* (or *isotropic*) *reflection*, when the diffused reflected radiance does not depend on the direction. *The mirror reflection*, when the maximum of the reflected radiance coincides with the direction of the mirror reflection (the reflection angle is equal to the incident angle) and *the backward reflection* when the maximum is situated along the direction opposite to the incident radiation direction. The mirror reflection evidently characterizes the surfaces close to the ideally smooth surface and otherwise the backward one characterizes the surfaces close to the strongly rough surface because it is formed by a large amount of the micro-grounds oriented perpendicular to the incident direction of radiation. The observations some of them we will consider in detail in Chap. 3 indicate that the cloud and snow are the closest to the orthotropic surface, the water is the most mirror surface and others are mainly backward reflected surfaces. However, the reference to the observation is excessive because of the mirror reflection of the banks from the water that everybody has seen and the backward reflection maximum is clearly observed from the airplane board.

The orthotropic surfaces are especially convenient for the theoretical analysis and practical calculations because they are characterized with only one parameter – the albedo and because of the simplicity of the mathematical description. We would like to point out that the assumption concerning the orthotropic reflection is an approximation and its accuracy is necessary to evaluate in a concrete problem. It is said that the anisotropic reflection from other surfaces needs some additional values for its description. The rather variable characteristics of the anisotropic reflection are considered in different studies, however here we are describing the general problem without its concretization. Note also that the reflection processes depend on the incident radiation polarization accompanying its change (Sivukhin 1980). Therefore, the consideration of the reflection without an account of polarization is an

---

into account the atmosphere, the other characteristics of the system "atmosphere plus surface" are analogously defined. For example, the incoming irradiance to the surface from the diffusing atmosphere depends itself on the surface albedo (true) because it contains the part of the reflected radiation that scattered back to the surface. Mark that on the one hand, only true values are used in formulas of the transfer theory for the reflection characteristics, and on the other hand, only characteristics of the system "atmosphere plus surface" are available for the observation.

approximation. Considering different orientations of the micro-roughness of the natural surfaces it is possible to assert that as reflection is closer to the orthotropic, then reflected radiation is less polarized. The homogeneous distribution of reflected radiation over directions is corresponded to the fully chaotic orientation of the micro-reflectors that causes the chaotic distribution of the polarization ellipses, i.e. the unpolarized light. Thus, the orthotropic reflection means also the absence of the dependence upon the polarization. Otherwise, when the anisotropy is stronger the dependence is clearer. The water surface is the most anisotropic surface, therefore, in this case the question about the exactness of the approximation of unpolarized radiation needs special study.

The function $R(\mu, \varphi, \mu', \varphi')$, defined from the relation between the radiances incoming on the surface $I(\tau_0, \mu, \zeta, \varphi')$, $(\mu' > 0)$ and reflected from the surface $I(\tau_0, \mu, \zeta, \varphi)$, $(\mu < 0)$, characterizes the radiation reflection from the surface:

$$I(\tau_0, \mu, \mu_0, \varphi) = \frac{1}{\pi} \int\limits_0^{2\pi} d\varphi' \int\limits_0^1 R(\mu, \varphi, \mu', \varphi') I(\tau_0, \mu', \mu_0, \varphi') \mu' d\mu' . \qquad (1.73)$$

It is easy to test that for the orthotropic surface (1.71) and (1.73) yield the equality $R(\mu, \varphi, \mu', \varphi') = A$ and just it defines the existence of the factor μ'/π for normalizing in (1.73). Equation (1.73) in the operator form is written as:

$$\mathbf{I}^\uparrow = \mathbf{R} \mathbf{I}^\downarrow , \qquad (1.74)$$

where: $\mathbf{I}^\uparrow = I(\tau_0, \mu, \mu_0, \varphi)$ is the reflected radiance, $\mathbf{I}^\downarrow = I(\tau_0, \mu, \mu_0, \varphi)$ is the incoming radiance, and $r = \frac{\mu'}{\pi} R(\mu, \varphi, \mu', \varphi')$ is the operator of the reflection from the surface.

The necessity of accounting the reflection from the surface in the radiative transfer theory is based on the evident assumption that the reflection is equal to *the illumination of the atmosphere from the bottom* (i.e. from the bottom boundary of the atmosphere $\tau = \tau_0$). Thus, it is enough to solve the radiative transfer problems for diffused radiation in the atmosphere first with the illumination from the top and then with the illumination from the bottom, and after all, it is necessary to add both results.

Introduce the following notation system:

1. the values related to the system "atmosphere plus surface" are specified with the upper line;

2. the values related to the atmosphere illuminated from the bottom without surface are specified with the symbol $\sim$;

3. the values related to the atmosphere illuminated from the top without surface are specified without special marks.

Then the solution of the radiative transfer problem, written in the operator form (1.57), will be the following: $\mathbf{I} = \mathbf{T} \mathbf{I}_0$ where $\mathbf{I}_0$ is the radiance incoming to the top of the atmosphere. Introduce operator $\mathbf{T}^{\downarrow}$, so that $\mathbf{I}^{\downarrow} = \mathbf{T}^{\downarrow}\mathbf{I}_0$ specifying that operator $\mathbf{T}^{\downarrow}$ is to describe the transfer of both diffused and *direct radiation* throughout the atmosphere. The latter has been excluded from the radiative transfer equation and it must be taken into account when we are considering the reflection from the surface. The solution of the radiative transfer problem with the illumination from the bottom is $\tilde{\mathbf{I}} = \tilde{\mathbf{T}}\tilde{\mathbf{I}}_1$ where $\tilde{\mathbf{I}}_1$ describes the radiation field coming from the bottom to the lower boundary, which is taken into account according to (1.58). Besides, operator $\tilde{\mathbf{T}}$ has to describe also the transfer of direct reflected radiation (i.e. the radiation transferring from the surface without scattering). Operator $\tilde{\mathbf{T}}^{\uparrow}$ describes the radiance incoming from above to the lower boundary illuminated from the bottom with radiance $\tilde{\mathbf{I}}_1$ so that $\tilde{\mathbf{I}}^{\uparrow} = \tilde{\mathbf{T}}^{\uparrow}\tilde{\mathbf{I}}_1$. Value $\tilde{\mathbf{I}}^{\uparrow}$ means the radiance reflected from the surface then scattered to the atmosphere and after all returned to the surface. Mathematically, the problem of constructing all operators $\mathbf{T}, \mathbf{T}^{\downarrow}, \tilde{\mathbf{T}}, \tilde{\mathbf{T}}^{\uparrow}$ is uniform as it follows from the previous section.

The radiance with a subject to the surface reflection is evidently obtained as a sum of the following components. Firstly, it is the radiance of direct solar radiation diffused to the atmosphere $\mathbf{T}\mathbf{I}_0$. Secondly, it is the radiance of direct and diffused radiation reflected from the surface $\tilde{\mathbf{T}}\tilde{\mathbf{I}}_1$ that is the combination $\tilde{\mathbf{T}}\mathbf{R}\mathbf{T}^{\downarrow}\mathbf{I}_0$ with taking into account (1.74). Further, it follows a subject to secondary reflected radiation $\tilde{\mathbf{T}}\tilde{\mathbf{I}}_2 = \tilde{\mathbf{T}}(r\tilde{\mathbf{T}}^{\uparrow}\tilde{\mathbf{I}}_1) = \tilde{\mathbf{T}}(r\tilde{\mathbf{T}}^{\uparrow}\mathbf{R}\mathbf{T}^{\downarrow}\mathbf{I}_0)$, etc. Finally, for the desired radiance calculation we are obtaining the following:

$$2\tilde{\mathbf{I}} = \mathbf{T}\mathbf{I}_0 + \tilde{\mathbf{T}}(1 + \mathbf{R}\tilde{\mathbf{T}}^{\uparrow} + (\mathbf{R}\tilde{\mathbf{T}}^{\uparrow})^2 + (\mathbf{R}\tilde{\mathbf{T}}^{\uparrow})^3 + \dots)\mathbf{R}\mathbf{T}^{\downarrow}\mathbf{I}_0 \ . \qquad (1.75)$$

Expression (1.75) is known as a radiance expansion over the reflection order. It is widely used in the algorithms of the numerical methods where it allows organizing the recurrent calculations of the desired radiance. Note that the series converges faster if the reflection is weaker. The operator approach is presented in particular in the books by Hulst (1980) and Lenoble (1985).

Consider a particular problem concerned with radiative transfer and reflection from the surface. Let us consider only the radiance at the boundaries $\tilde{I}(0, \mu, \mu_0, \varphi)$ ($\mu < 0$) and $\tilde{I}(\tau_0, \mu, \mu_0, \varphi)$ ($\mu > 0$) without consideration of it between the boundaries. The obvious examples are the problems of the interpretation of the satellite and ground-based observations of the diffused solar radiance. In these problems, the viewing angles are assumed to be in the range $[0, \pi/2]$, i.e. the value of μ is assumed positive. Then the desired values of the radiance are written as $\tilde{I}(0, -\mu, \mu_0, \varphi)$ and $\tilde{I}(\tau_0, \mu, \mu_0, \varphi)$ according to transfer geometry (In any case it is $\mu_0 > 0$).

Specify *the reflection and transmission functions* in accordance with Sobolev (1972) are shown as

$$I(0, -\mu, \mu_0, \varphi) = S\mu_0\varrho(\mu, \mu_0, \varphi) \ , \quad I(\tau_0, \mu, \mu_0, \varphi) = S\mu_0\sigma(\mu, \mu_0, \varphi) \ , \quad (1.76)$$

where the reflection from the surface is not taken into account. Specify the analogous function for the case of illumination from the bottom:

$$\tilde{I}(0, -\mu, \mu', \varphi) = \check{S}\mu'\tilde{\varrho}(\mu, \mu', \varphi) \ , \quad \tilde{I}(\tau_0, \mu, \mu', \varphi) = \check{S}\mu'\tilde{\sigma}(\mu, \mu', \varphi) \ , \quad (1.77)$$

Define these functions to the direction of the incoming irradiance πS from the bottom $(\mu', 0)$ and assume this back-standing geometry completely similar to the case of the illumination from the top then $\mu' > 0$ and consider τ_0 in this case as a top of the atmosphere.

The symmetry relations are the most important property of the reflection and transmission functions:

$$
\begin{aligned}
\varrho(\mu, \mu_0, \varphi) &= \varrho(\mu_0, \mu, \varphi) \,, \\
\tilde{\varrho}(\mu, \mu_0, \varphi) &= \tilde{\varrho}(\mu_0, \mu, \varphi) \,, \\
\sigma(\mu, \mu_0, \varphi) &= \tilde{\sigma}(\mu_0, \mu, \varphi) \,.
\end{aligned}
\qquad (1.78)
$$

In general case, the proof of (1.78) is complicated and presented e.g. in the book by Yanovitskij (1997). Specify the analogous functions for the system "atmosphere plus surface" $\bar{\varrho}(\mu, \mu_0, \varphi)$ and $\bar{\sigma}(\mu, \mu_0, \varphi)$.

It is possible to exclude the azimuthal dependence of the reflection and transmission functions presenting them as expansions over the azimuthal harmonics as follows:

$$
\varrho(\mu, \mu_0, \varphi) = \varrho^0(\mu, \mu_0) + 2 \sum_{m=1}^{N} \varrho^m(\mu, \mu_0) \cos m\varphi \,,
\qquad (1.79)
$$

and the analogous expressions for the functions $\sigma(\mu, \mu_0, \varphi)$, $\bar{\varrho}(\mu, \mu_0, \varphi)$ etc. Every harmonic satisfies relations of the symmetry $(\varrho^m(\mu, \mu_0,) = \varrho^m(\mu_0, \mu)$ etc.).

Now consider the simplest but widespread case of the orthotropic surface with albedo A. It is easy to demonstrate (Sobolev 1972) that the consideration of the only zeroth harmonics for the isotropic reflection is enough. Actually, if non-zeroth harmonics (1.79) varied it would mean the azimuthal dependence of the reflected radiation as per definitions (1.76)–(1.77) that contradicts the assumption about the orthotropness of the reflection.

Write the explicit form of the integral operators from (1.75). According to the definition of $\mathbf{T}$ operator (1.58) and to the expression for extraterrestrial radiance I_0 (1.41) we are getting the following:

$$
\mathbf{T}I_0 = \int\limits_0^{2\pi} d\varphi' \int\limits_0^1 d\eta' \, T(\mu, \mu', \varphi, \varphi') \pi S \delta(\mu' - \mu_0) \delta(\varphi' - 0) = \pi S T(\mu, \mu_0, \varphi, 0)
$$

comparing it with (1.76) and taking into account only the zeroth harmonics the following is inferred:

$$
T(\mu, \mu', \varphi, \varphi') = \frac{\mu_0}{\pi} \varrho^0(\mu, \mu_0) \quad \text{for the top of the atmosphere}
$$

$$
T(\mu, \mu', \varphi, \varphi') = \frac{\mu_0}{\pi} \sigma^0(\mu, \mu_0) \quad \text{for the bottom of the atmosphere}
$$
$$
(\mu' \equiv \mu_0 \quad \text{and} \quad \varphi' \equiv 0) \,.
$$

Direct radiation is necessary to take into consideration also for the description of the reflection because:

$$T^{\downarrow}(\mu, \mu', \varphi, \varphi') = \frac{\mu_0}{\pi}[\sigma^0(\mu, \mu_0) + \exp(-\tau_0/\mu_0)] \ .$$

For the case of the illumination from the bottom to direction $(\mu'\varphi')$ the analogous expressions are obviously deriving. Further, according to the definition of the operator (1.58) and with a subject to equality $\tilde{I}_0(\mu', \varphi') = \tilde{S}$ (due to the orthotropy of the reflection, the link between the radiance and irradiance (1.4) and equality of the irradiances in definitions (1.77)) we finally obtain for the bottom of the atmosphere:

$$\tilde{T}^{\uparrow}(\mu, \mu', \varphi, \varphi') = 2\int\limits_0^1 \mu'\bar{\varrho}^0(\mu', \mu)d\mu' \ ,$$

i. e. the $\tilde{T}^{\uparrow}$ depends only on μ.

The analogous expression is obtained for the top of the atmosphere

$$\tilde{T}(\mu, \mu', \varphi, \varphi') = 2\int\limits_0^1 [\bar{\sigma}^0(\mu, \mu') + \exp(-\tau_0/\mu)]\mu'd\mu' \ ,$$

where direct radiation and condition $\tilde{T}(\mu, \mu', \varphi, \varphi') = \tilde{T}^{\uparrow}(\mu, \mu', \varphi, \varphi')$ are taken into account.

The product of the integral operators is found by definitions (1.58) and (1.73)

$$R\tilde{T}^{\uparrow}(\mu, \mu', \varphi, \varphi') = \frac{1}{\pi}\int\limits_0^{2\pi} d\varphi'' \int\limits_0^1 R(\mu, \varphi, \mu'', \varphi'')\tilde{T}^{\uparrow}(\mu'', \mu', \varphi'', \varphi')\mu''d\mu'' \ ,$$

that yields after substituting the above-obtained expressions, in particular $R = A$, constant AC for $R\tilde{T}^{\uparrow}$, where

$$C = 4\int\limits_0^1\int\limits_0^1 \bar{\varrho}^0(\mu', \mu'')\mu'\mu''d\mu'd\mu'' \ . \tag{1.80}$$

Absolutely analogously $RT^{\downarrow}$ is found as $A\mu_0\phi(\mu_0)$, where

$$\phi(\mu_0) = 2\int\limits_0^1 \sigma^0(\mu', \mu_0)\mu'd\mu' + \exp(-\tau_0/\mu_0) \ .$$

As $R\tilde{T}^{\downarrow}$ is a constant relatively to the variable of the integration the following is deduced with a subject to symmetry relations (1.78):

$$\tilde{T}RT^{\downarrow} = A\mu_0\phi(\mu)\phi(\mu_0) \quad \text{for the top of the atmosphere ,}$$

$$\tilde{T}RT^{\downarrow} = A\mu_0\psi(\mu)\phi(\mu_0) \quad \text{for the bottom of the atmosphere ,}$$

where

$$\psi(\mu) = 2\int_0^1 \tilde{\varrho}^0(\mu,\mu')\mu'd\mu' . \tag{1.81}$$

After substituting the obtained expressions to series (1.75), summarizing the geometric progression, accounting definitions (1.76), and expansions (1.79) we are getting the known (Sobolev 1972) relations:

$$\tilde{\varrho}(\mu,\mu_0,\varphi) = \varrho(\mu,\mu_0,\varphi) + \frac{A\phi(\mu)\phi(\mu_0)}{1-AC} ,$$
$$\tilde{\sigma}(\mu,\mu_0,\varphi) = \sigma(\mu,\mu_0,\varphi) + \frac{A\psi(\mu)\phi(\mu_0)}{1-AC} . \tag{1.82}$$

Relations (1.82) together with (1.80)–(1.81) are expressing the simple links of the reflection and transmission functions in the case of the system "atmosphere plus surface" with the same functions without reflection from the surface. That is to say, a subject to the surface actually is not a complicated problem. Mark that the symmetry of the reflection function is also correct with the reflection from the surface $\tilde{\varrho}(\mu,\mu_0,\varphi) = \varrho(\mu_0,\mu,\varphi)$.

1.5
Cloud impact on the Radiative Transfer

Clouds are the most variable component of the climatic system and they play a key role in atmospheric energetics. The actual elaborations in the field of the numerical climate simulation are the following: the creation of the high-resolution models for the limited area scales which are suitable for the parameterization of cloud dynamics in the climate simulation; the evaluation of the yields of the cloud radiative forcing and microphysical processes (with taking into account the atmospheric aerosols) to the formation of the properties and structure of the cloud cover; the algorithms improvement for the cloud characteristics retrieval from the remote sounding data. The methods of the calculation of the characteristics of solar radiation (the semispherical irradiances, radiances and absorption) and derivation of the cloud optical characteristics from the radiation observational data will be considered below to solve the problems mentioned above.

The process of the radiative transfer within clouds is also described with radiative transfer equation (1.35), but the multiple scattering plays the main role unlike the clear atmosphere. Only the horizontally homogeneous atmosphere will be considered below. Applying to the cloudy atmosphere, it means the considering of the model of the infinitely extended and horizontally homogeneous cloud layer. In reality, it is the stratus cloudiness, which corresponds best of all to this case. Fix on the properties of the stratus clouds, which allow applying the considered theoretical methodologies to the real cloudiness.

Remember the classification of the stratiform clouds: the stratiform clouds of the lower level are Stratus (St), Stratus-cumulus (Sc), Nimbus stratus (Ns); the stratiform clouds of the medium level are Alto-stratus (As), Alto-cumulus (Ac); the high stratiform clouds are Cirrus-stratus (Cs); and also the Frontal cloud systems are Ns-As, As-Cs, Ns-As-Cs (Feigelson 1981; Matveev et al. 1986; Marchuk et al. 1986; Mazin and Khrgian 1989). The extended stratus clouds are important in the feedback chain of the climatic system influencing essentially the albedo, radiative balance of the "atmosphere plus surface", and total circulation of the atmosphere (Marchuk et al. 1986; Marchuk and Kondratyev 1988). The stratus clouds widening over vast regions impact on the Earth radiative balance not only in the regional but also in the global scale.

The cloud albedo is significantly higher than the ground or ocean albedo without snow cover. Basing on this and assuming that clouds prevent the heating of the surface and lower atmospheric layer in the low and middle latitudes, a negative yield to the Earth radiative balance is usually concluded. The clouds in the high latitudes don't increase the light reflection because the snow albedo is also high and, in this case, the clouds play a prevailing role in atmospheric heating.

However, it has been elucidated in the last few decades that the situation is more complicated: the clouds themselves absorb a certain part of incoming radiation providing the atmospheric heating in all latitudes. Thus, the problem of the interaction between the clouds and radiation comes to the foreground in the stratus clouds study. The climate simulation requires inputting the adequate optical models of the clouds and so it is necessary to obtain the real cloud optical parameters (volume scattering and absorption coefficients). The including of atmospheric aerosols in the processes of the interaction between *short-wave radiation (SWR)* and clouds affect equivocally the forming of the heat regime of the atmosphere and surface. *The direct* and *indirect* aerosol heating effects are indicated in the literature (Hobbs 1993; Charlson and Heitzenberg 1995; Twohy et al. 1995). The radiation absorption by the carbonaceous and silicate aerosols causes the direct effect. The indirect effect is attributed to the hydrophilic atmospheric aerosols necessary to the water vapor condensation and to the generating of the cloud droplets. Hence, the high concentration of these aerosols increases the droplets number and optical thickness of the cloud that, in turn, intensifies the reflection of solar radiation and reduces the radiation absorption in the atmosphere and on the surface. From the results of the airborne radiative observations of the last few decades, it has been revealed that direct and indirect effects of the aerosols influence differently the increasing or extinction of the solar radiation absorption in clouds of different origin in different geographical regions (Hobbs 1993; Harries 1996).

Table 1.1. The probability (%) of the conservation of the cloudiness with the cloud amount equal to 1 above the European territory of the former USSR

Probability, %	Duration of the existence of 1-amount cloudiness (h)				
	1	3	6	12	24
Winter	93	87	83	78	74
Summer	80	64	52	41	35

Table 1.2. The average altitudes of cloud z_b (bottom) and z_t (top) and geometrical thickness $\Delta H = z_t - z_b$ (km)

Type of cloudiness	z_b		z_t		ΔH	
	Winter	Summer	Winter	Summer	Winter	Summer
St	0.25	0.29	0.55	0.58	0.30	0.29
Sc	0.85	1.26	1.14	1.59	0.29	0.33
As	3.80	3.93	4.73	4.83	0.93	0.90

The detailed analysis accomplished in Harries (1996) with employing the previous results of the observations concerning the greenhouse effect in the global climate change has demonstrated an unfeasibility to evaluate accurately the influence of the clouds and aerosols radiative forcing on the global warming.

The data in the books by Feigelson (1981), Matveev (1984) and Matveev et al. (1986) obtained as a generalization of the airborne and satellite measurements and observations of the meteorological stations network illustrate the recurrence of the stratiform clouds and the average period of their conservation above Europe that is equal to 13–15 h in winter and about 5 h in summer time. The probability of the conservation of the cloud amounts equal to 1 above the European territory of Russia (over 10 stations) during different time intervals is presented in Table 1.1 according to Mazin and Khrgian (1989).

As has been mentioned in the book by Feigelson (1981) it is necessary to understand that the obtained cloud characteristics and parameters relate to the very given cloud in the given time period because of the strong variability of clouds. Nevertheless, the certain recurrences of some parameters of the stratus clouds for some geographical regions are marked. Thus, for example, the prevailing altitude of the stratus clouds in the polar and temperate zones is about 2 km and in the tropical zone is about 3 km. After the data averaging of the airborne and balloon observations the most typical values of the stratus cloud top and bottom altitudes have been obtained that Table 1.2 illustrates.

The stratus clouds, which are not farther than 200 km from the boundary of an atmospheric front when it is forming and passing the atmosphere, are called *frontal clouds*. The width extension of the frontal zone in central Europe could reach 1000 km according to the satellite observations (Marchuk et al. 1986,1988). The length of this zone is about 7000 km. The cloud zones are broken to the macro cells of a hundred kilometers in length, which, in turn, consist

Table 1.3. Recurrence (%) of the extension of the frontal upper cloudiness above the European territory of the former USSR

Width of zone (km)	Type of front Cold	Type of front Warm	Width of zone (km)	Type of front Cold	Type of front Warm
< 50	2.4	–	501– 600	–	25.0
51–100	12.2	–	601– 700	–	20.8
101–200	26.8	–	701– 800	–	11.7
201–300	29.3	6.5	801– 900	–	6.2
301–400	22.0	10.4	901–1000	–	2.6
401–500	7.3	12.9	1001–1500	–	2.6

Table 1.4. Recurrence (%) of the cloud areas with different extension of the frontal lower cloudiness

Extension (km)	Type of front Cold	Type of front Warm
<10	20	14
10– 20	28	19
20– 30	19	19
30– 50	18	21
50– 75	9	11
75–100	4	8
100–150	0	3
150–200	1	4
200–300	1	1

of the cloud strips or of the solid fields with the cloud cells inhomogeneous in the density of tens of kilometers in length (Feigelson 1981; Matveev 1984; Matveev et al. 1986).

The information about the recurrence of the frontal clouds zone width of the upper and lower levels presented in books (Feigelson 1981; Matveev 1984; Matveev et al. 1986) from the airborne observations is illustrated in Table 1.3. The main conclusion of Table 1.3 is that the high-level cloud fields of the length less than 200 km are typical for the cold front, and of the length 500–600 km are typical for the warm front.

Table 1.4 indicates that the most frequent frontal lower level clouds have a horizontal extension not exceeding 50 km for the cold front and 75 km for the warm front. Thus, it follows from the above that it is possible to simulate the stratus cloudiness with the homogeneous horizontally extended layer. Besides, the stratus cloudiness is rather stable hence, the methodology for the retrieval of cloud optical parameters based on the ground observations of the solar irradiance for different solar incident angles (i. e. different moments with the intervals 1–2 h) described below, could be applied to real data.

References

Bass AM, Paur RJ (1984) The ultraviolet cross section of ozone. In: Zerofs CS, Chazi AP (eds) The measurements of atmospheric ozone. Reidel, Dordrecht, pp 606–610

Bohren CF, Huffman RK (1983) Absorption and Scattering of Light by Small Particles. Wiley, New York

Charlson RJ, Heitzenberg J (eds) (1995) Aerosol Forcing of Climate. Wiley, New York

Dave JV (1970) Coefficients of the Legendre and Fourier series for the scattering function of spherical particles. Appl Opt 9:1888–1896

Deirmendjian D (1969) Electromagnetic Scattering on Spherical Polydispersions. Elsevier, Amsterdam

Feigelson EM (ed) (1981) Radiation in the cloudy atmosphere. Gidrometeoizdat, Leningrad (in Russian)

Goody RM, Yung UL (1996) Atmospheric Radiation: Theoretical Basis. Oxford University Press, New York

Harries JE (1996) The greenhouse Earth: A view from space. Quart J R Meteor Soc 122(Part B,532):799–818

Henyey L, Greenstein J (1941) Diffuse radiation in Galaxy. Astrophys J 93:70–83

Hobbs PV (ed) (1993) Aerosol-Cloud-Climate Interactions. Academic Press, New York

Hulst HC Van de (1957) Light scattering by small particles. Willey, New York

Hulst HC Van de (1980) Multiple Light Scattering. Tables, Formulas and Applications. Vol. 1 and 2. Academic Press, New York

Kolmogorov AN, Fomin SV (1999) Elements of the theory functions and the functional analysis. Dover Publications

Kondratyev KYa (1965) The actinometry. Gydrometeoizdat, Leningrad (in Russian)

Korn GA, Korn TM (2000) Mathematical handbook for scientists and engineers: definitions, theorems, and formulas for reference and review, 2nd edn. Dower, New York

Lenoble J (ed) (1985) Radiative transfer in scattering and absorbing atmospheres: standard computational procedures. DEEPAK Publishing, Hampton, Virginia

Makarova EA, Kharitonov AV, Kazachevskaya TV (1991) Solar irradiance. Nauka, Moscow (in Russian)

Marchuk GI, Mikhailov GA, Nazaraliev NA et al. (1980) The Monte-Carlo method in the atmospheric optics. Springer-Verlag, New York

Marchuk GI, Kondratyev KYa (1992) Priorities of the global ecology. Nauka, Moscow (in Russian)

Marchuk GI, Kondratyev KYa, Kozoderov VV, Khvorostyanov VI (1986) Clouds and climate. Gidrometeoizdat, Leningrad (in Russian)

Marchuk GI, Kondratyev KYa, Kozoderov VV (1988) Radiation balance of the Earth: Key aspect. Nauka, Moscow (in Russian)

Matveev LT, Matveev YuL, Soldatenko SA (1986) The global cloud field. Gidrometeoizdat, Leningrad (in Russian)

Matveev VS (1972) The approximate presentation of the absorption coefficient and equivalent linewidth with the Voigt function (Bilingual). J Appl Spectrosc 16:228–233

Matveev YuL (1984) Physics-statistical analysis of the global cloud field (Bilingual). Izv RAS Atmosphere and Ocean Physics 20:1205–1218

Minin IN (1988) The theory of radiation transfer in the planets atmospheres. Nauka, Moscow (in Russian)

Mazin IP, Khrgian AKH (eds) (1989) Clouds and cloudy atmosphere. Reference book. Gidrometeoizdat, Leningrad (in Russian)

Penner SS (1959) Quantitative molecular spectroscopy and gas emissivities. Addison-Wesley, Massachusetts

Rothman LS, Gamache RR, Tipping RH et al. (1992) The HITRAN molecular database: 1991 and 1992 edn. J Quant Spectr Rad Trans 48:469–597

Rozanov VV, Timofeyev YuM, Barrows G (1995) The informative content of measurements of outgoing UV, visual and near IR solar radiation. (Instruments of GOME) (Bilingual). Earth Observations and Remote Sensing 6:29–39

Sedunov YuS, Avdyushin SI, Borisenkov EP, Volkovitskiy OA Petrov NN, Reintbakh RG, Smirnov VI, Chernov AA (eds) (1991) The Atmosphere. Reference book. Gydrometeoizdat, Leningrad (in Russian)

Sivukhin DV (1980) The introductory survey of physics. Nauka, Moscow (in Russian)

Smirnov VI (1974) Course of the high mathematics. Vol. 3, part 2. Nauka, Moscow (in Russian)

Sobolev VV (1972) The light scattering in the planets atmospheres. Nauka, Moscow (in Russian)

Stepanov NN (1948) The spherical trigonometry. Gostehizdat, Moscow-Leningrad (in Russian)

Twohy CH, Durkee PA, Huebert BJ, Charlson RJ (1995) Effects of aerosol particles on the microphysics of coastal stratiform clouds. J Climate 8:773–783

Vasilyev AV (1996) Multipurpose algorithm for calculation of optical characteristics of homogeneous spherical aerosol particles I. Single particle. Bulletin of the St. Petersburg University, Series 4: Physics, Chemistry 4(25):3–11 (in Russian)

Vasilyev AV (1997) Multipurpose algorithm for calculation of optical characteristics of homogeneous spherical aerosol particles II. Ensemble of particles. Bulletin of the St. Petersburg University, Series 4: Physics, Chemistry 1:14–24 (in Russian)

Vasilyev AV, Ivlev LS (1996) Multipurpose algorithm for calculation of optical characteristics of two-layer spherical particles with the homogeneous core and shell. Atmosphere and Ocean Optics 9:1552–1561 (Bilingual)

Vasilyev AV, Ivlev LS (1997) Empirical models and optical characteristics of aerosol ensembles of two-layer spherical particles. Atmosphere and Ocean Optics 10:856–865 (Bilingual)

Vasilyev AV, Rozanov VV, Timofeyev YuM (1998) The analysis of the informative content of measurements of outgoing reflected and diffused solar radiation in the spectral ranges 240–700 nm. Earth Observations and Remote Sensing (2):51–58 (Bilingual)

Vasilyev OB, Vasilyev AV (1994) Two-parametric model of the phase function. Atmosphere and Ocean Optics 7:76–89 (Bilingual)

Yanovitskij EG (1997) Light scattering in inhomogeneous atmospheres. Springer, Berlin Heidelberg New York

Zuev VE, Belov VV, Veretennikov VV (1997) The theory of linear systems in the optics of diffuse medium. Printing House of Siberian Filial RAS, Tomsk (in Russian)

CHAPTER 2

Theoretical Base of Solar Irradiance and Radiance Calculation in the Earth Atmosphere

In this chapter, we are considering two concrete calculation methods for the solar radiances and irradiances needed for observational data interpretations. They are the Monte-Carlo method for clear atmosphere and the asymptotic method for overcast sky. The authors are not presenting or even reviewing all numerical and analytical methods of the radiative transfer theory, referring readers to the book by Lenoble (1985). The reasons of our choice among the diversity of methods will be explained during the consideration below.

2.1
Monte-Carlo Method for Solar Irradiance and Radiance Calculation

The Monte-Carlo method (the more strict name is the method of statistical modeling) is a most powerful method of radiative transfer theory. It allows us to solve the problems concerned with the radiance calculation while taking into account spherical geometry, polarization, heterogeneity of the atmosphere and surface, etc. (Marchuk et al. 1980). Here we are applying this method for solving the rather simple (compared with the above-mentioned) problem of the solar radiance and irradiance calculation in the horizontally homogeneous and plane parallel atmosphere.

There are two approaches to describe the Monte-Carlo method application in atmospheric optics (Kargin 1984; Marchuk et al. 1980). The first one describes it as a method of formal mathematical solving the problems (equations) of radiative transfer and the second one considers it as a method of simulation of the physical processes of radiative transfer in the atmosphere, when it is not needed to attract a body of the transfer theory. Concerning our problem the second way is easier for understanding so we are following it concurring with the author of book (Kargin 1984) and then we will show that the elaborated algorithm corresponds to the mathematical state of the problem considered in Sect. 1.3.

Unlike other approaches in the Monte-Carlo method, it is appropriate not to divide radiation to the direct, diffused and reflected from the surface. In addition, we are not considering the optical thickness τ, but the altitude z as a vertical coordinate. The reasons for this can be seen in Chap. 5.

Thus, we are solving the problem for the atmospheric model shown in Fig. 2.1. Let the parallel solar flux $F_0 = \pi S$ income from direction $(\mu_0, 0)$ to

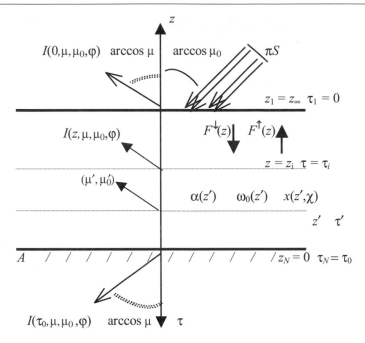

Fig. 2.1. Model of the atmosphere

the top of the plane parallel horizontally homogeneous atmosphere z_∞. The base of the atmosphere $z = 0$ is the orthotropic surface with albedo A (note that the condition of the orthotropic reflection is not essential for the Monte-Carlo method. The anisotropic reflection model will be considered in Chap. 5). The initial optical parameters of the atmosphere are provided as look-up tables over the altitude grid: volume extinction coefficient $\alpha(z_i)$, probability of the quantum surviving $\omega_0(z_i)$, phase function as a table over altitudes and cosines of the scattering angles $x(z_i, \chi_j)$, $j = 1, \ldots, M$, $\chi_1 = 1$, $\chi_M = -1$, where $i = 1, \ldots, N$, $z_1 = z_\infty$, $z_N = 0$. The physical atmospheric model (the vertical profiles of the temperature, pressure, concentrations of the absorbing gases and the aerosol model described in Sect. 1.2) defines all these parameters. It is necessary to find the numerical values of the semispherical fluxes – the downward one $F^\downarrow(z)$ and upward one $F^\uparrow(z)$ – at arbitrary altitude $0 \leq z \leq z_\infty$ or (and) radiance $I(z, \mu, \varphi)$ for arbitrary direction (μ, φ). All mentioned parameters and values are monochromatic for the chosen wavelength.

Let us express the optical thickness as a function of altitude by means of (1.39) before presenting the Monte-Carlo method. Using the trapezoidal quadrature, we obtain:

$$\tau(z) = \sum_{j=1}^{k-1} \frac{1}{2}[\alpha(z_j) + \alpha(z_{j+1})](z_j - z_{j+1}) + \frac{1}{2}[\alpha(z) + \alpha(z_k)](z_k - z) \qquad (2.1)$$

where number k is defined from condition $z_{k+1} < z \leq z_k$. The table of the optical thickness simply appeared from (2.1):

$$\tau_i \equiv \tau(z_i) = \sum_{j=1}^{i-1} \frac{1}{2} [\alpha(z_j) + \alpha(z_{j+1})](z_j - z_{j+1}) , \tag{2.2}$$

where $i = 1, \ldots, N$, moreover $\tau_1 = 0$, $\tau_N = \tau_0$ is the optical thickness of the atmosphere defined by (1.40). We are using the linear interpolation in accordance with the trapezoidal quadrature here and further. Then function $\alpha(z)$ in (2.1) is expressed as

$$\alpha(z) = \alpha(z_k) \frac{z - z_{k+1}}{z_k - z_{k+1}} + \alpha(z_{k+1}) \frac{z_k - z}{z_k - z_{k+1}} , \tag{2.3}$$

which with taking into account (2.2) gives the polynomial of the power equal to two

$$\tau(z) = \tau_k + \alpha(z_k)(z_k - z) + \frac{1}{2} \frac{u(z_{k+1}) - u(z_k)}{z_k - z_{k+1}} (z_k - z)^2 . \tag{2.4}$$

After obtaining function $\tau(z)$ according to (2.2) and (2.4) it is possible to use the altitude, as a coordinate because it is more appropriate in practice. The input tables of the initial atmospheric parameters are directly converted to $\alpha(\tau_i)$, $\omega_0(\tau_i)$ and $x(\tau_i, \chi_i)$, $i = 1, \ldots, N$, and for obtaining the intermediate values, for example $\omega_0(\tau)$, it is possible to use either the linear interpolation directly over τ or to find the altitude as function $z(\tau)$ and to interpolate over altitude z in (2.3) that is more correct. Inverse function $z(\tau)$ from (2.4) could be written:

$$z(\tau) = z_k + \frac{\alpha(\tau_k) - \sqrt{\alpha^2(\tau_k) + 2\Delta_k(\tau - \tau_k)}}{\Delta_k} , \tag{2.5}$$

where number k is deduced from condition $\tau_k \leq \tau < \tau_{k+1}$, and

$$\Delta_k = [\alpha(\tau_k) - \alpha(\tau_{k+1})] / [z_k - z_{k+1}] .$$

We should mention that as the procedure for the coordination of the altitude and optical depth is not linked with the specific of the Monte-Carlo method at all, it is possible to apply it in other numerical methods of radiative transfer theory.

Here we will give an account of the Monte-Carlo method. It is based on the modeling of radiative transfer in the atmosphere as a random process: the motion of the conditional particle of light called the "photon", the simulation of the process on computer, and the calculation of the desired characteristics as a mathematical expectation of random numbers appearing during the simulation (Kargin 1984; Marchuk et al. 1980).

For the statistical simulation on computer, it is necessary to reproduce a process that will play the role of the random event. Such an algorithm, called *a random number generator* or *randomizer*, is well known nowadays and we are not dwelling upon its specific pointing only that we have been using here the randomizer proposed in the study by Molchanov (1970). The totality of the random numbers uniformly distributed over the interval [0, 1] is the basis of the Monte-Carlo method. We are implying only these numbers using the term "the random number", specifying them by sign β, and at every appearance in the text we mean a new random number. For compact writing of the numerous recurrent relations occurring in the method we are using the operation of the assignment ":=" as it is accepted in the programming languages.

Let the probability of a certain discrete random event be equal to P. Choose the random number and if $\beta \leq P$, then assume that the event has happened, in the opposite case assume that it has not happened. The grounds of this approach are evident: if the quantity of the simulating acts tends to the infinity then the ratio of the quantity of the simulating acts when the event has happened to the quantity of all acts is equal to the probability of the event, i.e. to P due to the uniformity of the random numbers distribution. Note that for the continuous random value simulating characterized with probability density $\varrho(u)$ within the interval $[a, b]$ the probability value u within the interval $[a, u]$ is equal to $P(u) = \int_a^u \varrho(u')du'$ according to the definition. The application of the above-mentioned approach for the discrete random values leads directly to the following equation for values u simulating:

$$\int_a^u \varrho(u')du' = \beta \, . \tag{2.6}$$

As has been mentioned above, the process of radiative transfer in the Monte-Carlo method is simulated as a photon motion. Coming to the atmosphere the photon is moving along a certain trajectory, which finishes either with a photon outgoing from the atmosphere or with its absorption in the atmosphere or at the surface. The position of the photon in the atmosphere is determined completely with three coordinates: τ', μ', φ', hence, the simulation of the trajectory reduces to the consequent counting of the coordinate data. Therefore, it is enough to simulate only three processes: the free path of a photon (i. e. without interaction with the atmosphere), the interaction with the atmosphere (the absorption and scattering), and the interaction of a photon with the surface (the absorption and reflection).

A free photon path is analogous to the transfer of solar direct radiation throughout the atmosphere. Remember the formula of Beer's Law (1.42):

$$F_d(\tau) = F_0\mu_0 \exp(-\tau/\mu_0) \, .$$

Let K photons income to the top of the atmosphere, i. e. $F_0\mu_0 = KE$, where E is the energy of a single photon. Substituting KE to Beer's Law we obtain that the quantity of photons reaching optical depth τ is $K(\tau) = K\exp(-\tau/\mu_0)$. However, owing to the probability definition it means that the probability for a photon to reach optical depth τ is $\exp(-\tau/\mu_0)$. After replacing cosine of incident angle

μ_0 to cosine μ' that characterizes an arbitrary direction and specifying the free pass as $\Delta\tau'$, we obtain the following due to the method of the continuous random values simulating:

$$\Delta\tau' = -\mu' \ln\beta, \quad \tau' := \tau' + \Delta\tau'. \tag{2.7}$$

Mark that, firstly, (2.7) is corrected both for the photon moving downward ($\mu' > 0$ and τ' increases) and for the photon moving upward ($\mu' < 0$ and τ' decreases). Secondly, deriving (2.7) we don't need the explicit form of the probability density of the photon free path function, however we will need it later. As follows from the above-mentioned relations the probability of the photon path within the range of 0 to τ is $P(\tau) = 1 - \exp(-\tau/\mu_0)$. After differentiating with respect to τ the following is obtained:

$$\varrho(\Delta\tau') = \frac{1}{|\mu'|} \exp(-\Delta\tau'/\mu'). \tag{2.8}$$

The probability of the photon scattering in the atmosphere is $\omega_0(\tau')$ (directly to the terminology – the probability of the quantum surviving). Thus if $\beta \leq \omega_0(\tau')$, then the photon scattering is occurring in the opposite case the absorption is happening, i.e. at the end of the trajectory. The cosine of the scattering angle χ and the azimuth of the scattering ϕ are to be obtained in the scattering case. As the phase function does not depend on the azimuth it is uniformly distributed within the interval $[0, 2\pi]$ that gives $\phi = 2\pi\beta$. The density of the probability of the scattering to the angle with cosine χ is phase function $x(\chi)$ [according to the definition (1.2) in Sect. 1.1]. As this value is specified in the look-up tables, based on simulating rule (2.6) and repeating literally the reasons for (2.1)–(2.5) with accounting of 1/2 factor in normalizing relation (1.18) we obtain:

$$\chi = \chi_k + \frac{x(\tau', \chi_k) - \sqrt{x^2(\tau', \chi_k) + 2\Delta_k(2\beta - X_k(\tau'))}}{\Delta_k}, \tag{2.9}$$

where $X_i(\tau') = \sum_{j=1}^{i-1} \frac{1}{2}[x(\tau', \chi_{j+1}) + x(\tau', \chi_j)](\chi_j - \chi_{j+1})$, number k is derived from condition $X_k(\tau') \leq 2\beta < X_{k+1}(\tau')$, value β is the same as in (2.9) and:

$$\Delta_k = \frac{x(\tau', \chi_{k+1}) - x(\tau', \chi_k)}{\chi_k - \chi_{k+1}} \quad \left(\Delta_k = 0 \quad \text{provides} \quad \chi = \chi_k - \frac{2\beta - X_k(\tau')}{x(\tau', \chi_k)}\right).$$

Owing to linear relation $X_j(\tau')$ and $x(\tau', \chi_j)$ it is convenient to construct the preliminary table $X_j(z_i)$, where $j = 1, \ldots, M$, $i = 1, \ldots, N$ from the initial data and to use it for the interpolation of function $X_j(\tau')$ with (2.3) and (2.5). After scattering, the photon needs to determine the next direction of its motion. This not complicated problem is reduced to the solving of the spherical triangles (Stepanov 1948). Specifying the direction of the photon before the scattering by value $\mu_1' = \mu'$, we obtain:

$$\mu' = \mu_1'\chi + \sqrt{(1 - \mu_1'^2)}\sqrt{(1 - \chi^2)}\cos\phi \,,$$

$$\varphi' := \varphi' + \arccos\left(\frac{\chi - \mu'\mu_1'}{\sqrt{1 - \mu'^2}\sqrt{1 - \mu_1'^2}}\right). \tag{2.10}$$

It is possible to attribute an evident meaning of the reflection probability to the albedo in the description of the interaction with the surface: the reflection occurs if $\beta \leq A$ and the opposite case corresponds to the photon absorption by the surface and to the end of the photon trajectory. Due to the reflection orthotropness all possible directions of the photon are uniformly distributed because:

$$\mu' = -\cos\left(\frac{\pi}{2}\beta\right), \quad \varphi' = 2\pi\beta, \quad \tau' = \tau_0 . \tag{2.11}$$

After the repeating of the above reasoning about the photon free path to calculate the desired irradiances we are revealing that it is enough to count the number of the photons passing level $\tau(z)$ to the interested direction, i.e. downward ($\mu' > 0$) for $F^{\downarrow}(z)$ and upward for ($\mu' < 0$) for $F^{\uparrow}(z)$. For that we introduce functions $\xi_1^{\downarrow}(z)$ and $\xi_1^{\uparrow}(z)$ with the zeroth initial values. Let τ_1' be the photon coordinate before simulating of the free path (2.7) and $\tau_2' = \tau_1' + \Delta\tau'$ be the coordinate after the simulation. Then:

$$\xi_1^{\downarrow}(z) := \xi_1^{\downarrow}(z) + 1, \quad \text{provided} \quad \tau_1' \leq \tau(z) \leq \tau_2' ,$$
$$\xi_1^{\uparrow}(z) := \xi_1^{\uparrow}(z) + 1, \quad \text{provided} \quad \tau_2' \leq \tau(z) \leq \tau_1' . \tag{2.12}$$

Equation (2.12) is also used in the case of the photon going out of the atmosphere ($\tau_2' < 0$) or reaching the surface ($\tau_2' \geq \tau_0$). After the simulation of a certain number K of trajectories the desired irradiances are found as a number of the counting photons multiplied by energy of the single photon, which is equal to $F_0\mu_0/K$ (see above). Thus, we are inferring:

$$F^{\uparrow}(z) = \frac{1}{K}\xi_1^{\uparrow}(z)F_0\mu_0 , \quad F^{\downarrow}(z) = \frac{1}{K}\xi_1^{\downarrow}(z)F_0\mu_0 . \tag{2.13}$$

Values $\xi_1^{\downarrow}(z)$ and $\xi_1^{\uparrow}(z)$ will be called further *the counters* and the expressions analogous to (2.12) will be treated as *writing to the counters*.

Equation (2.12) seems to contradict the formula of a link of the radiance and irradiance (1.4) because they don't account for the cosine of the photon zenith angle. However, the photon is a carrier of the very irradiance and not of the radiance. It is easy to understand from the physical meaning: the real quantum of light as a photon has energy independent of the direction. The formal proof of the correctness of (2.12) is elementary. Consider the trajectory of a single photon ($K = 1$). Let the photon go out from the top of the atmosphere after the first scattering event. Then, the value of the upward irradiance at the top as per (2.13) is to be equal to the value of the downward irradiance due to the law of energy conservation (there is no absorption), and this condition wouldn't be broken only if the writing to the counters according to (2.12) does not depend on value μ'.

In result, the following algorithm of the irradiance calculation is obtained:

1. In the beginning of every trajectory it can be written $\tau' = 0$, $\mu' = \mu_0$, $\varphi' = 0$.

2. Further, the photon free path is simulated according to (2.7) with the writing to the counters (2.12).

3. If the photon is going out of the atmosphere ($\tau' \leq 0$), its trajectory will finish and the trajectory of the following photon will be simulated.

4. If the photon reaches the surface ($\tau' \geq \tau_0$), its interaction (reflection or absorption) with the surface will be simulated.

5. The absorption means the end of the trajectory and the trajectory of the following photon is simulated.

6. The reflection from the surface gives the new direction of the trajectory according to (2.11) and then the photon recurrent free path is simulated.

7. If the photon stays in the atmosphere ($0 < \tau' < \tau_0$), its interaction (scattering or absorption) with the atmosphere is simulated.

8. The absorption leads to the end of the trajectory.

9. In the case of the scattering, the new direction of the photon is determined using (2.9) and (2.10) and the photon following free path is simulated. The desired values of the irradiances are found with (2.13) after sufficient numbers K of the trajectories.

Mark that ratios $\xi_1^\downarrow(z)/K$ and $\xi_1^\uparrow(z)/K$ are the expectations (the arithmetic means) of photon numbers written to the counters as the result of the simulation of a single photon trajectory. Introduce counters $\xi^\downarrow(z)$ and $\xi^\uparrow(z)$ with zeroth energies in the beginning of every trajectory and write a relation analogous to (2.12) for a single trajectory. Then, (2.12) transforms to:

$$\xi_1^\downarrow(z) := \xi_1^\downarrow(z) + \xi^\downarrow(z)\,, \quad \xi_1^\uparrow(z) := \xi_1^\uparrow(z) + \xi^\uparrow(z)\,, \tag{2.14}$$

moreover, the writing to the counters as per (2.14) is carried out at the end of every trajectory.

Thus, the problem of the irradiance calculation by the Monte-Carlo method is reduced in fact to the calculation of the expectations of random values $\xi^\downarrow(z)$ and $\xi^\uparrow(z)$ (the number of photons written to the counters) over the finite sample from K trajectories by (2.13) and (2.14). It is possible to calculate not only the expectation but also other statistical estimations of random values $\xi^\downarrow(z)$ and $\xi^\uparrow(z)$. Further, obtain their variance. For that we introduce the counters of squares $\xi_2^\downarrow(z)$ and $\xi_2^\uparrow(z)$ with zeroth initial values and write together with (2.14) the following:

$$\xi_2^\downarrow(z) := \xi_2^\downarrow(z) + [\xi^\downarrow(z)]^2 , \quad \xi_2^\uparrow(z) := \xi_2^\uparrow(z) + [\xi^\uparrow(z)]^2 . \tag{2.15}$$

Using the known expression for variance $\mathbf{D}(\xi) = \mathbf{M}(\xi^2) - \mathbf{M}^2(\xi)$, where $\mathbf{M}(\dots)$ is expectation, we obtain:

$$\mathbf{D}(\xi^\downarrow) = \frac{1}{K}\xi_2^\downarrow(z) - \left(\frac{1}{K}\xi_1^\downarrow\right)^2 , \quad \mathbf{D}(\xi^\uparrow) = \frac{1}{K}\xi_2^\uparrow(z) - \left(\frac{1}{K}\xi_1^\uparrow\right)^2 . \tag{2.16}$$

The behavior of the distribution of random values $\xi^\downarrow(z)$ and $\xi^\uparrow(z)$ is unknown. However, the distribution of its expectations according to the central limit theorem tends to the normal distribution as $K \to \infty$. Hence, desired irradiances (2.13), which are also considered as random values, have the distributions asymptotically close to the normal distribution. It is known that the normal distribution is characterized with *the expectation* and *the variance* expressed by (2.16). For *the standard deviation (SD)* $(\mathbf{s}(\dots) = \sqrt{\mathbf{D}(\dots)})$ of the irradiances in accordance with the study by Marchuk et al. (1980) with taking into account the known rule for the variances addition the following is obtained:

$$\mathbf{s}(F^\downarrow(z)) = F_0\mu_0\sqrt{\mathbf{D}(\xi^\downarrow)/K} , \quad \mathbf{s}(F^\uparrow(z)) = F_0\mu_0\sqrt{\mathbf{D}(\xi^\uparrow)/K} . \tag{2.17}$$

As follows from (2.17), the increasing of the number of trajectories K leads to the minimization of the standard deviation (SD), i.e. of the random error of the irradiances calculation. Evaluating the SD with (2.15)–(2.17) is of great practical interest because it allows accomplishment of the calculations with the accuracy fixed in advance. Actually, the calculation of the SD gives the possibility of estimating the necessary number of photon trajectories and as soon as the SD is less than the fixed value, the simulating is finished.

The above-considered scheme of the simulating of photon trajectories is called *"direct modeling"* (Kargin 1984) as it directly reflects our implication concerning photon motion throughout the atmosphere. However, direct modeling is not enough for accelerating the calculation according to the algorithm of the Monte-Carlo method or for the radiance calculation (Kargin 1984). Consider two approaches to increase the calculation effectiveness that we have applied. It is possible to find detailed descriptions of other approaches in the books by Kargin (1984), and Marchuk et al. (1980).

The basis of optimizing the calculation with the Monte-Carlo method is an idea of decreasing the spread in the values written to the counters. Then the variance expressed by (2.16) decreases too and fewer trajectories are necessary for reaching the fixed accuracy according to (2.17).

Assume that the photon could be divided into parts (as it is a mathematical object and not a real quantum here). Then a part of the photon equal to $1 - \omega_0(\tau')$ is absorbed at every interaction with the atmosphere and the rest $\omega_0(\tau')$ is scattered and, then, continues the motion. During the interaction with the surface these parts are equal to $1 - A$ and to A (A is the surface albedo) correspondingly. We specify the value w' called *the weight of a photon* (Kargin 1984; Marchuk et al. 1980), which it is possible to formally consider as a fourth coordinate. Assume value $w' = 1$ in the beginning of every trajectory and while writing to the counters, (2.12) will be assigned not unity but value w'. Then the simulation of the interaction with the atmosphere is reducing to the assignment $w' := w'\omega_0(\tau')$ at every step, and the simulation of the interaction with the surface is reducing to the assignment $w' := w'A$. Now the photon trajectory can't break (the surviving part of the photon always remains), the break of the trajectory occurs only when the photon is outgoing from the atmosphere top. Usually for not driving the photon with too small weight within the atmosphere *parameter of the trajectory break* W is introduced: the trajectory is broken if $w' < W$. It is suitable to evaluate value W based on the accuracy needed for the calculation: $W = s\delta$, where s is the minimal (over all altitudes z for the downward and upward irradiances) needed relative error of the calculation; δ is the small value (we have used $\delta = 10^{-2}$). This approach of the photon "dividing" is known under the unsuccessful name *"the analytical averaging of the absorption"* (Kargin 1984) (the words "analytical averaging" are associated with a certain approximation, which is not used in reality).

Consider a photon at the beginning of the trajectory at the top of the atmosphere. In this case, before the simulation of the first free path ($\tau' = 0, \mu' = \mu_0$, $w' = 1$) using Beer's Law (1.42) it is possible to account direct radiation, i. e. radiation reaching level $\tau(z)$ without interaction with the atmosphere. For that it is necessary to write to all counters $\xi^\downarrow(z)$ the value depending on z instead unity:

$$\psi = w' \exp\left(-\frac{\tau - \tau'}{\mu'}\right), \tag{2.18}$$

and further writing to the counters is not implemented for the first free path (direct radiation). This approach is easy to extend to other parts of the trajectory: before the writing of the free path to the counter, which the photon can reach ($\xi^\downarrow(z)$, for $\mu' > 0$ and $\tau \geq \tau'$, or $\xi^\uparrow(z)$, for $\mu' < 0$ and $\tau \leq \tau'$) value ψ calculated with (2.18) is writing and the further photon flight through the counters is not registering. Note that as it has been shown above the exponent in (2.18) is a probability of the photon started from level τ' to reach level τ. This general approach of writing to the counter the probability of the photon to reach the counter is called *"a local estimation"* (Kargin 1984; Marchuk et al. 1980).

The analysis of the above-described algorithm of the irradiances calculation indicates that the irradiances are not depending on photon azimuth φ'. Actually, calculated only in two cases with (2.10) and (2.11), azimuth φ' does not influence other coordinates and hence, the values written to the counters. Thus, the "photon azimuth" coordinate is excessive in the task and it could be excluded for accelerating the calculations (but only in this task of the irradiances calculations above the orthotropic surface).

Consider the second of the problems described above: the problem of radiance $I(z, \mu, \varphi)$ calculation. It is obvious that the procedures either of the simulation of photon trajectories or of the calculation of the expectation and variance are depending on the desired value, and hence they wouldn't change. The difference is concerning the procedure of writing the values to the counters. Encircle the cone with small solid angle $\Delta\Omega(\mu, \varphi)$ around direction (μ, φ). We will be writing to the counter all photons, which have reached level z and have come to the cone for radiance I according to the equation analogous to (2.12). Moreover, in this case value $1/|\mu|$ has to be written to the counter instead of unity in the case of the irradiance calculation to satisfy the link of the radiance and irradiance (1.4). Pass further from the above-described (but not realized) scheme of the direct modeling of the radiances to the schemes of the *weight modeling* and *local estimation*. Let the photon have coordinates (τ', μ', φ'). According to the definition of the phase function as a density of the probability of the scattering (Sect. 1.2), the probability of the photon coming to solid angle $\Delta\Omega(\mu, \varphi)$ after scattering at level τ' is equal to the integral of the phase function over the angle intervals defined by (1.17) (i.e. $\Delta\Omega$ and scattering angle $\angle(\mu', \varphi')(\mu, \varphi)$) with taking into account normalizing factor $1/4\pi$. Let value $\Delta\Omega$ decrease toward zero. Then we are revealing that the density of the probability of the photon to reach direction (μ, φ) coincides with the value of the phase function for argument $\chi' = \cos(\angle(\mu', \varphi')(\mu, \varphi))$, which is computed with (1.46). This probability is necessary to multiply by factor ψ defined with (2.18), i.e. by the probability of the photon to reach level $\tau(z)$. Finally, the local estimation for the radiance is obtained according to the results of the books by (Kargin 1984, Marchuk et al. 1980).

$$\psi = \frac{w'}{4\pi |\mu|} x(\tau', \chi') \exp\left(-\frac{\tau - \tau'}{\mu'}\right)$$
$$\chi' = \mu\mu' + \sqrt{(1 - \mu^2)(1 - \mu'^2)} \cos(\varphi - \varphi') .$$

(2.19)

Thus, the considered algorithm of the radiance computation according to the Monte-Carlo method differs from the irradiance computation algorithm just with the other equation for the local estimation (2.19) instead of (2.18) and with other equations for the counters: for radiance over single trajectory $\xi(z, \mu, \varphi)$, for expectation $\xi_1(z, \mu, \varphi)$ and for the square of the expectation $\xi_2(z, \mu, \varphi)$. Both algorithms (for radiance and irradiance) could be carried out on computer with one computer code. It is pointed out that the condition of the clear atmosphere (the small optical thickness) has not been assumed so the Monte-Carlo method algorithms can be also applied for the cloudy atmosphere.

In conclusion, illustrate that the considered algorithms actually correspond to the solution of the equation of radiative transfer (1.47).

The desired radiation characteristic (radiance, irradiance) could be written in the operator form according to expressions of the radiance through the source function (1.52), and as per the link of the irradiance and the radiance (1.4):

$$\Psi\mathbf{B} = \int \Psi(u)\mathbf{B}(u)du ,$$

(2.20)

where function $\Psi(u)$ is a certain function allowing the desired value calculation through the source function [e.g. (1.52)]. Variable u specifies here and further coordinates τ', or (and) μ', φ' according to (1.52) and (1.6). The source function in its turn is defined by the Fredholm integral equation of the second kind (1.54) and (1.55) with kernel $\mathbf{K}$ and $\mathbf{q}$ as an absolute term.

The Monte-Carlo method has been primordially elaborated for computing the integrals analogous to (2.20):

$$\int \Psi(u)\mathbf{B}(u)du = \mathbf{M}_\xi(\Psi(\xi)) , \tag{2.21}$$

where $\mathbf{M}_\xi(\ldots)$ is the expectation of random value ξ simulated with probability density $\mathbf{B}(u)$ as per (2.6). Therefore, (2.20) and the equation for the source function (1.54) at the Monte-Carlo method are written for a single trajectory and the desired value is computed over the totality of the trajectories as an expectation according to (2.21). Applying (2.20) to the formal solution of the Fredholm equation, i.e. to the Neumann series (1.56) we obtain:

$$\Psi\mathbf{B} = \Psi\mathbf{q} + \Psi\mathbf{K}\mathbf{q} + \Psi\mathbf{K}^2\mathbf{q} + \Psi\mathbf{K}^3\mathbf{q} + \ldots . \tag{2.22}$$

The computer scheme of the Monte-Carlo method is reduced to consequent applying of (2.22). Term $\Psi\mathbf{q}$ is formed as follows: we are simulating random value $\xi^{(1)}$ corresponded to probability density $\mathbf{q}$ and value $\Psi(\xi^{(1)})$ is being written to the counter. Then the term $\Psi\mathbf{K}\mathbf{q}$ is forming: using value $\xi^{(1)}$ random value $\xi^{(2)}$ corresponded to density of the probability of the transition $\mathbf{K}(\xi^{(1)}, \xi^{(2)})$ is simulating, and value $\Psi(\xi^{(2)})$ is being written to the counter. The following procedures are simulating analogously. Finally, the absolute term $\Psi\mathbf{K}^n\mathbf{q}$ is forming: using value $\xi^{(n)}$ we are simulating random value $\xi^{(n+1)}$ corresponded to density of the probability of the transition $\mathbf{K}(\xi^{(n)}, \xi^{(n+1)})$ and value $\Psi(\xi^{(n+1)})$ is being written to the counter. The photon trajectory in the phase space is a chain of the pointed transitions, the simulation is accomplished over many trajectories, and, in accordance with (2.22) the desired value is mean value $\Psi(\xi)$ over all trajectories.

Now we are showing that the explicit form of operators $\mathbf{q}$, $\mathbf{K}$ and Ψ in the above-described algorithms corresponds to their form in the equations of radiative transfer theory presented in Sect. 1.3. Furthermore, as direct radiation is not included in (1.54)–(1.56), operator $\mathbf{K}\mathbf{q}$ corresponds to $\mathbf{q}$ in (1.55) and (1.56), the latter is specified as $\mathbf{q}'$. The phase space is specified with three coordinates (τ', μ', φ'). Operator $\mathbf{q}$ is evidently extraterrestrial solar radiation $q = F_0\mu_0\delta(\mu-\mu_0)\delta(\varphi)$ that corresponds to operator $\mu_0 I_0$ considered in Sect. 1.3 while (1.57) have been derived. Hence, to prove the correspondence of the Monte-Carlo method algorithms to (1.54)–(1.56) it is enough to demonstrate the correspondence of integral operators $\mathbf{K}$ to each other.

To begin with, consider the case without accounting for photon weights w', i.e. the radiation absorption is simulated explicitly. Let $w' \equiv 1$ in the local estimation expressed by (2.18) and (2.19). The $\mathbf{K}$ operator describes, as has been mentioned above, the probability density of the photon path between two points of the phase space, whose coordinates are specified as (τ', μ', φ') and

(τ, μ, φ) for the conformity with definitions (1.55). According to its meaning, the probability density is the product of three probability densities: the density of the photon free path of distance $\Delta \tau' = \tau - \tau'$ according to (2.8), density of the non-absorption of the photon in the atmosphere $\omega_0(\tau)$, and the density of the scattering of the photon with change of the direction from (μ', φ') to (μ, φ), which is equal to $x(\tau, \chi')/(4\pi)$ as per (2.19). However, this product is exactly equal to K according to (1.55)! Taking into account that as per (2.6) and (2.11) the photon probability within the directional interval $[-\mu', 0]$ is equal to $P(\mu') = 2\arccos(-\mu')/\pi$ the following condition is added to (1.55) for $\tau' = \tau_0$, $\mu' < 0$ to consider the surface albedo in the Monte-Carlo method:

$$\mathbf{K} = K(\tau, \mu, \varphi, \tau', \mu', \varphi') = -\frac{A}{\pi^2 \mu' \sqrt{1 - \mu'^2}} \exp\left(-\frac{\tau - \tau'}{\mu'}\right). \tag{2.23}$$

Now to find operator Ψ remember that the variables noted in the definition of the K operator (1.55) as (τ, μ, φ), later become the integration variables themselves when the desired values are calculated using (2.20). For example, during the calculation of the radiance according to source function (1.52) τ' is a variable noted in equations of the source function (1.53)–(1.55) as τ. Therefore, coordinates of the point (τ, μ, φ) are to be noted as (τ', μ', φ') at (2.10). After the radiance calculation with (1.52), the irradiance is computed according to relation (1.6) and factor $1/\mu'$ is canceled out. The integrating is accomplished over all three variables (τ', μ', φ') and for operator Ψ it yields the expression exactly equal to simple local estimation (2.18). When the radiance is computed with (1.52) the integration variable is τ' only, so there is no dependence of the source function upon coordinates (μ', φ'). Actually, the probability density of transition K is written accounting for the change of the notions for coordinates (τ', μ', φ') and for the radiance computation, using (1.52) coordinates (τ', μ, φ) are applied. Hence, the scattering angle, which the photon trajectory is simulated with, in the K operator according to the Monte-Carlo method, differs from the operator defined by (1.55) in the transfer equation. Therefore, the probability density of the scattering to direction (μ', φ') has not yet accounted for. To account for it we are accomplishing the multiplication by the phase function in the equation for local estimation (2.19). Thus, there is a complete correspondence between (2.19) and (1.52)–(1.55) also during the consideration of the radiance.

The case of simulating the photon trajectories with weights w' corresponds to the coordinated transformation of operators K and Ψ taking into account that they are used in solution (2.22) only as a convolution of K with Ψ. In this case, the multiplication by probability of the quantum surviving $\omega_0(\tau)$ is passing from operator K to Ψ. It corresponds to the changing of photon weight w' when the powers of the K operator are calculated in (2.22), and then to the multiplication of the local estimation to photon weight w' in (2.18) and (2.19) (Kargin 1984). Analogously it is concluded that the direct modeling of the irradiances otherwise corresponds to the passing from the exponential factor (the local estimation (2.18)) to the K operator. Similar transformations, many of which are difficult to present from the physical point of view, are the basis for various other approaches of the calculation optimization in the Monte-Carlo method (Kargin 1984; Marchuk et al. 1980), e.g. the computing of the derivatives of the irradiances that will be considered in Chap. 5. As has been shown using these methods, the same transfer equation (1.47) is solved with different versions of operators $\mathbf{K}$ and Ψ simulating. In practice, it is appropriate to use the following procedure. Assume that the probability density of transition $\mathbf{K}$ is always determined by the concrete scheme of the photons trajectories simulating, and operator Ψ is determined by the concrete writing to the counters (in other words, $\mathbf{K}$ is responsible for radiative transfer and Ψ answers for the model of its "observation").

2.2
Analytical Method for Radiation Field Calculation in a Cloudy Atmosphere

Let us consider the model of an extended and horizontally homogeneous cloud of large optical thickness $\tau_0 \gg 1$ as Fig. 2.1 illustrates. At the first stage, the cloud layer is assumed vertically homogeneous as well and the influence of the clear atmosphere layers above and below the cloud layer is not taken into account. The volume coefficients of scattering α and absorption κ, linked with the cloud characteristics as $\kappa + \alpha \equiv \tau_0/\Delta z$, $\alpha \equiv \omega_0 \tau_0/\Delta z$, $\kappa \equiv \tau_0(1 - \omega_0)/\Delta z$, are used for the cloud description. The optical properties of the cloud are described by the following parameters: single scattering albedo ω_0; optical thickness τ, and mean cosine of the scattering angle g, which characterizes a phase function. From the bottom the cloud layer adjoins the ground surface and its reflectance is described by ground albedo A. The underlying atmosphere could be taken into account if albedo A is implying as an albedo of the system "surface + atmosphere under the cloud". Parallel solar flux πS is falling on the cloud top at incident angle $\arccos \mu_0$. The reflected and transmitted radiance is observed at viewing angle $\arccos \mu$. The reflected radiance (in the units of incident extraterrestrial flux $\pi S \mu_0$) is expressed with reflection function $\varrho(\tau_0, \mu, \mu_0)$ and the transmitted radiance (in the same units) is expressed with transmission function $\sigma(\tau_0, \mu, \mu_0)$.

2.2.1
The Basic Formulas

At a sufficiently large optical depth within the cloud layer far enough from the top and bottom boundaries the asymptotic or diffusion regime set in owing to the multiple scattering. This regime permits a rather simple mathematical description (Sobolev 1972; Hulst 1980). The region within the cloud layer is called a diffusion domain. The physical meaning yields the following specific features of the diffusion domain:

1. the role of the direct radiation (transferred without scattering) is negligibly small compared to the role of the diffused radiation;

2. the radiance within the diffusion domain does not depend on the azimuth;

3. the relative angle distribution of the radiance does not depend on the optical depth (Sobolev 1972).

The name "diffusion" appears because the equation of radiative transfer is transformed to the diffusion equation in that case (Hulst 1980). In the scattering layer of a large optical thickness the analytical solution of the transfer equation is possible and it is expressed through the asymptotic formulas of the theory of radiative transfer (Sobolev 1972; Minin 1988), moreover the existence and uniqueness of the solution have been proved (Germogenova 1961). According to the books by Sobolev (1972), Hulst (1980), and Minin (1988), the solution of the transfer equation, expressed through reflection ρ and transmission σ functions, is the following:

$$\rho(0, \mu, \mu_0, \varphi) = \rho_\infty(\mu, \mu_0, \varphi) - \frac{m\bar{l}K(\mu)K(\mu_0)\exp(-2k\tau_0)}{1 - \bar{l}\bar{l}\exp(-2k\tau_0)}$$

$$\sigma(\tau_0, \mu, \mu_0) = \frac{m\bar{K}(\mu)K(\mu_0)\exp(-k\tau_0)}{1 - \bar{l}\bar{l}\exp(-2k\tau_0)}.$$

(2.24)

In these equations $\rho_\infty(\mu, \mu_0, \varphi)$ is the reflection function for a semi-infinite atmosphere; $K(\mu)$ is the escape function, which describes an angular dependence of the reflected and transmitted radiance; m, l, k are the constants, depending on the cloud optical properties, the formulas for its computing are presented below; $\bar{K}(\mu)$ and $\bar{l}$ depends on ground albedo A as well. The following expressions are taking into account the ground surface reflection according to Sobolev (1972), Ivanov (1976) and Minin (1988):

$$\bar{l} = l - \frac{Amn^2}{1 - Aa^\infty}, \quad \bar{K}(\mu) = K(\mu) + \frac{Aa(\mu)n}{1 - A}.$$

(2.25)

In these expressions $a(\mu)$ is the plane albedo and a^∞ is the spherical albedo of a semi-infinite atmosphere (the atmosphere of the infinite optical thickness).

$$\bar{K}(\mu) = K(\mu) + A\bar{Q}a(\mu), \quad \bar{n} = \frac{n}{1 - Aa^\infty}, \quad \bar{l} = l - Am\bar{Q}Q$$

(2.26)

where $a(\mu)$, a^∞, and value n are defined by the integrals:

$$a(\mu) = 2\int_0^1 \rho(\mu, \mu_0)\mu_0 d\mu_0, \quad a^\infty = 2\int_0^1 a(\mu)\mu d\mu$$

$$n = 2\int_0^1 K(\mu)\mu d\mu, \quad \bar{n} = 2\int_0^1 \bar{K}(\mu)\mu d\mu,$$

It is seen that (2.24) are the asymmetric formulas relatively to variables μ and μ_0, which are input with escape functions $K(\mu)$ and $\bar{K}(\mu)$. It links with different boundary conditions at the top and bottom of the layer. The top is free and it could be assumed as an absolutely absorbing one for the upward radiation and the bottom boundary reflects partly the downward radiation. Thus each of them generates its own light regime described by different escape functions $K(\mu)$ and $\bar{K}(\mu)$ and constants l and $\bar{l}$.

Consider the semispherical fluxes of diffused solar radiation (solar irradiances) in relative units of incident solar flux πS. Reflected irradiance $F^{\uparrow}(0, \mu_0)$ and transmitted irradiance $F^{\downarrow}(\tau, \mu_0)$ are described by the formulas similar to (2.24), where reflection function $\varrho_\infty(\mu, \mu_0)$ and escape function $K(\mu)$ are substituted with their integrals $a(\mu_0)$ and n, according to (1.6) and (2.26). As a result, the following formulas are inferred:

$$F^{\uparrow}(0, \mu_0) = a(\mu_0) - \frac{mn\bar{l}K(\mu_0)\exp(-2k\tau_0)}{1 - l\bar{l}\exp(-2k\tau_0)} \, ,$$

$$F^{\downarrow}(\tau_0, \mu_0) = \frac{m\bar{n}K(\mu_0)\exp(-k\tau_0)}{1 - l\bar{l}\exp(-2k\tau_0)} \, . \tag{2.27}$$

The radiation absorption within the cloud layer is determined by the radiative flux divergence (Sect. 1.1). It is computed with the obvious equation:

$$R = 1 - F^{\uparrow}(0, \mu_0) - (1 - A)F^{\downarrow}(\tau_0, \mu_0)$$

$$= 1 - a(\mu_0) + \frac{nK(\mu_0)m\exp(-k\tau_0)}{1 - l\bar{l}\exp(-2k\tau_0)}\left[\bar{l}\exp(-k\tau_0) - \frac{1 - A}{1 - Aa_\infty}\right] \, . \tag{2.28}$$

Mention that the term "asymptotic" specifies the light regime installed within the cloud and it does not point out any approximation. Equations (2.24), (2.27) and (2.28) are rigorous in the diffusion domain. Their accuracy will be studied below depending on the optical thickness.

2.2.2
The Case of the Weak True Absorption of Solar Radiation

In clouds, the absorption is extremely weak compared with scattering $(1-\omega_0 \ll 1)$ within the short-wavelength range. As has been shown in the books by Sobolev (1972), Hulst (1980), and Minin (1988) in this case both functions $\varrho_\infty(\mu, \mu_0)$ and $K(\mu)$ and constants m, l, k are expressed with the expansions over powers of small parameter $(1 - \omega_0)$. We consider here that parameter s, where $s^2 = (1 - \omega_0)/[3(1 - g)]$, is more convenient for the problem in question than parameter $(1 - \omega_0)$. Value g is a mean cosine of the scattering angle or, here, the asymmetry parameter of Henyey-Greenstein function (1.31). Then, these expansions over the powers of s for the constants in (2.24)–(2.28) look like:

$$k = 3(1-g)s\left[1 + s^2\left(1.5g - \frac{1.2}{1+g}\right)\right] + O(s^3),$$

$$m = 8s\left[1 + \left(6 - 7.5g + \frac{3.6}{1+g}\right)s^2\right] + O(s^4),$$

$$l = 1 - 6q's + 18q'^2 s^2 + O(s^3),\qquad(2.29)$$

$$a^\infty = 1 - 4s + 12q's^2 - \left(36q' - 6g - \frac{1.608}{1+g}\right)s^3 + O(s^4),$$

$$n = 1 - 3q's + \left(9q'^2 - 3(1-g) - \frac{2}{1+g}\right)s^2 + O(s^3).$$

For the functions in (2.24)–(2.28) the followings expansions are correct according to books by Sobolev (1972), Minin (1988), and Yanovitskij (1997):

$$K(\mu) = K_0(\mu)(1 - 3q's) + K_2(\mu)s^2 + O(s^3),$$

$$a(\mu) = 1 - 4K_0(\mu)s + a_2(\mu)s^2 + a_3(\mu)s^3 + O(s^4),\qquad(2.30)$$

$$\varrho_\infty(\mu,\mu_0) = \varrho_0(\mu,\mu_0) - 4K_0(\mu)K_0(\mu_0)s + \varrho_2(\mu,\mu_0)s^2 + \varrho_3(\mu,\mu_0)s^3 + O(s^4),$$

where the nomination is introduced:

$$q' = 2\int_0^1 K_0(\zeta)\zeta^2 d\zeta \cong 0.714.$$

In these expansions functions $\varrho_0(\mu,\mu_0)$ and $K_0(\mu)$ are functions $\varrho_\infty(\mu,\mu_0)$ and $K(\mu)$ for the conservative scattering ($\omega_0 = 1$) correspondingly, functions $a_2(\mu)$ and $K_2(\mu)$ are the coefficients by the item s^2. They are presented either in analytical or in table form (Sobolev 1972; Hulst 1980; Minin 1988; Yanovitskij 1997). Asymptotic expansions (2.29) and (2.30) have been mathematically rigorously derived, their errors are defined by items $\sim s^3$ or $\sim s^4$ omitting in the series.

The coefficients by items s^2 and s^3 in the expansion for reflection function $\varrho_\infty(\mu,\mu_0)$ have been derived in the study by Melnikova (1992) and look like:

$$\varrho_2(\mu,\mu_0) = \frac{a_2(\mu)a_2(\mu_0)}{a_2},\qquad \varrho_3(\mu,\mu_0) = \frac{a_3(\mu)a_3(\mu_0)}{a_3},\qquad(2.31)$$

where a_2, a_3, $a_2(\mu)$ and $a_3(\mu)$ are the coefficients by s^2 and s^3 in the series for spherical a^∞ albedo as per (2.29) and in series for plane $a(\mu)$ albedo as per (2.30) correspondingly.

According to the book by Minin (1988), where it has been shown that it is possible to neglect the dependence of escape function $K_0(\mu)$ upon the phase function for the conservative scattering and values $0.65 \le g \le 0.9$, we present the following table:

Table 2.1. Values of escape function $K_0(\mu)$ for cloud layers ($0.65 \leq g \leq 0.9$)

μ	1.0	0.9	0.8	0.7	0.6	0.5	0.4	0.3	0.2	0.1
$K_0(\mu)$	1.271	1.193	1.114	1.034	0.952	0.869	0.782	0.690	0.591	0.476

The approximation for function $K_0(\mu)$ with the error 3% for $\mu > 0.4$ has been proposed in the book by Sobolev (1972): $K_0(\mu) = 0.5 + 0.75\mu$. In the book by Yanovitskij (1997) and in the paper by Dlugach and Yanovitskij (1974) the results of escape function $K(\mu)$ have been presented for the set of values of phase function parameter g and single scattering albedo ω_0. The analysis of these numerical results yields the following approximation for function $K_0(\mu)$ with taking into account the phase function dependence:

$$K_0(\mu) = (0.7678 + 0.0875\,g)\mu + 0.5020 - 0.0840\,g \; . \tag{2.32}$$

The correlation coefficient of the formulas is about 0.99–0.93 depending on parameter g.

In the book by Minin (1988) it has been proposed to present the function $K_2(\mu)$ with the expression $K_2(\mu) = n_2 K_0(\mu) w(\mu)$, auxiliary function $w(\mu)$ is specified with the table.

The numerical analysis in Melnikova (1992) of the table presentation of escape function $K(\mu)$ according to the paper Dlugach and Yanovitskij (1974) gives the analytical approximation of function $K_2(\mu)$:

$$K_2(\mu) = n_2 K_0(\mu) w(\mu) = 1.667\, n_2 (\mu^2 + 0.1) \; . \tag{2.33}$$

This approximation after the integration with respect of variable μ yields value n_2 with an error less than 0.02%.

In the study by Yanovitskji (1995) the rigorous expression for the function $a_2(\mu)$ has been derived, and the simple approximation for $a_3(\mu)$ accounting for the formula from the book by Minin (1988) (4.55, p. 155) has been deduced (Melnikova 1992):

$$
\begin{aligned}
a_2(\mu) &= 3K_0(\mu) \left[\frac{3}{1+g}(1.271\mu - 0.9) + 4q' \right] , \\
a_3(\mu) &= 4K_0(\mu) \left[4.5g - \frac{1.6}{1+g} - 3 - n_2 w(\mu) \right] .
\end{aligned}
\tag{2.34}
$$

The integration of the expressions for functions $a_2(\mu)$ and $a_3(\mu)$ with respect to μ leads to values

$$a_2 = 12q' + \frac{9}{1+g}(1.271\,q' - 0.9) = 12\,q' + 0.007$$

Table 2.2. Values of second coefficient $a_2(\mu)$ of the plane albedo expansion for the semi-infinite layer and parameter $0.75 \leq g \leq 0.9$

g	0	0.1	0.2	0.3	0.4	μ 0.5	0.6	0.7	0.8	0.9	1.0
0.75	1.310	2.220	3.118	4.078	5.126	6.256	7.475	8.786	10.19	11.70	13.29
0.80	1.267	2.236	3.151	4.117	5.163	6.289	7.494	8.796	10.18	11.66	13.23
0.85	1.201	2.242	3.181	4.148	5.198	6.320	7.512	8.798	10.17	11.63	13.18
0.90	1.092	2.244	3.208	4.193	5.237	6.350	7.529	8.808	10.16	11.60	13.12

and

$$a_3 = 36\,q' - 6g - \frac{1.6}{1+g}$$

that gives errors 0.04 and 0.004% correspondingly. The values of function $a_2(\mu)$ computing for four values of parameter g are presented in Table 2.2.

Surface albedo A is assumed by the formulas:

$$\bar{K}_0(\mu) = K_0(\mu) + A/(1-A) \,,$$

$$\bar{K}_2(\mu) = K_2(\mu) + \frac{A}{1-A}\left[3K_0(\mu)\frac{3.8\mu - 2.7}{1+g} + n_2\right] \,, \qquad (2.35)$$

2.2.3
The Analytical Presentation of the Reflection Function

The following group of formulas is the approximations obtained from the analysis of the numerical values of the reflection function. As is usually done (Sobolev 1972; King 1983; Minin 1988; Yanovitskij 1997), let us describe the reflection function with the above-mentioned expansion over the azimuth angle cosine to separate the item independent of the azimuth angle:

$$\varrho(\varphi, \mu, \mu_0) = \varrho^0(\mu, \mu_0) + 2\sum_{m=1}^{\infty} \varrho^m(\mu, \mu_0)\cos m\varphi \,, \qquad (2.36)$$

where functions $\varrho^m(\mu, \mu_0)$ are the harmonics of the reflection function of order m. Superscripts specify here the number of the azimuthal harmonics. As has been mentioned above, we are using here the phase function described by the Henyey-Greenstein formula (1.31).

The analysis of the numerical calculations (Yanovitskij 1972; King 1983; King 1987; Yanovitskij 1997) shows that for the accurate description of function $\varrho(\varphi, \mu, \mu_0)$ it is enough to know the zeroth and first 6 harmonics if either of cosines μ and μ_0 are greater than 0.15 even for value $g = 0.9$, unfavorable for computing accuracy. This limitation does not restrict our consideration

Table 2.3. Linear approximation for coefficients a^m, b^m, c^m in formula (2.37) for zero, first and second azimuthal harmonics of the reflection function

m	a^m	b^m	c^m	μ_{limit}
0	$2.051\,g + 0.508$	$-1.420\,g + 0.831$	$0.930\,g + 0.023$	–
1	$1.821\,g - 0.558$	$-1.413\,g + 0.387$	$1.150\,g - 0.239$	0.80
2	$2.227\,g - 0.669$	$-1.564\,g + 0.481$	$1.042\,g - 0.293$	0.55

Table 2.4. Power approximation for the coefficients a^m, b^m, c^m in (2.37) for 3rd, 4th, 5th and 6th azimuthal harmonics of the reflection function

m	a^m	$0.3 \leq g \leq 0.9$ b^m	c^m	μ_{limit}
3	$62.00\,g^3 - 90.28\,g^2 + 42.42\,g - 6.26$	$-15.24\,g^3 + 19.70\,g^2 - 8.73\,g + 1.25$	$2.75\,g^2 - 2.03\,g + 0.39$	0.50
4	$105.26\,g^3 - 155.06\,g^2 + 72.93\,g - 10.76$	$-30.30\,g^3 + 43.04\,g^2 - 19.83\,g + 2.89$	$3.70\,g^2 - 3.20\,g + 0.65$	0.45
5	$120.63\,g^3 - 177.60\,g^2 + 83.48\,g - 12.32$	$-25.84\,g^3 + 35.15\,g^2 - 15.61\,g + 2.22$	$3.23\,g^2 - 2.75\,g + 0.55$	0.35
6	$144.92\,g^3 - 202.16\,g^2 + 90.48\,g - 12.85$	$-32.60\,g^3 + 43.88\,g^2 - 19.15\,g + 2.67$	$3.90\,g^2 - 3.41\,g + 0.70$	0.35

because it is also necessary to use a complicated model of the spherical atmosphere and to take into account the refraction of solar rays for the small cosines of zenith solar and viewing angles. These cases are not studied here.

The values of $\varrho^m(\mu, \mu_0)$ for $m = 0, \ldots, 6$ have been analyzed in the study by Melnikova et al. (2000). The following expression, which is similar to the formula for the zeroth harmonic in the book by Sobolev (1972), is used for the description of high harmonics $\varrho^m(\mu, \mu_0)$:

$$\varrho^m(\mu, \mu_0) = [a^m \mu \mu_0 + b^m(\mu + \mu_0) + c^m]/(\mu + \mu_0) \,. \tag{2.37}$$

This presentation provides the reciprocity of the reflection function relative to both zenith viewing and zenith solar angles.

The approximation of coefficients a^m, b^m and c^m in the range of parameter g $0.3 \leq g \leq 0.9$ is presented in Tables 2.3 and 2.4.

The well-known relation of the rigorous theory (Sobolev 1972; Minin 1988; Yanovitskij 1997) is assumed for the isotropic and conservative scattering ($g = 0$, $\omega_0 = 1$), namely:

$$\varrho^0(\mu, \mu_0) = \frac{\varphi(\mu)\varphi(\mu_0)}{4(\mu + \mu_0)} \,, \tag{2.38}$$

where $\varphi(\mu)$ is Ambartsumyan's function (Sobolev 1972). In this case the following approximation is correct: $\varphi(\mu) = 1.874\,\mu + 1.058$ and it has been obtained that $a^0 = 0.88$, $b^0 = 0.47$, and $c^0 = 0.28$ (Melnikova 1992). It is known that the reflection function for the isotropic scattering does not differ very much from the anisotropic values of $\varrho^0(\mu, \mu_0)$ if $\mu, \mu_0 > 0.25$ (Minin 1988; Melnikova et al. 2000), so it is possible to improve this approach for the enlarged angle ranges. The formula for the isotropic scattering (2.38) could be corrected approximately with the linear dependence upon the asymmetry parameter as follows (Melnikova 1992; Melnikova et al. 2000):

$$\varrho^0(\mu, \mu_0) = \frac{\varphi(\mu)\varphi(\mu_0) + g[4.8\,\mu_0\mu - 3.0(\mu_0 + \mu) + 1.9]}{4(\mu_0 + \mu)}. \tag{2.39}$$

In the case of the Henyey-Greenstein phase function the high harmonics are close to zero ($\varrho^m(\mu, \mu_0) \approx 0$, $m > 0$) if either of zenith angle cosines μ and μ_0 are greater than μ_{limit}. The values of μ_{limit} distinguish for different harmonics and they are shown in Tables 2.3 and 2.4. The approximation by (2.37) with coefficients a^m, b^m and c^m in Tables 2.3–2.4 gives an acceptable presentation for all the harmonics of the reflection function considered here. The errors of this approximation have been shown to depend on the values of the zenith solar and viewing angles cosines, on the number of the harmonic m, and on phase function parameter g (Melnikova et al. 2000). Some details of the error analysis will be presented in Sect. 2.4.

The presented totality of rigorous asymptotic formulas (2.24)–(2.28), expansions (2.29)–(2.31) and approximations (2.32)–(2.35) allows computing the reflected and transmitted radiance and irradiance together with the radiative flux divergence for the cloud layer if the layer properties and the geometry of the problem are known. The considered model has to satisfy the applicability ranges of the presented formulas: *large optical thickness* and *weak true absorption*. These ranges will be analyzed in Sect. 2.4 in detail. However, it is necessary to point out that for the application of (2.24)–(2.28) the large optical thickness is a condition with known asymptotic functions and constants. The using of expansions (2.29)–(2.31) needs the weak absorption condition.

We would like to mention that the approximation formula for the reflection function, for which needs to be known the whole phase function, has also been proposed in the study by Konovalov (1997).

2.2.4
Diffused Radiation Field Within the Cloud Layer

Radiation within the cloud layer (in the diffusion domain: $\tau_0 - \tau_{N-1} \gg 1$ and $\tau_1 \gg 1$) is described with formulas different from those presented above. The correspondent analysis could be found in Minin (1988) and Ivanov (1976). Here we are offering the results useful for further consideration.

The diffused radiance in energetic units in the diffusion domain at optical depth τ satisfied conditions $\tau \gg 1$, $\tau_0 - \tau \gg 1$ and is expressed with the equation, derived in the book by Minin (1988)

$$I(\tau, \mu, \mu_0, \tau_0) = SK(\mu_0)\mu_0 \exp(-k\tau)$$
$$\times \frac{i(\mu)\exp(k(\tau_0 - \tau))/i(-\mu)\bar{l}\exp(-k(\tau_0 - \tau))}{1 - l\bar{l}\exp(-2k\tau_0)}, \tag{2.40}$$

where S is the solar constant, function $i(\mu)$ characterizes the angular dependence of the radiance in deep levels of the semi-infinite atmosphere. The behavior of function $i(\mu)$ relative to the phase function shape and absorption in the medium has been studied in the book by Yanovitskij (1997). The expansion for function $i(\mu)$ has been derived in the paper by Yanovitskij (1972) in the case of weak true absorption, which is presented here in terms of parameter s:

$$i(\mu) = 1 + 3s\mu + 3\frac{1 - g^2 + 2P_2(\mu)}{1 + g}s^2$$

$$+ \left[9(1 - 1.5g)\mu + \frac{10.8\,P_3(\mu)}{(1 + g)(1 + g + g^2)} + \frac{3.6\,\mu}{1 + g}\right]s^3 + O(s^4).$$

(2.41)

Functions $P_i(\mu)$ for $i = 1, 2, \ldots$ are Legendre polynomials of power i.

The diffused irradiance in relative units of πS within the optically cloud layer is described with the following:

$$F^\downarrow(\mu_0, \tau, \tau_0) = \frac{K(\mu_0)\exp(-2k\tau_0)}{1 - \bar{l}\bar{l}\exp(-2k\tau_0)}[i^\downarrow \exp(k(\tau_0 - \tau)) - i^\uparrow \bar{l}\exp(-k(\tau_0 - \tau))],$$

$$F^\uparrow(\mu_0, \tau, \tau_0) = \frac{K(\mu_0)\exp(-k\tau_0)}{1 - \bar{l}\bar{l}\exp(-2k\tau_0)}[i^\uparrow \exp(k(\tau_0 - \tau)) - i^\downarrow \bar{l}\exp(-k(\tau_0 - \tau))],$$

(2.42)

where

$$i^\downarrow = 2\int_0^1 i(\mu)\mu d\mu, \quad i^\uparrow = 2\int_0^1 i(-\mu)\mu d\mu.$$

Expansions for values $i^\downarrow$ and $i^\uparrow$ have been derived in the book by Minin (1988) after integrating (2.41):

$$i^{\downarrow\uparrow} = 1 \pm 2s + 3s^2\frac{1.5 - g^2}{1 + g} \pm 3s^3\left[2 - 3g + \frac{0.8}{1 + g}\right] + O(s^4).$$

(2.43)

It is also convenient to describe the internal radiation field with the values of *internal albedo* $b(\tau_i) = F^\uparrow(\tau_i)/F^\downarrow(\tau_i)$ and net flux $F(\tau_i) = F^\downarrow(\tau_i) - F^\uparrow(\tau_i)$, according to Minin (1988) and Ivanov (1976)

$$F(\tau, \mu_0) = F^\downarrow(\tau, \mu_0) - F^\uparrow(\tau, \mu_0) = \frac{4sK(\mu_0)\exp(-k\tau)}{1 - \bar{l}\bar{l}\exp(-2k\tau_0)}[1 + \bar{l}\exp(-k(\tau_0 - \tau))]$$

$$\frac{F^\uparrow(\tau, \mu_0)}{F^\downarrow(\tau, \mu_0)} = b(\tau) = \frac{b^\infty - \bar{l}\exp(-2k(\tau_0 - \tau))}{1 - b^\infty\bar{l}\exp(-2k(\tau_0 - \tau))}.$$

(2.44)

Value b^∞ and function $b(\tau)$ are called the internal albedo of the infinite atmosphere and the internal albedo of the atmosphere of the large optical thickness correspondingly, moreover $b^\infty = 1 - 4s + s^2$ and the values of function $b(\tau)$ could be obtained from the observations or from the calculations of the semispherical irradiances at level τ.

2.2.5
The Case of the Conservative Scattering

In the absence of the true absorption, according to the definition, we have $\omega_0 = 1$ and the expressions for the radiative characteristics are particularly simple (Sobolev 1972; Minin 1988).
For the reflection and diffusion functions:

$$\varrho(0, \mu, \mu_0, \varphi) = \varrho_0(\mu, \mu_0, \varphi) - \frac{4K_0(\mu_0)K_0(\mu)}{3\left[(1-g)\tau_0 + \delta + \frac{4A}{3(1-A)}\right]},$$

$$\sigma(\tau_0, \mu, \mu_0) = \frac{4K_0(\mu_0)\bar{K}_0(\mu)}{3\left[(1-g)\tau_0 + \delta + \frac{4A}{3(1-A)}\right]};$$

(2.45)

for the semispherical fluxes in relative units of πS

$$F^{\uparrow}(0, \mu_0) = 1 - \frac{4K_0(\mu_0)}{3\left[(1-g)\tau_0 + \delta + \frac{4A}{3(1-A)}\right]},$$

$$F^{\downarrow}(\tau_0, \mu_0) = \frac{4K_0(\mu_0)}{3(1-A)\left[(1-g)\tau_0 + \delta + \frac{4A}{3(1-A)}\right]},$$

(2.46)

and, finally, the simple expression for the net flux that summarizes both equations (2.46) is feasible at any level in the conservative medium because the net flux is constant without absorption (Minin 1988)

$$F(\tau, \mu_0) = \frac{4K_0(\mu_0)(1-A)}{3(1-A)[(1-g)\tau_0 + \delta] + 4A}.$$

(2.47)

It should be emphasized that equality $F^{\uparrow}(\tau, \mu_0) = F^{\downarrow}(\tau, \mu_0) = K_0(\mu_0)$ is correct in the semi-infinite conservatively scattered atmosphere with a thick optical depth, where the sense of escape function $K_0(\mu)$ frequently met in our consideration is clear from. The case of the conservative scattering becomes true in a certain cloud layer at the single wavelengths within the visual spectral range. Equations (2.45)–(2.47) are correct in the wider interval of the optical depth ($\tau_0 \geq 3$) than (2.24), (2.26), (2.28) derived with taking into account the absorption. Corresponding relations of the characteristics of the inner radiation field are written as:
For the radiance:

$$I(\tau, \mu) = \frac{S\mu_0 K_0(\mu_0)\{(1-A)[3(1-g)(\tau_0 - \tau) + 1.5\delta + 3\mu] + 4A\}}{(1-A)[3(1-g)\tau_0 + 3\delta] + 4A},$$

(2.48)

for the upward and downward semispherical solar fluxes:

$$F^{\uparrow}(\tau, \mu_0) = K_0(\mu_0)\frac{(1 - A)[3(1 - g)(\tau_0 - \tau) + 1.5\delta - 2] + 4A}{(1 - A)[3(1 - g)\tau_0 + 3\delta] + 4A}$$

$$F^{\downarrow}(\tau, \mu_0) = K_0(\mu_0)\frac{(1 - A)[3(1 - g)(\tau_0 - \tau) + 1.5\delta + 2] + 4A}{(1 - A)[3(1 - g)\tau_0 + 3\delta] + 4A} \, .$$

(2.49)

It is possible to apply the formulas of the radiative characteristics in the case of conservative scattering for a rough estimation even for very weak absorption but the computational errors increase fast when the absorption grows and it is necessary to use the equations for the absorption medium to reach a certain accuracy.

2.2.6
Case of the Cloud Layer of an Arbitrary Optical Thickness

The optical thickness of certain cloud layers is not sufficient in some cases for making use of the above-presented equations and their application leads to significant errors and it causes the necessity of different approaches. We would like to mention the two-flux Eddington and delta-Eddington methods among all analytical approaches (Josef et al. 1980). These methods are notable for the simple expressions and they provide sufficient accuracy of the calculations, however, they are approximations. In addition, they are awkward enough and hence, are not convenient for the inverse problem transforming. A mathematically rigorous method has been developed for the calculation of the irradiances at the boundaries of the layer of arbitrary optical thickness in Yanovitskij (1991,1997) and Dlugach and Yanovitskij (1983). The restrictions to the true absorption are more rigorous than above and the optical thickness is accepted in the range $0.1 < \tau_0 < 5.0$ The irradiance outgoing from the layer is described with the following:

$$F^{\uparrow} = 1 - f[u(\mu, \tau_0)\,\text{ch}\,k\tau_0 - v(\mu, \tau_0)] \, ,$$

$$F^{\downarrow} = f[u(\mu, \tau_0) - v(\mu, \tau_0)\,\text{ch}\,k\tau_0] \, ,$$

$$f = \frac{4s}{\text{sh}\,k\tau_0} \, .$$

(2.50)

Functions $\text{sh}\,k\tau_0$ and $\text{ch}\,k\tau_0$ specify the hyperbolic sine and cosine, functions $u(\mu_0, \tau_0)$ and $v(\mu_0, \tau_0)$ are defined in several studies (Dlugach and Yanovitskij 1983; Yanovitskij 1991, 1997) and they are similar to the escape function. Here we are not adducing these definitions. It should be emphasized only that they depend on the optical thickness as well. Besides, these functions depend inexplicitly on the phase function. The tables containing the values of functions $u(\mu_0, \tau_0)$ and $v(\mu_0, \tau_0)$ for the wide set of the arguments and several values of phase function parameter g have been calculated and presented in Yanovitskij (1991) and Dlugach and Yanovitskij (1983)

According to Yanovitskij (1991, 1997) and Dlugach and Yanovitskij (1983) functions $p(\tau_0)$ and $q(\tau_0)$ have been specified for accounting the surface reflection

$$p(\tau_0) = 2 \int_0^1 u(\mu, \tau_0)\mu d\mu , \quad q(\tau_0) = 2 \int_0^1 v(\mu, \tau_0)\mu d\mu , \qquad (2.51)$$

moreover, relation $p(\tau_0) + q(\tau_0) = 1$ is correct for these functions. The irradiances outgoing from the layer at the boundaries with the reflecting surface at the bottom are described as

$$\bar{F}^\uparrow(\mu_0, \tau_0) = 1 - f\{[u(\mu_0, \tau_0)\, \mathrm{ch}\, k\tau_0 - v(\mu_0, \tau_0)] + A\bar{F}^\downarrow[p(\mathrm{ch}k\tau_0 + 1) - 1]\} ,$$
$$\bar{F}^\downarrow(\mu_0, \tau_0) = f[u(\mu_0, \tau_0) - v(\mu_0, \tau_0)\, \mathrm{ch}k\tau_0]/\{(1 - A) + Af[p(\mathrm{ch}\, k\tau_0 + 1) - 1]\} .$$
$$(2.52)$$

The above-presented expressions may be useful for computing the solar irradiances in the case of lower optical thickness (cirrus clouds or cloudless atmosphere with a heavy gaze).

2.3
Calculation of Solar Irradiance and Radiance in the Case of the Multilayer Cloudiness

The radiation field in the multilayer cloudiness was considered in many studies (e. g. Sobolev 1974; Germogenova and Konovalov 1974; Ivanov 1976). Applying the approaches developed in these studies, numerical difficulties arise connecting with the necessity of accounting the total interrelation of all layers. While solving the inverse problems, these difficulties are intensifying. However, it is possible to neglect the whole totality of the interrelations and consider every layer independently with taking into account the approximate influence of the neighbor layers in real problems concerning cloud layers of rather large optical thickness. Just such an approach was elaborated in a study (Melnikova and Minin 1977) for computing the downwelling and upwelling solar irradiances in the vertically heterogeneous medium consisting of two optically thick layers with different optical properties. It has been assumed that the irradiance transmitted by the upper layer accepted as an incident flux for the lower layer. The influence of the lower layer on the upper radiation field is determined by its spherical albedo, i. e. the lower layer is accepted as a reflecting surface for the upper layer. That is to say, the angle distribution of diffused radiation incoming from the bottom to the upper layer and from the top to the lower layer is accounted for approximately. The test of the approach has indicated the relative error of such approximation to be less than 1%.

Let the total optical thickness of the system of N cloud layers be $\tau_0 = \Sigma\tau_i \gg 1$, where τ_i is the optical thickness of i-th sublayer. The single scattering albedo of the i-th sublayer is ω_{0i}, moreover the true absorption is weak compared to the scattering, $1 - \omega_{0i} \ll 1$. The volume extinction coefficient is specified as ε_i, the absorption coefficient of the i-th layer is $\kappa_i = \varepsilon_i(1 - \omega_{0i})$, and the scattering coefficient is $\alpha_i = \varepsilon_i\omega_{0i}$. We are neglecting the radiation scattering in the optically thin clear atmosphere between the cloud layers and in the underlying clear layer and assuming that the lower layer adjoins the ground surface with albedo A.

Remember that *the diffused irradiances* outgoing from the optically thick layer are described in relative units πS by (2.27). The albedo for the upper layer is accepted as the value of the spherical albedo of the second layer (counting from above):

$$A_1 = a_2^\infty - \frac{n_2^2 m_2 \bar{l}_2 \exp(-2k_2\tau_2)}{1 - l_2\bar{l}_2\exp(-2k_2\tau_2)} .$$

Value a_2^∞ specifies the spherical albedo of the infinite atmosphere with properties of the second layer: $a_2^\infty = 1 - 4s_2 + 6\delta s_2^2$. The subscripts indicate for which layer the values are calculated. In the system of N layers the escape function $K(\mu_0)$ of the layer with number $i > 1$ is replaced with the integral of the function with respect to the value μ_0 (with value n_i) and multiplied by the irradiance transmitted by the upper layer $n_iF^\downarrow(\tau_{i-1})$. The following specifications have been accepted in the study by Melnikova and Zhanabaeva (1996):

$$\bar{f}_\sigma(\tau_i) = \frac{m_i \exp(-k_i\tau_i)}{1 - l_i\bar{l}_i \exp(-k_i\tau_i)} ,$$

$$\bar{f}_\rho(\tau_i) = l_i e^{-k_i\tau_i}\bar{f}_\sigma(\tau_i) = \frac{m_i\bar{l}_i \exp(-2k_i\tau_i)}{1 - l_i\bar{l}_i \exp(-2k_i\tau_i)} , \qquad (2.53)$$

$$\bar{n}_i = \frac{n_i}{1 - A_i a_i^\infty} .$$

Finally the expressions of the diffused irradiances at its boundaries are derived for the layer with number $k > 1$:

$$F_k^\downarrow = \frac{K(\mu_0)}{n_1} \prod_{i=1}^{k} \bar{n}_i n_i \bar{f}_\sigma(\tau_i) = F_{k-1}^\downarrow \bar{n}_k n_k \bar{f}_\sigma(\tau_k) , \qquad (2.54)$$

$$F_k^\uparrow = [a_k^\infty - n_k^2\bar{f}_\rho(\tau_k)]F_{k-1}^\downarrow .$$

and

$$A_i = a_{i+1}^\infty - Q_{i+1}f_\rho(\tau_{i+1}) .$$

The formulas for computing the solar diffused radiance are derived as above by substituting the product that described diffused radiation incoming to layer:

$n_i\sigma_{i-1}(\tau_{i-1},\mu_0,\mu)$. The expressions for the radiance are obtained as:

$$\sigma_k = \frac{K_1(\mu_0)}{n_1} \prod_{i=1}^{k} \bar{K}_i(\mu)n_i\bar{f}_\sigma(\tau_i) = \sigma_{k-1}\bar{K}_k(\eta)n_k\bar{f}_\sigma(\tau_k) , \tag{2.55}$$

$$\varrho_k = [a_k(\mu) - K_k(\mu)n_k\bar{f}_\varrho(\tau_k)]\sigma_{k-1} .$$

The subscripts in these expressions are related to the layer with the correspondent number and optical parameters g_i, ω_{0i} and τ_i. If there is a conservative scattering in the layer with number i, escape function $K(\mu)$ converts to function $K_0(\mu)$, values n_i and a_i^∞ are equal to unity, $\bar{n}_i$ accepts value $1/(1 - A_i)$ and functions $f_\varrho(\tau_i)$ and $f_\sigma(\tau_i)$, defining the dependence upon the optical thickness are expressed with the formula:

$$\bar{f}_\varrho(\tau_i) = \bar{f}_\sigma(\tau_i) = \frac{4(1 - A_i)}{(1 - A_i)[3(1 - g_i)\tau_i + 3q'] + 4A_i} . \tag{2.56}$$

For the case of the layers of the arbitrary optical thickness, (2.52) is used for the derivation of the expressions for multilayer clouds analogous to the thick layers. The formulas for the irradiances for the upper layer coincide with (2.52). The irradiances for the lower layers are expressed with:

$$F_i^\uparrow = F_{i-1}^\downarrow \bar{A}_i = F_{i-1}^\downarrow \left[A_i + \frac{A_{i+1}V_i^2}{1 - A_{i+1}A_i} \right] ,$$

$$F_i^\downarrow = F_{i-1}^\downarrow \bar{V}_i = F_{i-1}^\downarrow \frac{V_i}{1 - A_{i+1}A_i} , \tag{2.57}$$

where values A_i and V_i are computed with the relations:

$$A_i = 2 \int_0^1 F_i^\uparrow \mu d\mu = 1 - f_i[p(\tau_i)(\mathrm{ch}\, k\tau_i + 1) - 1] , \tag{2.58}$$

$$V_i = f_i[p(\tau_i)(\mathrm{ch}\, k\tau_i + 1) - \mathrm{ch}\, k\tau_i] .$$

2.4
Uncertainties and Applicability Ranges of the Asymptotic Formulas

The asymptotic formulas of the transfer theory presented in this chapter are obtained rigorously. It is necessary to take into consideration that they are describing the radiation field within the boundaries and at them the more exact, the bigger the optical thickness is and the less true absorption is. In addition, there is a strong relationship between the accuracy and the degree of the scattering anisotropy (the extension of the phase function forward or the magnitude of parameter g). Certain mathematical aspects concerning the estimation of the applicability ranges for the asymptotic formulas of reflection

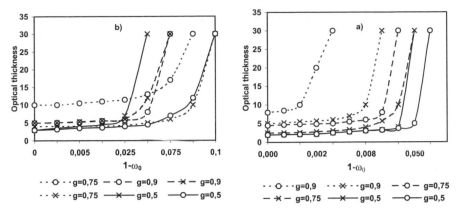

Fig. 2.2a,b. The applicability ranges of the asymptotic formulas of radiative transfer theory in the case of calculation reflected irradiance (**a**) and radiative flux divergence (**b**) for the cloudy layer. Curves correspond to the relative uncertainty equal to 5%. The *solid curve* is for phase function parameter $g = 0.5$; *dashed line* – $g = 0.75$ and the *dashed dotted line* – $g = 0.9$; *curves with circles* correspond to $\mu = 1$, with *crosses* – to $\mu = 0.5$

and transmission functions $\rho(\mu, \mu_0, \tau_0)$ and $\sigma(\mu, \mu_0, \tau_0)$ were analyzed in studies by Konovalov (1974,1975). The accuracy of the formulas for $\rho(\mu, \mu_0, \tau_0)$ and $\sigma(\mu, \mu_0, \tau_0)$ has turned out roughly equal to each other. The uncertainties of the formulas for reflected and transmitted radiation are about 2% beginning from optical thickness $\tau_0 \geq 4/(1 - k)$. The numerical analysis of the formulas of the spherical albedo and transmittance (the values of reflected and transmitted irradiances integrated with respect to the cosine of the solar zenith angle) for the wide set of parameters has been accomplished in the study by Harshvardhan and King (1986). It has been shown there that their uncertainty does not exceed 5% by values $\tau \geq 2.0$ and $\omega_0 \geq 0.7$.

The accuracy of the formulas for irradiances was tested to provide the relative errors less than 5% in the region plotted in coordinates "$\tau - \omega_0$" in Fig. 2.2 (Demyanikov and Melnikova 1986). The curves in Fig. 2.2a,b correspond to the level of 5% error of the reflected irradiance (a) and radiative flux divergence (b) calculated for asymmetry parameters $g = 0.5, 0.75$ and 0.9 and for two values of cosine $\mu_0 = 1$ and 0.5. The result for the transmitted irradiance is similar to the result shown in Fig. 2.2 for the reflected irradiance.

The numerical analysis of the accuracy of the radiance calculation in the optically thick layer has been accomplished in the book by Yanovitskij (1997). According to this book the applicability region of the radiance ($\tau > 15$; $\omega_0 > 0.99$) is more strongly restricted than of the irradiance ($\tau > 7$; $\omega_0 > 0.9$), which in their turn is narrower than for the integral over the zenith angle characteristics ($\tau > 2$; $\omega_0 > 0.8$).

The accuracy of asymptotic expansions (2.29) and (2.30) is defined by the omitted items proportional to s^3 or s^4. The accuracy of the approximations was tested by comparison with the function values computed by the numeri-

Table 2.5. Uncertainty of the calculation of escape function $K(\mu)$, %

ω_0	0.999		0.995		0.990		0.980		
g	0.5	0.9	0.5	0.9	0.5	0.9	0.5	0.75	0.9
s	0.02580	0.05774	0.05774	0.12910	0.08165	0.18257	0.11550	0.16330	0.25820
$\mu = 0.1$ 0.1	0.2	0.4	1.0	0.5	2.0	10	33	127	
$\mu = 0.5$ 0.1	0.4	0.1	2.0	0.1	4.0	6.0	29	79	
$\mu = 0.7$ 0.3	0.5	0.3	0.8	0.4	3.0	5.0	25	64	
$\mu = 1.0$ 0.2	0.6	0.6	2.0	1.0	4.0	2.5	12	45	

Table 2.6. Uncertainty of the calculation of reflection function $\varrho_\infty(\mu, \mu)$ of the semispherical layer

ω_0	0.999		0.995		0.990	
g	0.5	0.9	0.5	0.9	0.5	0.9
s	0.02580	0.05774	0.05774	0.12910	0.08165	0.18257
$\mu = 0.1$	0.2	0.6	0.2	1.0	0.3	2.6
$\mu = 0.5$	0.2	0.3	0.4	1.0	1.0	3.0
$\mu = 1.0$	0.2	0.3	0.5	1.0	0.7	3.0

cal methods and presented in Yanovitskij (1997) and Dlugach and Yanovitskij (1974). The relative uncertainties of the escape function computed with approximations (2.31) are presented in Table 2.5. It has been found that uncertainties are rather small as far as $\omega_0 = 0.98$ for magnitudes $g = 0.5$ and $\mu > 0.2$. Table 2.5 illustrates that the errors of the escape function calculation do not exceed 6% for value $s < 0.12$.

Comparison of the results of the reflection function $\varrho_\infty(\mu, \mu_0)$ calculation accounting for coefficients $\varrho_2(\mu, \mu_0)$ and $\varrho_3(\mu, \mu_0)$ of expansion (2.30) with the numerical computing results of studies by Yanovitskij (1997) and Dlugach and Yanovitskij (1974) yields the errors shown in Table 2.6. Equation (2.31) for functions $\varrho_2(\mu, \mu_0)$ and $\varrho_3(\mu, \mu_0)$ allow the computing of corresponding values with a rather small error as far as $\omega_0 = 0.9$. Therefore, it is possible to calculate the solar radiance reflected from the cloud layer in the shortwave spectral range with the analytical formulas, and this fact is useful for the interpretation of the satellite radiation data.

The accuracy of the formulas in the case of an arbitrary optical thickness has been tested by comparison with the results of the numerical calculations using the following methods: double and adding method, delta-Eddington method and Monte-Carlo method. A wide set of parameters has been analyzed: $\tau_0 = 0.1 - 5.0$; $\omega_0 = 0.99 - 0.9999$ and $g = 0.25 - 0.75$ (Melnikova and Solovjeva 2000). The results of all four methods have turned out to coincide with the variations from 0.1 to 5% independent of the magnitudes of τ_0, ω_0 and g. Thus, it can be thought that all tested magnitudes of the parameters are in the applicability region of the formulas derived in the work of Yanovitskij (1991) and Dlugach and Yanovitskij (1983). For the weakly extended phase function ($g \leq 0.5$) calculations, the errors are not exceeding 1%.

The accuracy of the calculations of the radiative characteristics with (2.53)–(2.56) for multilayer cloudiness has been tested for the following cases: $\tau_i = 5, 7, 10$; $g_i = 0.65, 0.75, 0.85$ and $\omega_{0i} = 0.99, 0.995, 0.999$ by comparison with the values calculated with the doubling and adding method. Equations (2.53)–(2.56) for the irradiance and radiance are accurate for all values of ω_{0i} and g_i when $\tau_i \geq 7$, the errors are less than 1–2%, and when $\tau_i \sim 5$ the error reaches 10% (Melnikova and Zhanabaeva 1996).

2.5
Conclusion

Specific features of two methods are considered in Chap. 2: the first is the Monte-Carlo method, one of the most widely used numerical methods for the calculation of radiative characteristics; the other is a method of asymptotic formulas from transfer theory applied to the calculation of radiative characteristics in the case of the overcast sky.

The Monte-Carlo method allows for all features of the interaction of radiation with the atmosphere and surface with high accuracy that makes it indispensable for the standard calculations of the radiative characteristics of the atmosphere. Besides, the Monte-Carlo method makes possible the simulation of the processes of the real radiation measurements, which is especially important for problems of observational data interpretation (Fomin et al. 1994). This is the main reason for the application of the method in our analysis of airborne observational data of the solar irradiances that will be considered in Chaps. 3–5. Finally, we would like to mention that the Monte-Carlo method is rather simple and flexible, which allows easy realization of computing algorithms on PC and the application of these to different problems of the theory of radiative transfer. Further dissemination of the method could be expected in the near future taking into account the appearance of modern computer systems with the ability to perform parallel calculations (Sushkevitch et al. 2002). The main and rather serious disadvantage of this method is the random error contained in its results (i. e. the method is a full analog of the observations). An increase in computing time and modern computing systems can lead to a decrease in this error.

The approach for the calculation of the reflection function of the semi-infinite atmosphere with the analytical formulas is proposed for the Henyey-Greenstein phase function. We would like to point out that on the one hand the phase function for real clouds could be more complicated than the Henyey-Greenstein formula. However, on the other hand Raleigh scattering together with the influence of multiple scattering could turn out to be rather significant and smooth the shape of the real cloud phase function. Thus, the proposed approach can provide less computational error for the real cloudiness than is to be expected according to the theory. We would like to stress that the analytical method is especially convenient for inverse problem solving namely for the retrieval of optical parameters from the solar radiance and irradiance observations. Analytical formulas presented here will be used later to derive the correspondent inverse formulas, and to express the optical parameters of cloud layers through the measured values of the solar radiance and irradiance (Chap. 6).

References

Demyanikov AI, Melnikova IN (1986) On the applicability region determination for asymptotic formulas of monochromatic radiation transfer theory. Izv Acad Sci USSR Atmosphere and Ocean Physics 22:652–655 (Bilingual)

Dlugach JM, Yanovitskij EG (1974) The optical properties of Venus and Jovian planets. II. Methods and results of calculations of the intensity of radiation diffusely reflected from semi-infinite homogeneous atmospheres. Icarus 22:66–81

Dlugach ZhM, Yanovitskij EG (1983) Illumination of surface and albedo of planet atmosphere by quasy-conservative scattering. Izv AN USSR Atmosphere and Ocean Physics 19:813–823 (Bilingual)

Fomin VA, Rublev AN, Trotsenko AN (1994) Standard calculations of fluxes and radiative flux divergences in aerosol and cloudy atmosphere. Izv Acad Sci USSR, Atmosphere and Ocean Physics 30:301–308 (Bilingual)

Germogenova TA (1961) On character of transfer equation for plane layer. J Comput Math Math Physics 1:1001–1019 (in Russian)

Germogenova TA, Konovalov NV (1974) Asymptotic characteristics of solution of transfer equation in inhomogeneous layer problem. J Comput Math Math Physics 14:928–946 (in Russian)

Harshvardhan, King MD (1993) Comparative accuracy of diffuse radiative properties computed using selected multiple scattering approximations. J Atmos Sci 50:247–259

Hulst HC Van de (1980) Multiple Light Scattering. Tables, Formulas and Applications. Vol. 1 and 2, Academic Press, New York

Ivanov VV (1976) Radiation field in the optically thick planet atmosphere adjacent to reflected surface. Astronomical J 53:589–595 (in Russian)

Ivanov VV (1976) Radiation transfer in multi layer optically thick atmosphere. II. Studies of Leningrad University Astronomic Observatory 32:23–39 (in Russian)

Joseph JH, Wiscombe WJ, Weiman JA (1980) The delta-Eddington approximation for radiative flux transfer. J Atm Sci 33:2452–2459

Kargin BA (1984) Statistical modeling for solar radiation field in the atmosphere. Printing house of SO AN USSR, Novosibirsk (in Russian)

King MD (1983) Number of terms required in the Fourier expansion of the reflection function for optically thick atmospheres. J Quant Spectrosc Radiat Transfer 30:143–161

King MD (1987) Determination of the scaled optical thickness of cloud from reflected solar radiation measurements. J Atmos Sci 44:1734–1751

King MD, Harshvardhan (1985) Comparative accuracy of selected multiple scattering approximations. J Atmos Sci 43:784–801

King MD, Harshvardhan (1986) Comparative accuracy of the albedo, transmission and absorption for selected radiative transfer approximations. NASA Reference Publication, 1160, January

King MD, Radke LF, Hobbs PV (1990) Determination of the spectral absorption of solar radiation by marine straticumulus clouds from airborne measurements within clouds. J Atmos Sci 47:894–907

Kokhanovsky A, Nakajima T, Zege E (1998) Physically based parameterizations of the short-wave radiative characteristics of weakly absorbing optically media: application to liquid-water clouds. Appl Opt 37:4750–4757

Konovalov NV (1974) Asymptotic properties of solution of one-velocity transfer equation in homogeneous plane layer. Problem with azimuthal dependence. Preprint of Keldysh Institute for Applied Mathematics (KIAM) RAS, Moscow (in Russian)

Konovalov NV (1975) On applicability region of asymptotic formulas for calculation of monochromatic radiation in inhomogeneous optically thick plane layer. Izv RAS Atmosphere and Ocean Physics 11:1263–1271 (Bilingual)

Konovalov NV (1997) Certain properties of the reflection function of optically thick layers. Preprint of Keldysh Institute for Applied Mathematics (KIAM) RAS, Moscow (in Russian)

Lenoble J (ed) (1985) Radiative transfer in scattering and absorbing atmospheres: standard computational procedures. A. DEEPAK Publishing, Hampton, Virginia

Marchuk GI, Mikhailov GA, Nazaraliev NA et al. (1980) The Monte-Carlo method in the atmosphere optics. Springer-Verlag, New York

Melnikova IN (1992) Analytical formulas for determination of optical parameters of cloud layer on the basis of measuring characteristics of solar radiation field. 1. Theory. Atmosphere and Ocean Optics 5:169–177 (Bilingual)

Melnikova IN, Minin IN (1977) To the transfer theory of monochromatic radiation in cloud layers. Izv Acad Sci USSR Atmosphere and Ocean Physics 13:254–263 (Bilingual)

Melnikova IN, Zshanabaeva SS (1996) Evaluation of uncertainty of approximate methodology of accounting the vertical stratus structure in direct and inverse problems of atmospheric optics. International Aerosol Conference, December, Moscow (in Russian)

Melnikova IN, Solovjeva SV (2000) Solution of direct and inverse problem in case of cloud layers of arbitrary optical thickness and quasi-conservative scattering. In: Ivlev LS (ed) Natural and anthropogenic aerosols. St. Petersburg, pp 86–90 (in Russian)

Melnikova IN, Dlugach ZhM, Nakajima T, Kawamoto K (2000) On reflected function calculation simplification in case of cloud layers. Appl Optics 39:541–551

Minin IN (1988) The theory of radiation transfer in the planets atmospheres. Nauka, Moscow (in Russian)

Mishchenko MI, Dlugach JM, Yanovitskij EG, Zakharova NT (1999) Bi-directional reflectance of flat, high-albedo particulate surfaces: an efficient radiative transfer solution and applications to snow and soil surfaces. J Quant Spectrosc Rad Transfer 63:409–432

Molchanov NI (ed) (1970) Set of computer codes of small electronic-digital computer "Mir". Vol 1, Methods of calculations. Naukova dumka, Kiev (in Russian)

Nakajima T, King MD (1992) Asymptotic theory for optically thick layers: Application to the discrete ordinates method. Appl Opt 31:7669–7683

Sobolev VV (1972) The light scattering in the planets atmospheres. Nauka, Moscow (in Russian)

Sobolev VV (1974) Light scattering in inhomogeneous atmosphere. Astronomical J 51:50–55 (in Russian)

Stepanov NN (1948) The spherical trigonometry. Moscow–Leningrad Gostehizdat (in Russian)

Sushkevitch TA, Strelkov SA, Kulikov AK, Maksakova CA, Vladimirova EV, Ignatyeva EI (2002) On perspectives of modeling of Earth radiation with supercomputers accounting for aerosols and clouds. Thesis presented at IV International Conference "Natural and anthropogenic aerosols", September, 2002, St. Petersburg. Chemistry Institute, St. Petersburg University Press, St. Petersburg

Yanovitskij EG (1972) Spherical albedo of planet atmosphere. Astronomical J 49:844–849 (in Russian)

Yanovitskij EG (1991) Radiation flux in the weakly absorbing atmosphere of an arbitrary optical thickness. Izv Acad Sci USSR Atmosphere and Ocean Physics 27:1241–1246 (Bilingual)

Yanovitskij EG (1997) Light scattering in inhomogeneous atmospheres. Springer, Berlin Heidelberg New York

Spectral Measurements of Solar Irradiance and Radiance in Clear and Cloudy Atmospheres

Certain information concerning the instruments and methodologies of the experimental study together with some results of the observations of solar radiative characteristics in the atmosphere is presented in this chapter. The data of detached radiative spectral airborne measurements obtained in the Laboratory for the Shortwave Radiation of Atmospheric Department of Physics of the Institute of Leningrad (now St. Petersburg) State University are considered here as well.

Ground-based, airborne and ship observations of spectral solar radiances and irradiances have been accomplished during more than 20 years (1970–1990) under the guidance of Kirill Ya Kondratyev, Vladimir S Grishechkin and Oleg B Vasilyev. The results of these observations and their interpretation were described in numerous articles, annals and books (Kondratyev et al. 1972, 1973, 1974, 1975, 1976, 1978, 1987, 1987, 1990; Berlyand et al. 1974; Chapurskiy and Chernenko 1975; Chapurskiy et al. 1975; Kondratyev and Ter-Markaryants 1976; Vasilyev O et al. 1982, 1987, 1995; Grishechkin and Melnikova 1989; Grishechkin et al. 1989; Vasilyev A et al. 1994, 1997a, 1997b, 1997c). The review of all or even of part of these results would be too bulky so here we are dwelling on the results of the airborne radiance and irradiance measurements carried out during 1983–1988, which have been processed recently and have not yet published. We would like to mention that the airborne observations of solar radiances and irradiances, which are aimed at studying atmospheric energetics and the reflectance of surfaces, have been carried out rather intensively before the middle of the 80-th in Russia (Kondratyev et al. 1972, 1973; Kondratyev and Ter-Markaryants 1976). Later the interest in them decreased which is bound up with the accumulation of a significant volume of data, the wide development of satellite observations, and with well-known economic reasons as well. However, single experiments including the airborne ones have still been accomplished (Skuratov et al. 1999).

3.1
Complex of Instruments for Spectral Measurements of Solar Irradiance and Radiance

The measuring complex of the instruments was created at the beginning of the 80-th in St. Petersburg University and regretfully is not used nowadays.

Therefore, together with the modern level of optical instrument making, the features of the old measuring complex are not of great interest nowadays. However, the methodological experience of many-years exploitation of the measuring complex on board aircraft has not lost its actuality. Thus, we will concentrate on it.

By the beginning of the 1980th, instruments for radiation observations were united in a complex: an information-measuring system (Vasilyev O et al. 1987). The measuring part of the complex was provided with the K-3 spectrometer (Mikhailov and Voitov 1969). It was a diffractive mirror spectrometer with a grating of 600 lines per mm as a dispersing element. The operating spectral range of the instrument was 330–978 nm, registration time of the single spectrum was 7 s and the spectral scanning was a mechanical one. The spectrometer had three overlapped operating ranges: the ultraviolet (UV), visual (VD) and near infrared (NIR), with corresponding photomultipliers used as a light receiver. The digital output signal was recorded to the magnetic tape using an ordinary tape-recorder. It allowed the carrying out of all consequent processing of the information on computer.

The spectral graduation, i. e. the determination of wavelengths λ'_i, attributed to the discrete points with numbers i, which are defined by the time moments in the process of mechanical scanning, was accomplished in the laboratory by measuring of the mercury lamp spectrum and by identification of the known spectral lines of the atmospheric gases at the output recording. The oxygen band at 760 nm in the spectrum of diffused radiation of the sky was also used for the graduation in the NIR range. As the mechanical scanning led to the displacement of the registration points, graduating values λ'_i were determined with a rather large uncertainty (both random and systematic). This uncertainty was equated to the root-mean-square (RMS) deviation of the same series of measurements of the mercury lamp spectrum and was equal to 1 nm.

The spectral instrumental function of K-3 spectrometer $f_\lambda(\lambda)$ (look at Sect. 1.1) has been presented in the study by Vasilyev O et al. (1987). This function has been obtained in the laboratory through the registration of laser line (in visual range) and can be approximated by the triangle function with halfwidth $\Delta\lambda$ equal to 3 nm, namely:

$$f_\lambda(\lambda) = 1 - \left| \frac{2\lambda - \lambda'_i}{\Delta\lambda} \right| , \qquad (3.1)$$

where $\Delta\lambda = 3$ nm. It is obvious, that the mentioned graduating uncertainty influences halfwidth $\Delta\lambda$ according to the conditions of the signal registration. The accuracy of approximation $f_\lambda(\lambda)$ by the triangle function is about 1%. It is important to mention that the measured value of the signal varies weakly with wavelength change within the majority of the spectral range of the K-3 instrument, so it is possible not to take into account the instrumental function and the uncertainties of the graduation for halfwidth $\Delta\lambda = 3$ nm as has been mentioned in Sect. 1.1. The exception is the comparatively narrow and deep band of oxygen absorption 760 nm together with some deep Fraunhofer lines of the solar spectrum in the UV region (Fig. 1.3). Therefore, the attention

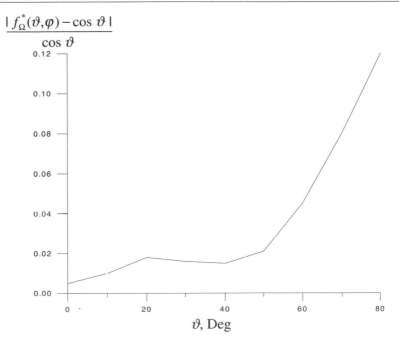

$$\frac{|f_\Omega^*(\vartheta,\varphi) - \cos\vartheta|}{\cos\vartheta}$$

Fig. 3.1. Deviation of real instrumental function $f_\Omega^*(\vartheta, \varphi)$ of the opaque-skinned integrator of light conductor from the theoretical cosine dependence

granted to the instrumental function of the K-3 instrument was not enough at those times. Regretfully, this lack of information makes the application of modern complex approaches difficult for interpretation of the observational results considered here.

The first concern of the accomplished airborne observations was to study the spectral reflectance properties of natural surfaces. For this purpose the upwelling radiance and downwelling irradiances were measured. Another concern was the elucidation of the radiative regime of the atmosphere from measurements of the upwelling and downwelling solar irradiances in different atmospheric layers. All mentioned observations were accomplished on board the IL-14 aircraft.

A K-3 spectrometer fixed on a special rotation device was set to the hatch, and allowed registration of nadir angles in the range 0–45° (the angle is counted off the vertical to the airplane flight direction). The varying of the viewing angle azimuth was reached with the change of flight direction, the azimuth scale was set as follows: 0° was the direction to the Sun, 180° was the direction opposite to the Sun. The instrument angle of view was 2° during the radiance registration.

For airborne observations of the upwelling and downwelling irradiances, a special light conductor was used. There was a metallic tube passing through the aircraft fuselage and provided with the device allowing the directing of the measured upward or downward light fluxes to the instrument receiver. The ends of the light conductor jutting out from the fuselage for 30 cm were provided with opaque-skinned integrators (over all directions of the hemisphere) made of opal glass MS-23. The edges of glasses were specially manufactured to provide cosine dependence (Sect. 1.1). Figure 3.1 demonstrates the relative deviation curve of the real light conductor instrumental function $f_\Omega^*(\vartheta, \varphi)$ from the desired cosine dependence obtained in the laboratory. As the direct radiation provides the essential part of the total flux under clear sky conditions, Fig. 3.1 illustrates that the systematic uncertainty of the downwelling irradiance measurements does not exceed 2% for solar zenith angles less than 50°, but for higher angles the uncertainty are increasing. These uncertainties were taken into account either during the observations or during the data processing accomplished over two stages.

The first stage of processing of the observational results is called *an initial processing* and obtained the radiance and irradiance spectra on the basis of the output signal of the instrument. During the initial processing, the beginning point of the spectrum λ_1' was defined through the logical search of the special benchmark (i. e. the square pulse formed by the mechanical system of scanning). The first point after the benchmark was assumed as a spectrum beginning point. Then the background (the value of the dark current) was defined by the mean value of the signal at several points after the benchmark. This value was subtracted from the signal magnitude at every wavelength. Note that the constancy of the background was ascertained during the repeated laboratory measurements. Further, the joining of the spectrum parts (UV, VD, NIR) was accomplished by excluding the overlap regions while making use of the known numbers of points at the beginning and at the end of every part. As graduating scales λ_i' of different samples of the instrument varied, it was necessary to carry out the linear interpolation over the spectrum from the initial scale λ_i' to the united scale λ_i to pool the data. In the capacity of the united scale the following set of wavelengths was chosen: 330–410 nm with the step 1 nm and 412–978 nm with the step 2 nm (365 points in whole), moreover the joining regions were 410–412 nm (the end of UV and the beginning of the VD regions) and 698–700 nm (the end of the VD and the beginning of the NIR regions).

The transformation of the obtained spectrum to energetic units: $mW\,cm^{-2}\,\mu m^{-1}$ for the irradiance and $mW\,cm^{-2}\,\mu m^{-1}\,sterad^{-1}$ for the radiance was the concluding step of the initial processing. The calibration was conducted in the laboratory by measuring the signal from the standard lamp SI-8 (Kondratyev et al. 1973a, 1974), whose spectrum was known in energetic units. The special source of the high level stabilization of the electric current and voltage was used for the lamp supply. During the calibration of the instrument, the registration of the lamp SI-8 irradiance was carrying out using the light conductor. The calibration result was the ratio of the calculated lamp energy (incoming to the instrument input slit) to the output signal of the instrument. This ratio was the factor, by which the output signal was multiplied during the measurements (look at the theoretical normalizing of the instrumental functions in Sect. 1.1). The accuracy of the calibration was defined through the accuracy of the lamp SI-8 signal measurements. It was 15% in the UV and 10% in the VD and NIR spectral regions.

To control the stability of the instrument sensitivity during the observation the special *internal standard* was installed inside the instrument – an incandescent bulb with a stabilized supply. During the observation, *the internal standard spectra* were monitored (about every 5 min). To correct the possible deviations of the sensitivity, these spectra were compared with the average instrument spectrum, which had been registered during the calibration. For this purpose, after every internal standard spectrum registration the mean ratio of this spectrum to the average instrument spectrum was obtained over the specially chosen intervals (each interval contained 10 wavelengths) within every spectral region of the instrument (UV, VD, NIR) to provide *three correcting factors*. Further, all measured spectra of the solar irradiances or radiances, which followed this internal standard spectrum, were divided by these correcting factors.

The *external standard spectra* similar to the internal ones (the spectra of the incandescent bulb installed outside of the instrument) were registered to correct for a factor of the transparency variations of the opal glass (due to contamination) during the measurement process together with the difference of the transparencies of the upper (for downward radiation) and lower (for upward radiation) opal glasses. During the airborne observational process, the recordings of the external standard through the opal glasses of the light conductor were being accomplished on board of a landed aircraft. Analogously to the above-considered procedure of obtaining the correcting factors, the curve to correct for a factor of the transparency variations was plotted from the ratio of the external standard spectra to the SI-8 lamp spectrum, registered through the light conductor.

The computer code of the initial processing, whose last version was realized on PC, allowed the fulfilling of all initial procedures (including the graduation and calibration) in an interactive regime, with different levels of researcher interference to the process: from step-by-step control to an automatic regime directly outputting the desired spectra (Vasilyev O et al. 1995).

Nowadays all the obtained spectra forms a computer database of the Airborne Spectral Radiative Observations (ASRO product) containing the observational results since autumn 1983 and including about 30,000 spectra. Every spectral file contains all necessary information concerning the observational process (date and time of the registration, altitude of flight, solar zenith angle, geographical coordinates, etc.) essentially simplifying and accelerating further data processing. This database is also provided with a flexible interface, allowing different procedures either with a separate spectrum or with groups of spectra, e. g. table output, the examination of plots, interactive regime correction, arithmetic operations with spectra (addition, subtraction, multiplying, division), smoothing, approximation with polynomial, elementary statistics (computing the mean value and the variance), etc.

Additional secondary processing of the obtained radiances and irradiances is necessary for determining the spectral reflectance characteristics of the surfaces and the values of radiative flux divergence in the atmospheric layers, which the observations were accomplished for. The specific features of secondary processing are defined by concrete methodologies of the observations and result production. These points will be considered further. Now we will continue discussing the common features of the observations and estimate the uncertainties of the results, extremely important for further interpretation.

The random error of the measurements with the K-3 instrument was estimated in the laboratory over the spectra series registered from the lamp SI-8 and was equal to 5% of the standard deviation in the UV, and to 1% in the VD and NIR regions (this uncertainty is not to be taken for a systematic uncertainty of the calibration, considered above). However, the real random uncertainty of the airborne observations was significantly higher because flight conditions influence the measurement results too. Certainly, the accuracy of the instrument on board was increasingly worse than the accuracy of the same instrument in the laboratory. The deviation of the receiving surface of the opal glasses from the horizontal plane, the deviations of the instrument optical axis from the fixed direction during the radiance registration, the unevenness of the ground surface illumination and heterogeneity were the additional "airborne" factors worsening the observation accuracy.

The aircraft lengthwise axis was disposed not horizontally but at a certain angle even during the constant flight altitude. This angle is called *a pitch* and it defines the slope of the light conductor. For compensation of the influence of the pitch, the light conductor was put at a specially chosen angle with a vertical direction but the influence couldn't be excluded completely because the pitch depended on the aircraft charging changing during the flight due to fuel depletion. As has been mentioned above direct radiation is the main part of the solar downwelling irradiance in a clear sky. The pitch influence on the accounting of direct radiation in the measured irradiance was obviously owing to the deviation of the angle of the solar beam incoming to the light conductor glass from the solar incident angle. As follows from the elementary geometrical consideration, this deviation would be maximum for the flight azimuth 0 and 180°, and minimum for 90 and 270°. Thus, the observations were mostly accomplished for the flight azimuth 90 and 270°.

It is easy to estimate the systematic uncertainty of the downwelling irradiance caused by the pitch, fixing the atmospheric model and using Beer's law (1.42). We would like to mention that this uncertainty is higher if the Sun is lower and the upper atmosphere is optically thinner, i.e. if the flight altitude is higher and the wavelength is longer (1.25)[1]. It is not complicated to estimate this uncertainty experimentally conducting the observations at different azimuth angles. Both estimations have given similar results: the uncertainty is less than 1% for the observations under azimuth angles 90 and 270° and increases up to 5% in the UV and VD and up to 10% in the NIR spectral regions. These values increase up to 10 and 15% for a solar zenith angle exceeding 60°.

Aircraft flight conditions called "*bumps*" appear owing to atmospheric turbulence. All parameters of the flight – altitude, pitch angle, *roll* angle (the angle

---

[1] Remember that the optical thickness of molecular scattering varies inversely with λ^4 as per Rayleigh law

between the aircraft transverse axis and the horizon) and the *yawing* angle (the angle between the aircraft lengthwise axis and the horizon) – randomly vary under the bumps conditions. These random variations lead to corresponding variations of the observation geometric parameters yielding the additional random error. The bumps influence mostly the roll angle, its variations can reach 10°. Moreover, the bumps become significant during a flight below a certain altitude. The experience of the IL-14 aircraft flights has shown that this altitude is 300 m for a flight above a water surface and 500 m for a flight above other surfaces. The experiments described here have been accomplished under conditions of weak bumps (according to the aircraft classification).

An analysis analogous to the above-mentioned one indicates that the influence of the roll variations on the observational accuracy is maximum just for the flight azimuths 90 and 270° that leads to similar estimations of the uncertainty (not systematic uncertainty but the random one).

It is evident that it is possible to neglect the influences of the pitch and bumps during the measurement of the upwelling irradiance because there is no principal direction of the upward radiation. The case of observations above the water surfaces with the mirror reflection could be a certain exclusion, however the maximum peak of the mirror reflection will be shown to turn out (Sect. 3.4) rather "smearing" due to the ripple. Hence, this peak weakly influences the upwelling irradiance when the opal glass slightly deviates from the horizon. The random variations of the pitch, roll and yawing angles lead to the random variations of the viewing angle during the radiance observations. However, with taking into account the weak dependence of the surface reflection properties upon the viewing angle it is possible to neglect the corresponding uncertainties. In some cases, the maximum peak of the mirror reflection could cause the uncertainty to increase. We should point out that all the above-considered uncertainties could be neglected under overcast conditions because there is no direct solar radiation in this case.

The influence of the illumination unevenness together with the ground surface heterogeneity links with the time interval of the spectrum registration (7 s). While the observations were made, the aircraft was flying at about 400 m and appearing above the other surface areas. It is obvious that in a clear atmosphere, the illumination unevenness of the upper opal glass and of the ground surface is negligible at such a distance. However, while flying below the heterogeneous clouds, the downwelling irradiance transmitted by clouds can vary. Moreover, if the surface type varies within the distance of 400 m (forest, tillage, marsh, etc.) the ground surface unevenness influences the accuracy of the upwelling radiance. Thus, observations of upward radiation can be accomplished only above areas with a large extent of homogeneous surfaces. In view of the observational experience, there are only three kinds of such surfaces: sand desert, water and snow. Practically, the observations were carried out only in cases when the influence of the mentioned factors did not exceed 10%. It was controlled by the visual estimation of the output signal variations at fixed wavelength within the VD spectral region.

We would like to mention that the uncertainty linking with surface heterogeneity decreases fast with flight altitude because the increasing of the surface

Table 3.1. Evaluation of the uncertainty (standard deviation) of airborne measurements of the radiative characteristics

Uncertainty source	Uncertainty type	Observations, which the uncertainty influences	Uncertainty estimation
Displacement of the wavelength scale	Systematic Random	All observations All observations	1 nm 1 nm
Deviation from the cosine dependence	Systematic	The irradiance observations	Look at Fig. 3.1
Calibration	Systematic	All observations	15% within UV, 10% within VD and NIR
K-3 spectrometer	Random	All observations	5% within UV, 1% within VD and NIR
Aircraft pitch	Systematic	Observations of the downwelling irradiance in the clear atmosphere	5% within UV, 10% within VD and NIR for the azimuths 0 and 180°
Aircraft bumps	Random	Observations of the downwelling irradiance in the clear atmosphere below the bumps level	5% within UV, 10% within VD and NIR for the azimuths 90 and 270°
Illumination heterogeneity	Random	Observations below the inhomogeneous clouds	10%
Surface heterogeneity	Random	Observations of the upwelling radiance and irradiance below the bumps level	10%

area in the field of view of the instrument is smoothing the surface heterogeneity. It is especially distinct during the upwelling irradiance observations: the corresponding estimations indicated that the surface heterogeneity could be neglected if the flight altitude was higher than the bumps level. Table 3.1 concludes the reasons and estimations of the uncertainties of the airborne observations with the information-measuring system based on the K-3 instrument.

3.2
Airborne Observation of Vertical Profiles
of Solar Irradiance and Data Processing

The concern of the spectral observations of solar irradiances was to calculate radiative flux divergences and it conditions both the observational scheme and the methodology of data processing. It is necessary to distinguish two different cases: observations under overcast and clear sky conditions. The observations either of upwelling or of downwelling irradiance were accomplished using *one instrument* through the upper and lower opal glasses in turn.

The observations of the solar irradiances in the overcast sky were accomplished out of the cloud (above the cloud top and below the cloud bottom) and within the cloud layer at every 100 m. As the implementation of the experiment under the overcast conditions needed both a horizontal homogeneity of the cloud and its stability in time, the observations were accomplished as fast as possible with measuring of only one pair of the irradiances (upwelling and downwelling) at every altitude level. Besides, only one circle of observations was needed as usual. We need to stress that cases of homogeneous and stable cloudiness are rare so the quantity of observations for the overcast sky are less than in the clear sky.

The main component of the uncertainty during irradiance observations under overcast conditions is the random error due to the heterogeneity of illumination (Table 3.1). It leads to distortions of the vertical profiles of the spectrum, as Fig. 3.2 demonstrates. The filtration of these distortions was possible using the smooth procedures, but the standard algorithms (Anderson 1971; Otnes and Enochson 1978) turned out to be ineffective in this case. Thus, it was necessary to elaborate the special one (Vasilyev A et al. 1994).

The smooth procedure of distortions of the spectral downwelling and up-welling irradiances provides the replacement of the irradiance value at every altitude level with the weighted mean value over this level and two neighbor (upper and below) levels:

$$F^{\downarrow}(z_i) = \sum_{j=-1}^{1} \beta_j f^{\downarrow}(z_{i+j}) , \quad F^{\uparrow}(z_i) = \sum_{j=-1}^{1} \beta_j f^{\uparrow}(z_{i+j}) , \quad \sum_{j=-1}^{1} \beta_j = 1 , \quad (3.2)$$

where β_j are the weights of smoothing (common for all wavelengths, altitudes and types of the irradiances); $f^{\downarrow}(z_i), f^{\uparrow}(z_i)$ are the observational results of the downwelling and upwelling irradiances at level z_i; $F^{\downarrow}(z_i), F^{\uparrow}(z_i)$ are the values of the irradiances calculated during the secondary data processing. Weights β_j in (3.2) have been obtained from the demands of the physical laws.

As the radiative flux divergence has to be positive, the net radiant flux does not increase with the optical thickness increasing (from the top to the bottom of the cloud) according to Sect. 1.1. That is to say, the following condition has to be fulfilled for the results of (3.2):

$$F^{\downarrow}(z_i) - F^{\uparrow}(z_i) \geq F^{\downarrow}(z_{i-1}) - F^{\uparrow}(z_{i-1}) \quad (3.3)$$

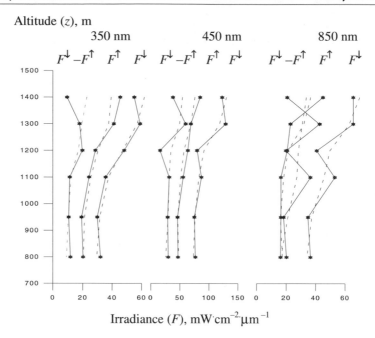

Fig. 3.2. Vertical profile of net, downward, and upward fluxes of solar radiation in the cloud for three wavelengths. *Solid lines* are the original measurements; *dashed lines* are the smoothed values. Observation 20th April 1985, overcast stratus cloudiness. Cloud top 1400 m, cloud bottom – 900 m, solar incident zenith angle $\vartheta_0 = 49°$ ($\mu_0 = 0.647$), snow surface

The substituting of (3.3) to (3.2) provided the conditions for obtaining weights β_j

$$\sum_{j=-1}^{1} \beta_j(f^{\downarrow}(z_{i+j}) - f^{\downarrow}(z_{i-1+j})) \geq \sum_{j=-1}^{1} \beta_j(f^{\uparrow}(z_{i+j}) - f^{\uparrow}(z_{i-1+j})), \quad \sum_{j=-1}^{1} \beta_j = 1.$$
(3.4)

The equation system (3.4) was solved with the iteration method. Firstly, weights β_j for measured values $f^{\downarrow}(z_i), f^{\uparrow}(z_i)$ were obtained after the conversion of the inequality to the equality in (3.4). Only three spectral points in the interval centers (UV – 370 nm, VD – 550 nm, NIR – 850 nm) were considered as a smoothing condition for all other wavelengths. Equation system (3.4) was solved using the Least-Squares Technique (LST) (Anderson 1971; Kalinkin 1978). The formulas and features of the LST in applying to atmospheric optics will be considered in Chap. 4 and here we are presenting the results only.

Then values $F^{\downarrow}(z_i), F^{\uparrow}(z_i)$ were calculated using (3.2), and conditions (3.3) were verified for all wavelengths and altitudes. The iterations were broken in the case of satisfying the conditions, otherwise the above-described procedure was repeated after substituting values $F^{\downarrow}(z_i), F^{\uparrow}(z_i)$ to $f^{\downarrow}(z_i), f^{\uparrow}(z_i)$ in (3.4). One other physical restriction was added in this case: the deviations of values $F^{\downarrow}(z_i)$, $F^{\uparrow}(z_i)$ from measured results $f^{\downarrow}(z_i), f^{\uparrow}(z_i)$ at any iteration can't exceed the root-mean-square random uncertainty of the measurements (10%, Table 3.1). Mark that two-three iterations were enough to obtain final values $F^{\downarrow}(z_i)$, $F^{\uparrow}(z_i)$. Figure 3.2 illustrates an example of the considered procedure.

Obtained values of the irradiances under the overcast condition $F^{\downarrow}(z_i)$, $F^{\uparrow}(z_i)$ were the results of the secondary processing. The root-mean-square deviation of the smoothed profile from the initial ones was accepted as a random uncertainty of the result. Note that the systematic error of calibration brought a considerable yield to the total uncertainty (Table 3.1), however the irradiances were considered as non-dimension combinations for further processing and interpretation, hence it was possible to ignore the calibration uncertainty. Note that the solar zenith angle varies negligibly (1–2°) owing to the fast accomplishment of the experiment, and during processing, the single value of the solar zenith angle was attributed to all spectra of the experiment.

The comparison of the measured irradiances with the extraterrestrial solar spectrum in the case of a clear atmosphere is of special interest. Beer's Law is the simplest ground of this approach if for example the optical thickness of the atmosphere is retrieved from the observational data. It is impossible to measure the solar extraterrestrial flux directly from the aircraft, thus the yield of the systematic uncertainty is essential during observations in a clear atmosphere.

The values of spectral radiative flux divergence are rather small in clear sky, and the random uncertainties of the results of the irradiance observations corresponding to the aircraft factors are extremely large. Thus, the main problem of experiment planning and data processing was the minimization of the random uncertainty of the results and correction of the systematic uncertainty during instrument calibration.

Increasing the measurement accuracy of the spectrometer is important itself but the measurement uncertainty onboard the aircraft due to flight factors, atmospheric conditions, and surface heterogeneity does not depend on an instrument and can reach high values. Therefore, the only method of getting the highly accurate experimental results is applying the most adequate approaches to the statistical data processing. It would be necessary to register several spectra at every level if we meant to perform the statistical processing at its simplest level – the data averaging. However, in this case, observations would have taken a lot of time and the irradiances at different levels would have been measured at essentially different solar zenith angles, complicating further the interpretation.

According to the above-mentioned difficulty, a special scheme of observations called *sounding* was elaborated (Kondratyev and Ter-Markaryants 1976; Vasilyev O et al. 1987). Corresponding to this scheme, two or three *preliminary ascents and descents* were carried out in a range from 50 m (1000 mbar) to 5600 m (500 mbar) with registrations every 100 mbar and *the detailed descent* was accomplished from 5600 m to 50 m at midday (during the period when the solar zenith angle is weakly varying) with registrations every 100 m (Fig. 3.3a). The registration of the numerous irradiance spectra with the minimal

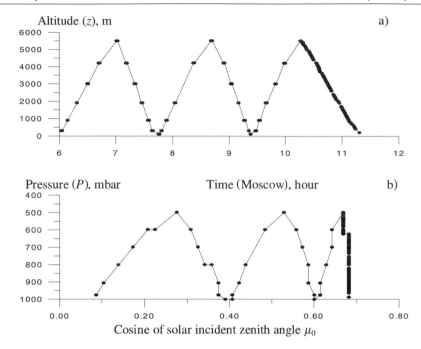

Fig. 3.3a,b. Scheme of the airborne sounding: **a** in the coordinates "time-altitude", **b** in the coordinates "cosine of the solar incident angle – atmospheric pressure". Observation 14th October 1983 above the Kara-Kum Desert, the *points* show the altitudes of the measurements

variation of the solar zenith angle during the detailed descent for obtaining the altitudinal dependence of the irradiance and the application of the irradiance values registered during the preliminary ascent and descent for correction of the solar zenith angle variations during the detailed descent were the main ideas of sounding. The minimal altitude 50 m was taken due to the special demands of flight safety; the maximal altitude 5600 m was taken due to the technical abilities of the IL-14 aircraft. While flying with the optimal regime, we succeeded in only two ascents and descents during one experiment, however, the crew gladly assisted during the observations allowing us to carry out three ascents and descents.

The flight altitude has been changed during the sounding but the scale of pressure has been used instead of the altitude scale during further data processing as Fig. 3.3b demonstrates. It was connected with the following: at altitudes higher than 500 m *the aircraft absolute scale of altitudes* was used, i.e. the altitude registered by the altimeter related to the level 1013 mbar or the atmospheric pressure was expressed in altitude units according to *the standard atmospheric model* (Standards 1981). The accuracy of the instrumental measurement of the altitude according to the absolute scale was about 50 m but it was difficult for the crew to set a concrete altitude level exactly while working under the conditions of time shortage so the real uncertainty of the altitude registration was assumed equal to 100 m. At altitudes below 500 m *the true aircraft altitude* was used because the distance between the aircraft and surface was measured with high accuracy with the radio altimeter. There was a gap between these two scales caused by the Earth's surface altitude above sea level and by the variations of pressure profile of the real atmosphere compared with the standard model (Standards 1981). This gap was determined through the comparison of the altimeter and radio altimeter registrations and was accounted for while forming the common altitude scale (by the pressure) for the irradiance profiles.

For accomplishment of the soundings, the areas of the Ladoga Lake, the Kara-Kum Desert (Turkmenistan, near the town of Chardjou) were chosen. This choice was conditioned by the demands of surface uniformity mentioned in the previous section and by the airports situated in the neighborhood as well. Correspondingly the soundings were carried out above three types of surface: snow (on the ice of the Ladoga Lake), water (the Ladoga Lake) and sand (the Kara-Kum Desert).

The most complicated stage of the secondary data processing was the initial one, i.e. the preliminary analysis and correction of the irradiance spectra. First, it was connected with the rather complicated conditions of the flights, which caused the malfunctions of the equipment on board and the errors of the registered spectra at some wavelengths. However, owing to the high scientific value of the data (and owing to the high price of the airborne experiments) it was inappropriate to exclude the whole spectrum because of the errors at one or several wavelengths. Hence, careful analysis of the errors together with the spectra correction was needed. Besides, the flight conditions did not allow us to realize the ideal sounding scheme as a whole; it caused the necessity of data correction while taking into account the deviation of the measuring procedure from the ideal scheme.

The attempts to create the universal algorithm of error correction of the measured spectra failed because of a huge variety of concrete errors. They were revealed and removed by hand, using the visual interface of the database described in the previous section. This algorithm was applied to observations in an overcast sky. However, applying this approach to the spectra of the clear atmosphere needed too much time because there were many more of these spectra. Just this obstacle was the reason why a significant volume of the data measured in 1983–1985 was processed only at the end of 1990th when a system for fast processing was created. The basis of the system was the idea of *the semiautomatic regime*. The data analysis was accomplished without an operator but after the error was revealed the passage to hand processing in the interactive regime occurred. In addition, the program code suggested different solutions to the operator.

The brief description of the proposed system of spectra processing with the detailed consideration of the approaches and schemes that could find a wide application in the preliminary analysis of the results of solar radiances and irradiances measurements are presented below.

At the first stage, the errors like *an overshoot* together with *breaks* of the spectrum parts are revealed using the logical analysis of every spectrum. The overshoot is an error where the values of the radiative characteristics at one or several spectral points are sharply distinct by a magnitude from the neighboring ones. If the relative difference of two neighbor values (following each other) of the spectral points exceeds the fixed level (e. g. 10%) the consequent point will be assumed as an overshoot. Note that a detailed logical analysis is necessary lest a strong absorption band is attributed to an overshoot, either it is necessary to account for all possible variants of the overshoot positions in the beginning or end of the spectrum and the nearby overshoots as well. An overshoot correction consists of the substitution of the point interpolated over the neighbor sure points to the error point. After the removal of the errors, the procedure is repeated (because the strongest overshoots can mask the weaker ones) until there is no overshoot at the recurrent iteration. The breaks at the boundaries of the UV–VD and VD–NIR regions of the spectrum are caused by the measurements with different photomultipliers at different spectrum regions (Sect. 3.1). These breaks are likely owing to the deviation of the dynamical characteristic of the photomultiplier from the linear one. The removal of the breaks is accomplished by the adding of the corresponding constant correcting values to the break spectrum region.

The elucidating of the errors using logical analysis is not effective enough. Usually, the operator easily identifies the errors visually just because he knows in advance, what the "right" spectrum looks like. Scientifically speaking he uses the a priori information about the spectrum shape accumulated from experience. The following stage of the elucidating and correcting of the errors is based just on that comparison of the spectrum shape with the certain *a priori spectrum*. The spectrum under processing and the a priori one are compared in relative units (they are reduced to the interval from 1 to 2) for excluding the relationship between the spectrum shape and the signal magnitude. If the modulus of the comparison result exceeds the standard deviation of the a priori spectrum multiplied by a certain factor the spectrum will be identified as an erroneous one. The factor is selected during the process of the system tuning. We have used the factor equal to 4.2 that differs from the traditional magnitude for the statistical interval equal to three standard deviations. There is an apparent dependence between the spectrum and atmospheric pressure together with solar zenith angle, so the distribution of the resulting error is rather different from Gaussian distribution that explains the deviation of the factor from 3. Two stages of the system provide the calculation of the standards and of their standard deviations. At the first stage, the a priori information is absent and the block of comparing with the standard is turned off. The standard (as an arithmetic mean over processed spectra) and its standard deviation are calculated from the results of the first stage (standards are being obtained separately for upwelling and downwelling irradiances and for different surfaces). At the second stage, all spectra are processed again with the block of comparing with the standard turned on. This system of algorithms, which are accumulating the a priori information, is a self-educating system as per the theory of the pattern recognition and selection (Gorelik and Skripkin 1989).

The practice of the data processing demonstrates that the application of self-educating systems in algorithms of the preliminary analysis of spectropho-

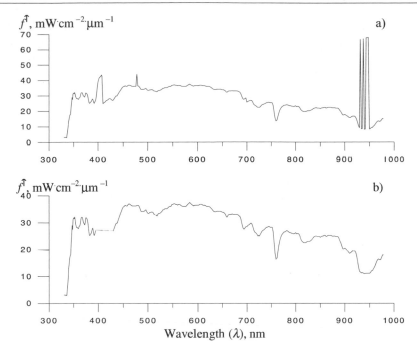

Fig. 3.4a,b. The example of the spectrum correction of the results of upward flux measurements 14th October 1983, time (Moscow) 7:12, altitude 4200 m: **a** the initial spectrum; **b** the corrected spectrum

tometer information is rather effective. Figure 3.4 illustrates an example of the error removal. The above-considered stages of the observational data processing deal with the analysis of the spectra shape.

Regretfully the errors were also revealed when the spectrum had a correct shape but differed from the "right" spectrum with the signal magnitude. To elucidate such situations, the dependence of the irradiance upon the atmospheric pressure and solar zenith angle was studied. The approximation of the dependence using the quadratic form gave an approximating curve rather close to single spectrums. If there had been some deviations, it would have been the reason to test the spectra for errors. For every wavelength the approximation of the dependence of the irradiance upon pressure P and the cosine of the solar zenith angle μ_0 was calculated (separately of the upwelling and downwelling irradiances).

Here is the example of the approximation of the downwelling irradiance:

$$f^{\downarrow}(P, \mu_0) = a_1 + a_2 P + a_3 \mu_0 + a_4 P^2 + a_5 \mu_0^2 + a_6 P \mu_0 . \tag{3.5}$$

Desired coefficients of the approximation $a_1, \ldots, a_6$ are obtained from the totality of registered irradiances $f^{\downarrow}(P_i, \mu_{0i})$ over every ascent and descent of the sounding. Equation system (3.5) is solved with the LST, where the inverse squares of the random standard deviation of the irradiances (Table 3.1) are taken as weights, for irradiances registered at the high solar zenith angles having a smaller weight, the uncertainty caused by the deviation from the cosine law is also included to the standard deviation as a random error.

The last stage of the preliminary analysis system is an accounting of individual specific features of the flight scheme. Solar zenith angle ϑ_0 ($\mu_0 = \cos\vartheta_0$) and a set of the atmospheric pressure values P_i, $i = 1,\ldots,N_i$ are chosen at this stage, which the final magnitudes of the irradiances will be obtained for as a result of the secondary processing of the sounding data. There are six levels in the ordinary flight scheme $N_i = 6$ and the irradiances magnitudes are output for the pressure levels from 1000 to 500 mbar through every 100 mbar.

After the above-described preliminary analysis, N_j downwelling irradiances $f^\downarrow(P_j, \mu_{0,j})$ and N_k upwelling irradiances $f^\uparrow(P_k, \mu_{0,k})$ are registered, from which it is necessary to obtain N_i values $F^\downarrow(P_i, \mu_0)$ and $F^\uparrow(P_i, \mu_0)$. The algorithm of this problem solution was described in Vasilyev O et al. (1987). However, this algorithm was based on several physically poor assumptions, e.g. on the supposition about the linear dependence of the irradiances upon solar zenith angle, on the square approximation of the dependence of the irradiances upon the atmospheric pressure, on the supposition about the monotonic increasing of the upwelling irradiance with altitude. Thus, the new algorithm has been elaborated for processing the results of soundings accomplished in the years 1983–1985. It is also based on certain assumptions but not so severe as before.

Let us present the dependence of the irradiance upon the solar zenith angle cosine and atmospheric pressure using Taylor series limiting by the items of second power:

$$F_i^\downarrow - Df_j^\downarrow = a_1 x_j + a_2 y_{ij} + a_3 x_j^2 + a_4 y_{ij}^2 + a_5 x_j y_{ij} \,,$$

$$F_i^\uparrow - Df_k^\uparrow = b_1 x_k + b_2 y_{ik} + b_3 x_k^2 + b_4 y_{ik}^2 + b_5 x_k y_{ik} \,,$$

$$(3.6)$$

where D is the correcting coefficient for the compensation of the systematic calibration uncertainty (the calibration factor). Specifications

$$F_i^\downarrow \equiv F^\downarrow(P_i, \mu_0) \,, \quad F_i^\uparrow \equiv F^\uparrow(P_i, \mu_0) \,,$$

$$f_j^\downarrow \equiv f^\downarrow(P_j, \mu_j) \,, \quad f_k^\uparrow \equiv f^\uparrow(P_k, \mu_k) \,,$$

$$x_j = \mu_0 - \mu_j \,, \quad x_k = \mu_0 - \mu_k \,,$$

$$y_{ij} = P_i - P_j \,, \quad y_{ik} = P_i - P_k$$

are introduced for a brevity. The desired values are $F_i^\downarrow$, $F_i^\uparrow$, D, $a_1,\ldots,a_5$, $b_1,\ldots,b_5$.

The conditions for determining calibration factor D are to be added to solve equation system (3.6). The extrapolation of the downwelling irradiance to the level $P_i = 0$ mbar and its comparison with known extraterrestrial flux $\delta F_0 \mu_0$, where correction factor δ accounts for the deviations of the Sun–Earth distance from the mean value for the date of the observation. The spectral magnitudes of

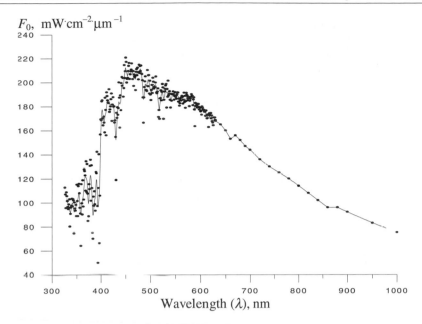

Fig. 3.5. Spectral solar extraterrestrial flux F_0, taking into account the instrumental (K-3) function (*solid curve*). *Points* show the initial values of F_0 of the high spectral resolution from the data according to Makarova et al. (1991)

F_0 have been taken from the book by Makarova et al. (1991, Fig. 1.3) where the recent data averaged over several original studies were presented. These values were recalculated with (1.12) while accounting for the spectral instrumental function expressed by (3.1) for a correct comparison with the data of the K-3 instrument. Figure 3.5 illustrates obtained curve $F_0(\lambda)$. The magnitudes of correction factor δ are presented in the book by Danishevskiy (1957). The system of linear equations is finally obtained:

$$a_1 x_j + a_2 y_{ij} + a_3 x_j^2 + a_4 y_{ij}^2 + a_5 x_j y_{ij} + Df_j^{\downarrow} - F_i^{\downarrow} = 0 \,,$$

$$b_1 x_k + b_2 y_{ik} + b_3 x_k^2 + b_4 y_{ik}^2 + b_5 x_k y_{ik} + Df_k^{\uparrow} - F_i^{\uparrow} = 0 \,, \qquad (3.7)$$

$$a_1 x_j + a_2(-P_j) + a_3 x_j^2 + a_4(-P_j)^2 + a_5 x_j(-P_j) + Df_j^{\downarrow} = \delta F_0 \mu_0 \,.$$

System (3.7) consists of $(N_j + N_k)N_i + N_j$ equations relative to $11 + 2N_i$ desired values. Levels P_i have been chosen for the equation quantity exceeding the number of the desired values not less than twice. System (3.7) is solved with the LST independently for every wavelength, where the inverse squares of the random standard deviation (Table 3.1) while accounting for the uncertainty of the deviation from the cosine law are taken as weights. This is to impose that the additional conditions of the formal mathematical solution do not contradict physical laws. Here they are: the non-negativity of the radiative flux divergences and the a priori restraints to the albedo:

$$F_{i+1}^{\downarrow} + F_i^{\uparrow} - F_i^{\downarrow} - F_{i+1}^{\uparrow} \geq 0 , \quad i = 1, \ldots, N_i - 1 ,$$

$$F^{\uparrow}(P_i = 1000\,\text{mbar})/F^{\downarrow}(P_i = 1000\,\text{mbar}) \geq A_{(-)} ,$$

$$F^{\uparrow}(P_i = 1000\,\text{mbar})/F^{\downarrow}(P_i = 1000\,\text{mbar}) \leq A_{(+)} ,$$

$$F_i^{\uparrow}/F_i^{\downarrow} \leq A_{(\text{max})} , \quad i = 1 \ldots, N_i .$$

(3.8)

The second and third lines in the set of restraints (3.8) account for the known range of the spectral albedo of the surface: $A_{(-)}$ is a minimal possible magnitude, $A_{(+)}$ is a maximal possible magnitude. These magnitudes $A_{(-)}$ and $A_{(+)}$ have been chosen from the spectral reflectivity data of similar surfaces (Chapurskiy 1986; Vasilyev A et al. 1997a, 1997b, 1997c) (spectral brightness coefficients to nadir with the approximation of the orthotropic surface equal to the albedo of sand, snow and pure lake water). These data will be considered in Sect. 3.4. The maximal albedo of the system "atmosphere plus surface" is assumed as $A_{(\text{max})} = 0.95$.

The solution of equation system (3.7) together with restraints (3.8) was accomplished with the iteration technique. Firstly, (3.7) was solved with the LST without accounting for restraints (3.8), and the fulfilling of restraints (3.8) was tested for the obtained solution. The iterations were broken when all these conditions had been fulfilled. Otherwise, the solution of system (3.7) was searched with restraints (3.8). Restraints (3.8) were transformed to the rigorous equality and the variables were excluded from system (3.7) by substitution of these equalities. The corresponding formulas expressing this solution will be presented in Chap. 4. The iteration scheme was constructed as follows. Firstly, values $F_{i+1}^{\downarrow}$ were excluded from the restraints for the irradiances and values $F_i^{\downarrow}$ – from the restraints for the albedo. The solution of system (3.7) was inferred for every excluded variable separately ($2N_i$ solutions as a whole) with the LST, and the one with the smallest error was chosen. For this solution, restraints (3.8) were tested again. If they failed the iterations were continued, and the couple of restraints were excluded, then three restraints, and so on. As the worse variant it was to examine $3 \cdot 2^{2N_i-2}$ solutions and it was the appropriate number for modern computers as in our experiments $N_i = 6$.

The final result of the secondary processing of the sounding data are the desired values of irradiances $F_i^{\downarrow}$ and $F_i^{\uparrow}$ for $i = 1, \ldots, N_i$ together with the covariance matrix of the errors. It should be emphasized that further interpretation of the results is to obtain the matrix as a whole and not only its diagonal (the variance of the irradiances values). If the solution has been obtained using restraints (3.8), the part of the irradiances is linearly dependent and hence non-informative. The indicator of the linear dependence has also been written to the output file of the secondary processing. We would like to point out that owing to the individual solution of system (3.7) while accounting for (3.8) for every wavelength the number of the independent (informative) irradiance values are essentially different for different wavelength. Coefficients $D, a_1, \ldots, a_5$, $b_1, \ldots, b_5$ and their standard deviations are additional result of the secondary processing.

We would like to point out that the three main sources of the systematic uncertainties of the obtained results are: the uncertainty of extraterrestrial solar flux F_0; the non-adequacy of (3.7) for dependence of the solar irradiance upon the pressure and solar zenith angle; and the atmospheric parameter variations during the observation. The first uncertainty is rather large (about several percents) according to the estimations of Makarova et al. (1991). However, if the same magnitude of F_0 as in (3.7) is used for further interpretation, this uncertainty will not influence the results. The second systematic uncertainty, as has been shown in Vasilyev O et al. (1987) for the old system of the equation, which is less exact than (3.7), does not exceed the random error of the observations and could be neglected. To an even greater degree, this conclusion may be applied to the more exact equation system (3.7). Finally, consider the third uncertainty. The solution of (3.7) is mean-weighted values over all observed spectra from the essence of the LST. Hence, they could be attributed to the atmospheric and surface parameters averaged over time and space. The spectra measured during the detailed descent give the maximal yield (just because there are more of these spectra than other ones) during the averaging. The detailed descent continues a bit longer than one hour (Fig. 3.3a) during the sounding that coincides with the time of a balloon flight. The space scale of the airborne observations is about 30 km that is also analogical to the horizontal distance of a balloon route. Thus, it is safe to say that the airborne data are not worse than any radio sounding data from the point of the space and time averaging of the atmospheric parameters.

3.3
Results of Irradiance Observation

The examples of the observational results and calculations according to the above-described technique are presented here for a clear and an overcast sky. The typical profiles of the downwelling and upwelling spectral irradiances are demonstrated in Figs. 3.6–3.8 and in Tables A.1–A.3 of Appendix A. The figures illustrate the vertical profiles of the downwelling (the upper group of the curves) and upwelling (the lower group) irradiance – 6 curves in every group from 500 mbar to 1000 mbar through 100 mbar from the upper curve to the lower one. These results were obtained from the sounding data above three kinds of surface: sand, snow and water. It is important to point out that the uncertainty of the results is rather significant at the boundaries of the spectral regions, where the sensitivity of the photomultiplier is weak.

The analysis of the observational results indicates the decreasing of both upwelling and downwelling irradiances with the increasing of the atmospheric pressure in all cases. This behavior is evident for the downwelling irradiance: solar radiation decreases owing to the radiation extinction in the atmosphere. For the upwelling irradiance this effect points to the predominance of scattering processes over absorption processes in the short wavelength range, i.e. the

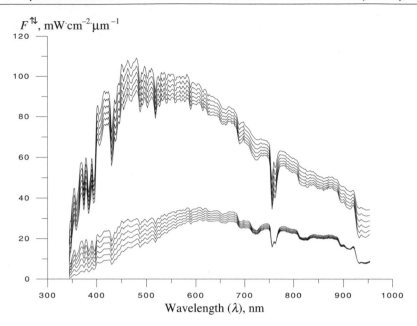

Fig. 3.6. Vertical profile of the spectral dependence of the solar semispherical irradiances from the results of the airborne sounding 16th October 1983. Sand surface, solar zenith angle 51°

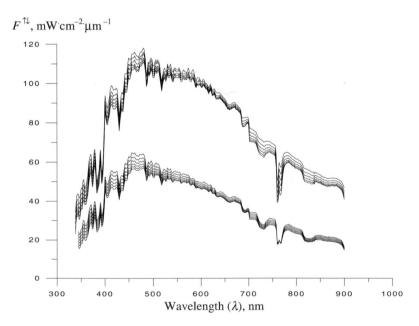

Fig. 3.7. Vertical profile of the spectral semispherical solar irradiance from the results of the airborne sounding 29th April 1985. Snow surface, solar zenith incident angle 48°

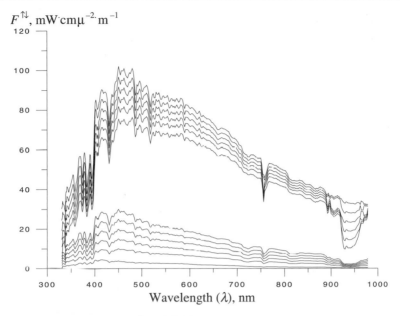

$F^{\uparrow\downarrow}$, $\mathrm{mW\cdot cm\mu^{-2}\cdot m^{-1}}$

Fig. 3.8. Vertical profile of the spectral semispherical solar irradiance from the results of the airborne sounding 16th May 1984. Water surface, solar zenith incident angle 43°

extinction of the upward radiation is weaker than its increasing caused by backscattering of the downward radiation.

As has been mentioned in the previous section not all spectrum points are independent and hence informative after the secondary processing. Figure 3.9 illustrates only the informative points of the same spectra as Fig. 3.6 does. In practice, the real number of the informative points differs very much for different spectra that seems to link with non-ideal weather conditions together with the errors during the registrations.

The spectral region is excluded from the further processing when there are less informative points in it. Thus, Fig. 3.9 demonstrates a sounding of high quality. An example of a "bad" sounding is shown in Fig. 3.10 that is analogous to Fig. 3.8 excluding the non-informative points.

The uncertainty of measurements is the most important characteristic varying strongly in different soundings. Figure 3.11 shows the minimal relative standard deviation over all realizations for downwelling and upwelling irradiances. It is easily seen from the comparison of the relative standard deviation with the initial values (Table 3.1) that the statistical processing significantly improves the accuracy of the results.

The vertical profiles of the spectral albedo of the "atmosphere plus surface" system characterizing three types of the surface are presented in Fig. 3.12. The figure demonstrates the results of the soundings above the sand surface (16 October 1983) – solid lines, above the snow surface (29 April 1985) – upper group of dashed lines, and above the water surface (16 May 1984) – lower group

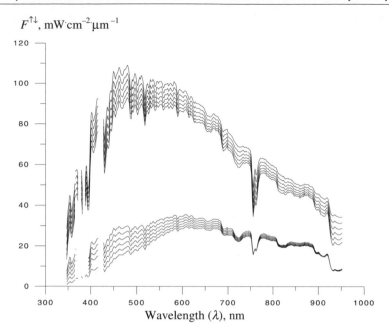

Fig. 3.9. Informative points of irradiance spectra obtained 16th October 1983. The figure is analogous to Fig. 3.6 excluding non-informative points

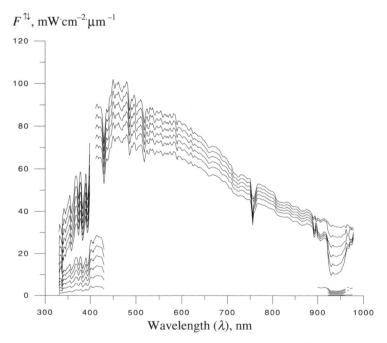

Fig. 3.10. Informative points of the irradiance spectra obtained 16th May 1984. The figure is analogous to Fig. 3.8, excluding non-informative points

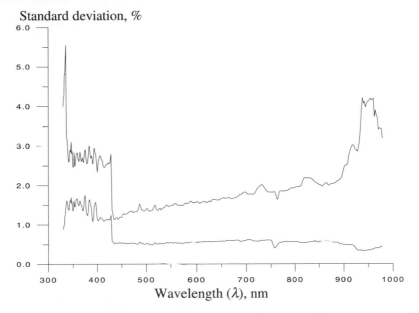

Fig. 3.11. Minimal value of the standard deviation over all data set of the airborne sounding. *Upper curve* – upwelling irradiance, *lower curve* – downwelling irradiance

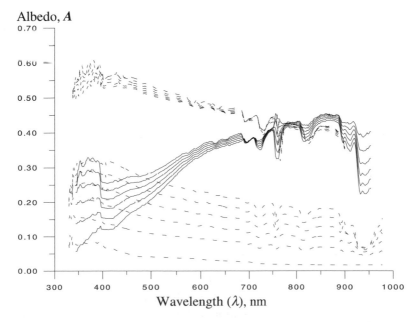

Fig. 3.12. Vertical profiles of the spectral albedo of the system "atmosphere plus surface"

of dashed lines. All values of the albedo increase when the atmospheric pressure decreases (with the altitude) especially if the surface is dark. It confirms the above conclusion about the predominance of scattering over absorption within the short-wavelength range excluding the absorption bands in the NIR region above the sand surface.

Figure 3.12 apparently indicates spectral transformation of the albedo in molecular absorption bands with the increasing of atmospheric thickness especially in the example of the water surface (Vasilyev A et al. 1997a, 1997b, 1997c). The figure also demonstrates that the magnitudes of the snow albedo of the Ladoga Lake surface are not very high compared with other observations (Chapurskiy 1986) that could be explained with the destruction and pollution of ice in spring (April). Carrying out the observations in winter is complicated owing to the low Sun. The standard deviation of the albedo is calculated with the covariance matrix of the couple of corresponded irradiances. The calculation methodology will be described in Chap. 4. The average uncertainty of the albedo is about 5%.

3.3.1
Results of Airborne Observations Under Overcast Conditions

The experiments on the overcast sky were carried out in the field by companies and conducted as components of CAENEX, GAAREX, GARP and GATE scientific programs. The results of these programs are considered in several books (Kondratyev 1972; Kondratyev and Ter-Markaryants 1976; Kondratyev et al. 1977; Kondratyev and Binenko 1981, 1984) and in several studies (Kondratyev et al. 1976; Vasilyev A et al. 1994; Kondratyev et al. 1996, 1997a, 1997b). The observations were carried out with K-2 and K-3 instruments (Mikhailov and Voitov 1969) and each experiment in the cloudy atmosphere was accompanied with the measurement for the same region under the clear sky conditions at the same height levels and at the close time. Only optically thick stratus clouds of large extension were studied during the overcast-sky experiments. The experimental results in different latitudinal zones in different time during 1971–1985 were analyzed using the uniform observational data sets. The geographical latitudes of the observations were changing from 15°N (the East part of the Atlantic Ocean close to the African coast) to 75°N (above the Cara Sea). All aircraft observations were accomplished above the homogeneous surfaces (sea and snow surface, deserts). Under these conditions, it was possible to exclude such factors as a horizontal heterogeneity of clouds and surface, broken cloudiness, radiation escape through the cloud sides. To estimate the cloud radiative forcing the data of the pyranometric (total SWR) and spectral observations were used simultaneously.

The surface albedo was calculated as a ratio of the upwelling to downwelling irradiances at the lowest level under the cloud layer. The information about the cloudy experiments, which will be further interpreted in Chap. 7, is presented in Table 3.2. The thickness of the cloud layer, the cosine of the solar zenith angle, the latitudes, the surface type and albedo, the total values of the radiative flux divergence over the spectral region in cases of the cloud and clear atmosphere

Table 3.2. Results of the airborne radiative observation in a cloudy atmosphere

No.	Experiment	μ_0	φ, °N	Data	A_s	Other condition / Total	f_s Sw	f_s Cloud	R, W m^{-2} Clear	R, W m^{-2}
GATE										
1	The Atlantic Ocean, cloud	0.966	16	12 Jul. 74	0.1	Above the Atlantic after	1.74	3.2	18.9	
2	The Atlantic Ocean, cloud	0.966	17	4 August 74	0.06	dust intrusion from the	1.45	2.9	26.1	
	The Atlantic Ocean, clear	0.966	17	13 August 74	0.02	Sahara Desert				2.43
CAENEX										
3	The Black Sea, cloud	0.819	44	10 April 71	0.06	Above sea surface	1.11	1.2	2.86	
4	Azov Sea, cloud	0.616	47	5 October 72	0.06	Above sea surface	1.16	2.5	12.8	
	Azov Sea, clear	0.616	47	6 October 72	0.08	Industrial pollution				3.60
5	City Rustavy, cloud	0.438	42	5 December 72	0.18	Above the ground	1.07	1.3	15.0	
	City Rustavy, clear	0.438	42	4 December 72	0.22	Industrial pollution				2.35
6	The Ladoga Lake, cloud	0.440	60	24 September 72	0.20	Above water surface	1.13	1.8	3.61	
	Ladoga Lake, clear	0.440	60	20 September 72	0.10	Above water surface				3.73
7	Ladoga Lake cloud	0.647	60	20 April 85	0.64	Above ice with snow	1.10	1.5	4.5	
	Ladoga Lake, clear	0.669	60	24 April 85	0.55	Above ice with snow				0.40
GARP										
8	Kara Sea, cloud	0.276	75	01 October 72	0.40	Above water with ice	1.00	1.1	4.63	
	Kara Sea, clear	0.276	75	30 September 72	0.40	Industrial pollution				1.97
9	Kara Sea, cloud	0.483	75	29 May 76	0.40		0.90	0.95	7.25	
10	Kara Sea, cloud	0.483	75	30 May 76	0.40	Above water with ice	1.00	1.2	1.1	
	Kara Sea, clear	0.460	75	21 April 76	0.05	Above water surface				1.87

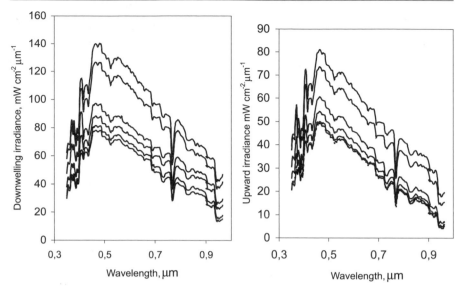

Fig. 3.13. The results of the airborne sounding in the overcast sky, experiment 7 in Table 3.2

are demonstrated. Value f_s characterizing the variations of solar radiation absorbed in the system "cloudy atmosphere plus surface" comparing with the system "clear atmosphere plus surface" is presented in Table 3.2 as well. We will describe value f_s in detail in the following section.

The data of the spectral radiation measurements accomplished on the 20th April 1985 above the Ladoga Lake and processed in accordance with the methodology described in Sect. 3.2 are presented in Fig. 3.13 and in Table A.3 of Appendix A (experiment 7 in Table 3.2). The comparison with the data of the observations carried out on 24 April 1985 in the clear atmosphere (Table A.2 of Appendix A) also above the Ladoga Lake indicate higher values of solar radiation absorption in the cloud layer. Besides, the values of the downwelling irradiance at level 1.4 km ($\sim$ 850 mbar) of the second observation are lower than the values of the first one. This might be caused by the extinction of radiation in thin cirrus clouds or aerosol layers in the upper troposphere and in the stratosphere.

3.3.2
The Radiation Absorption in the Atmosphere

Now we will pay attention to the estimation of the radiative flux divergence as a main aim of the radiative observations. To provide the possibility of comparison between the obtained results, the radiative flux divergence is normalized to the thickness of the atmospheric layer and then it computes according to (1.8).

The magnitude of the radiative flux divergence in the shortwave spectral region is close to zero and its uncertainty is rather high. The magnitude of the standard deviation of the radiative flux divergence is close to the radiative flux divergence mean value while calculating the uncertainty with the usual methodology. However, the radiative flux divergence is a non-negative value because it is a bounded value and its distribution differs from the Gaussian one. Thus, the values of the mean radiative flux divergence and their standard deviation obtained with the usual methodology do not correctly reflect the distribution of the radiative flux divergence as a random value. The application of the specially elaborated procedure of empirical simulation of the radiative flux divergence with computing its mean value together with the standard deviation removes this difficulty.

Let us consider one layer from P_{i+1} to P_i for the appropriate determination of the mean value and standard deviation of the radiative flux divergence. We use the randomizer described in the book by Molchanov (1970) with the expectation and variance equal to the correspondent values for the irradiance (Sect. 3.2). Irradiances $F_{i+1}^{\downarrow}$, $F_i^{\downarrow}$, $F_{i+1}^{\uparrow}$, $F_i^{\uparrow}$ are simulated as random values. The mean value of the radiative flux divergence and its standard deviation over the layer are computed by their concrete realizations with (1.7) and (1.8), excluding physically impossible cases of the negative radiative flux divergence values. Then, after accumulating enough statistics we get the estimation of the radiative flux divergence and its standard deviation. Moving on to the radiative flux divergence simulation for all layers the demand of the physical property of the radiative flux divergence *additivity* is necessary: the total radiative flux divergence has to be calculated as a sum of the radiative flux divergences of all layers during the layers merging. Hence, the multilayer situation is to be rejected if either of one layer has the negative value of the radiative flux divergence. It is also necessary to account that after the secondary processing the irradiance values correlate with each other so all irradiances are to be simulated at once as a randomly distributed vector with the fixed mean value and with the covariance matrix according to the methodology described in the book by Ermakov and Mikhailov (1976).

According to the results of soundings accomplished in 1970–1980th, the authors of various studies (Kondratyev and Ter-Markaryants 1976; Vasilyev O 1986; Vasilyev O et al. 1987) have revealed that it is possible to obtain the radiative flux divergence with the appropriate accuracy for the atmospheric layer of 100 mbar thickness if only the following set of conditions coincides:

- strong aerosol absorption;

- stability of the atmospheric parameters during the observations;

- stable functioning of the instruments.

All these conditions are hardly realized in practice. Thus, it has been proposed to consider the averaged irradiances in the atmospheric layer 1000–500 mbar, which are obtained as an arithmetic mean over the layers (with the corresponded recalculation of standard deviations).

Figure 3.14 illustrates the typical values of the radiative flux divergence above the Kara-Kum Desert and above Ladoga Lake. The molecular absorption bands of the atmospheric gases (ozone, oxygen and water vapor) are

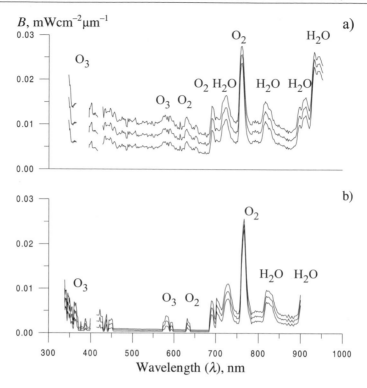

Fig. 3.14a,b. Examples of typical values of the radiative flux divergences in the atmospheric layer 1000–500 mbar; **a** above the Kara-Kum Desert, the airborne sounding 16th October 1983, solar zenith incident angle 51°, sand surface; **b** above Ladoga Lake, the airborne sounding 29th April 1985, solar zenith incident angle 48°, snow surface. There are three curves in every plot, average values and ranges of the 1 SD interval

specified in Fig. 3.14. These results completely agree with values obtained before (Kondratyev and Ter-Markaryants 1976; Vasilyev O et al. 1987).

It is important to mention that the clearer the atmosphere the less the radiative flux divergence and the more complicated is satisfying the conditions of its non-negativity. A large number of non-informative points in the spectrum of the sounding above Ladoga Lake is the usual situation. The best data are the sounding results presented in the article by Vasilyev O et al. (1987). It can be thought that the certain transformation of the molecular absorption bands in the spectrum of the sounding above Ladoga Lake (Fig. 3.14b) is caused by the same reasons.

The non-selective part (the constant level) in the irradiance spectra is to be attributed to the aerosol absorption essentially varying in the atmosphere. Aerosol absorption above the desert is about an order of magnitude higher than absorption above the water surface. In addition, it is possible to trace the specific features of aerosol absorption in the spectral dependence of the radiative flux divergences above the desert. Figure 3.15 illustrates the radiative

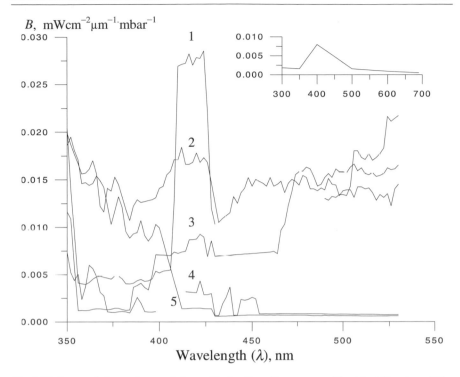

Fig. 3.15. Spectral dependence of the radiative flux divergences. The identification of the hematite absorption band in spectra. Above the Kara-Kum Desert: 1 – the airborne sounding 12th October 1983 under dust storm conditions; 2 – 10th October 1983 under dust gaze conditions; 3 – 23rd October 1984, the pure atmosphere. Above Ladoga Lake: 4 – airborne sounding 29th April 1985 (snow surface); 5 – the airborne sounding 16th May 1984 (water surface). Spectral dependence of the imaginary part of the complex refraction index of the hematite according to Ivlev and Popova (1975) is in the *right-hand upper corner*

flux divergences above the desert obtained during the beginning of the dust storm (12 October 1983), with the dust gaze (10 October 1983), and in the pure atmosphere (23 October 1983). The band of the selective aerosol absorption is apparent in corresponding curves, and it is possible to attribute this band to the ferrous oxides mixture (the component of the sand) called "*hematite*". The radiative flux divergences of the soundings above Ladoga Lake (29 April 1985 above the snow and 16 May 1985 above the water), where the mentioned band is absent, are presented for comparison.

Concerning the "hematite", it is important to point out that the concrete substance Fe_2O_3 usually implied under this term has the maximum of its absorption in the UV spectral region and does not have the apparent absorption selectivity as per the results of Ivlev and Popova (1975), Shettle (1996), and Ivlev and Andreev (1986). However, the authors of the study by Ivlev and Andreev (1986) have mentioned other ferrous oxides and hydroxides demonstrating absorption bands similar to the one shown in Fig. 3.15. Not only Fe_2O_3 but

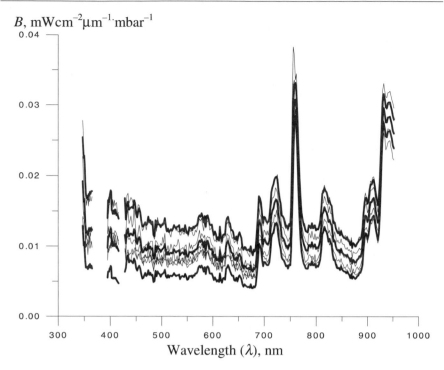

B, mWcm^{-2}μm^{-1}·mbar^{-1}

Fig. 3.16. Spectral dependence of the radiative flux divergences for every layer of the 100 mbar thickness from the results of the airborne sounding 16th October 1983 above the Kara-Kum Desert: *thin lines*; the average value for the layer of 1000–500 mbar and the ranges of 1 standard deviation interval – *thick lines*

also other ferrous oxides are evidently included in the sand composition, and we will symbolically call it "hematite". Thus, we are following the book by Zuev and Krekov (1986) where the complex mixture of ferrous oxides and hydroxides is implied as a hematite and where the data concerning its complex refractive index are taken from. The analysis of the radiative flux divergences in Fig. 3.15 shows that the hematite absorption band is rather narrow and has its maximum near 420 nm. Note that the high content of ferrous oxides in the sand's composition is typical for the Kara-Kum Desert (this is reflected in the name "Black Sands").

Analyzing the radiative flux divergences in separate layers it should be noted that only three soundings among all processed spectra are exact enough for identification of the atmospheric aerosols. Regretfully, there is no statistically significant altitudinal dependence: the radiative flux divergences are approximately equal to each other and close to the average radiative flux divergence for the whole layer 1000–500 mbar as Fig. 3.16 demonstrates.

In addition to that considered above, while processing the soundings data, complementary results have been obtained, namely: calibration curves D and coefficients of the relationship between the irradiance, pressure and solar zenith angle $a_1, \ldots, a_5, b_1, \ldots, b_5$ from (3.6). These parameters were supposed to be used during the correspondent correction of the irradiance spectra with other observation schemes (not soundings). However, the analysis indicated that the calibration curve D turns out to be highly dependent on the experiment series (i. e. linked with the laboratory calibration), which their use impossible for other experiments. The accomplished estimations affirm the standard deviations of calibration curve D to be equal to 2–3% in average, i. e. the calibration accuracy has been successfully improved by applying the above-described approach. However, even this error is too high, and it creates difficulties in applying the modern complex approaches of observational result interpretation, as will be shown in Chap. 5.

3.4
Results of Solar Radiance Observation.
Spectral Reflection Characteristics of Ground Surface

The main aim of the accomplished airborne observations of the solar radiance in the atmosphere was studying spectral reflectance properties of the surfaces. As has been shown in Sect. 1.4 the reflected characteristics of the surface described with function $R(\mu, \varphi, \mu', \varphi')$ are defined from the relation between the income and reflected radiation with (1.73). The simplest characteristic of the surface, the albedo is defined as a ratio of the upwelling to downwelling solar irradiance (1.72) (see the footnote on page 33) (Sivukhin 1980).

Nevertheless, taking into account the insignificant yield of the multiple scattered radiation to downwelling irradiance in the clear atmosphere, the observed reflection characteristics are assumed to correspond to the theoretical ones. However, the relationship between the observed reflection characteristics and ratio of direct and scattered radiation in the downwelling irradiance (Vasilyev O 1986) is particularly essential during comparison of the results obtained in the clear and cloudy atmosphere.

Owing to the diffused reflection (Sect. 1.4) function of four arguments $R(\mu, \varphi, \mu', \varphi')$ it is impossible to measure for the solar radiation field because the radiance really measured from direction (μ_0, φ) depends on the whole field of the income radiation [look at the definition of the reflection operator in (1.74)]. Therefore, the maximally informative characteristic of the reflection available from the observation is a spectral brightness coefficient (SBC) defined as follows:

$$r(\vartheta, \varphi) = I(\vartheta, \varphi)/I_0 , \tag{3.9}$$

where ϑ is the viewing angle (direction $\vartheta = 0$ is the nadir), φ is the viewing azimuth ($\varphi = 0$ corresponds to the Sun's direction), $I(\vartheta, \varphi)$ is the solar radiance reflected from the surface and I_0 is the radiance reflected from the absolutely white orthotropic surface.

The direct measurements of value I_0 were technically impossible during the flight so following the scheme of the SBC observations was used. Radiance I_0 was measured using the same instrument and simultaneously downwelling irradiance $F^\downarrow$ was measured using a second instrument on the ground. The aluminum plate covered with magnesium oxide (manufactured from the burning of magnesium shavings immediately before the airborne observation) was used as an absolutely white orthotropic surface. The albedo of this plate was assumed to be equal to 0.97. Calibration curve $\varrho = 0.97\,F^\downarrow/I_0$ was a result of the ground measurements. Mark that relation $\varrho = \pi$ for the absolutely white orthotropic surface follows from albedo definition (1.71) and from the expression for the upwelling irradiance through radiance I_0 (1.4). The values of downwelling irradiance $F^\downarrow$ and reflected radiance $I(\vartheta, \varphi)$ were registered simultaneously by two instruments during the airborne observations. The SBC was computed according to (3.9) as follows:

$$r(\vartheta, \varphi) = \varrho I(\vartheta, \varphi)/F^\downarrow \ . \tag{3.10}$$

The formula of the theoretical link between the SBC and surface albedo is obtained by expressing value $I(\vartheta, \varphi)$ from (3.10) with relations (1.4) and (1.71):

$$A = \frac{1}{\pi} \int\limits_0^{2\pi} d\varphi \int\limits_0^{\pi/2} r(\vartheta, \varphi) \cos\vartheta \sin\vartheta d\vartheta \ . \tag{3.11}$$

The instruments for the observations and the accuracy of the estimations have been described in Sect. 3.1. However, two different instruments measured the radiance and irradiance and the division (3.10) leads to the additional uncertainty connected with the random displacement of wavelength scales of the instruments relative to each other (Table 3.1). When the signal magnitude is weakly varying with wavelength, the effect of displacement is insignificant, but within the spectral regions, with the fast signal variations (e. g. within the oxygen absorption band 760 nm) the uncertainty of the SBC could strongly increase.

The random uncertainties of the SBC values calculated with (3.10) are caused by the flight factors, especially by the surface heterogeneity, and indicated in the SBC spectra as fast random oscillations. For its filtration the smooth procedure with the triangle weight function (Otnes and Enochson 1978) has been used that leads to the formulas:

$$R_i = d_0 r_i + \sum_{j=1}^{m} d_j(r_{i-j} + r_{i+j}) \ , \quad d_j = \frac{1}{m+1}\left(1 - \frac{j}{m+1}\right) \tag{3.12}$$

where r_i is the initial spectrum of the SBC and R_i is the smoothed one, subscript i corresponds to the point number of the spectrum, m is the smoothing halfwidth (we have used the value $m = 9$). We should mention that halfwidth m in (3.12) is a parameter of the frequency filtration of the data as pointed out in the book by Otnes and Enochson (1978) and it does not link with the instrumental function halfwidth expressed by (3.1). We should emphasize that only smoothed SBC spectra have been used for further analysis. The SBC spectra were considered in the spectral range 350–850 nm with a step of 10 nm accounting for the smoothing halfwidth and narrowing the spectral interval of the smoothed data (Vasilyev A et al. 1997a, 1997b, 1997c).

If the SBC is considered as a reflective characteristic of the surface only (and not of the system "atmosphere plus surface"), the influence of the atmospheric layer between the aircraft and surface on the SBC is to be maximally diminished. For this purpose, the observations seem to be conducted at minimal altitudes. However, the bumps become stronger with the altitude decreasing and the yield of the random item to the total uncertainty caused by the bumps increases. Thus, it is necessary to choose the optimal altitudes for the observations so that the influence of the atmospheric layer below the aircraft would be insignificant and the bumps would not be the main factor determining the random uncertainty. The experience of the flights on board of the IL-14 aircraft has shown that the optimal altitudes above the water surface are 200–300 m and above the ground are 300–500 m. However, sometimes the observations had an occasion to be accomplished at non-optimal altitudes. The approach excluding the influence of the atmospheric layer below the aircraft on the SBC has been proposed in Kondratyev and Markaryants (1976), and Vasilyev O (1986), with carrying out the measurement of the vertical profiles of the SBC. Such observations were also conducted but their amount in the total quantity of the spectra was not large thus, the authors have confined themselves to the analysis of the SBC, measured at the altitudes from 500 m and lower (Vasilyev A et al. 1997a, 1997b, 1997c). It is necessary to point out that the atmospheric influence on the SBC is impossible to exclude as a whole even while observing at low altitudes. It could be displayed as an overstating of the SBC values in the UV spectral range caused by strong Rayleigh scattering and as an understating of the SBC values within the oxygen and water vapor absorption bands in the NIR spectral region.

Studying the SBC spectra dependence upon the surface type, viewing direction, solar zenith angle etc. is of greatest interest while analyzing the obtained values SBC. The elucidating of the mentioned dependence is possible only after statistical processing of the SBC data array with taking into account the significance of the random item of the observational uncertainty. As there are numerous and strongly varying factors (for example solar zenith angle varies constantly) influencing the result, the application of the usual and easiest statistical methodology is ineffective in that case because *the equal observational conditions* are impossible to attribute to large groups of spectra.

To overcome the difficulties mentioned above, cluster analysis (the method of the formal classification of the SBC spectra) was applied during the data analysis. Its essence is to divide the whole totality of the SBC spectra into classes (or groups, clusters; Duran and Odell 1974; Gorelik and Skripkin 1989), each of them forming without any a priori information but only using the principle of the spectra "closeness". Thus, the cluster analysis is reducing the problem to a description of only a few numbers of classes. The metric function, i. e. the distance between two classified objects is used as a numerical characteristic of the spectra closeness (Duran and Odell 1974; Kolmogorov and Fomin 1989). The algorithm of the classification has been constructed recurrently: let certain spectra (for the recursion start it would be one spectrum) to be selected in the class. The following spectrum is added to the same class if the distance between it and the selected class is less than a certain fixed value. After all spectra are tested, the procedure of the classification will be returned for all rested spectra, etc., till the whole totality will have been divided.

There is a sufficient amount of different functions in available books (Duran and Odell 1974; Kolmogorov and Fomin 1989) satisfying the axiomatic demands to the metric introduced in the book by Kolmogorov and Fomin (1989). We should mention that the application of different metrics leads to different results of the classification, so the metric is to be chosen based on the concrete problem conditions. As has been mentioned in the books by Duran and Odell (1974), these conditions sometimes don't allow finding the mathematically correct metrics. Thus, different heuristic metrics (Duran and Odell 1974) are used which are not metrics in the strict sense of the word. We have had to follow the latter way and to use the function below as a measure of the distance between spectra $R^{(1)}$ and $R^{(2)}$:

$$\varrho(R^{(1)}, R^{(2)}) = \max_i \frac{2|R_i^{(1)} - R_i^{(2)}|}{s_i(R_i^{(1)} + R_i^{(2)})}, \tag{3.13}$$

where s_i is the relative random standard deviation of the measured SBC (concrete values are in the articles by Vasilyev A et al. 1997a, 1997b, 1997c). The differences between spectra at every wavelength (not at all wavelengths in average) are accounted in (3.13) because the spectra difference even within the narrow spectral region could turn out the essential one for classification. It is especially important for revealing the erroneous spectra, as will be considered further. For transformation to the relative values, the difference of the spectrum values in (3.13) is normalized to their mean arithmetic and to the standard deviation to take into account the uncertainty variations over the spectrum. As a distance between the spectrum and the class, the distance to the starting spectrum of the class is used. The spectrum will be attributed to the class if the distance is less then 3, i. e. according to the known statistical rule "the allowed deviation from the average does not exceed three standard deviations".

The choice of the starting spectrum (also, the starting spectrum of every following class recurrently) is the indefinite point of the cluster analysis. The problem is that the spectrum is to correspond to the maximum of the distribution of the multi-dimensional function (of the histogram) of the classified objects totality and the search for the maximum is a complicated problem either from the calculating or mathematical point of view. We have analyzed the applicability of different approaches of the choice of starting spectrum to our problem as per the book by Duran and Odell (1974) and we finally decided in favor of the following algorithm. Every spectrum of the classified totality has been tested for the possibility of using it as a starting one. For this spectrum, the number of the spectra of the same class is determined together with the average distance to the spectra of this class expressed by (3.13) and the ratio of the average distance to the number of the spectra of the same class. The latter evidently has a sense of the spectra density in this class. Hence, the spectrum providing the maximal spectra density in the class has been chosen as starting.

The formal computer classification has to be followed by a stage of hand analysis. At this stage, the information concerning measuring factors is analyzed for every class. Some classes are united after this analysis. The important result of the cluster analysis is the automatic revealing of the erroneous spectra that can find a wide application in the operative processing of the atmospheric and surface radiative characteristics. Actually the spectrum could be assumed an erroneous one if there is only one spectrum in the class.

After the classification procedure the mean value and standard deviation is calculated for every class over all contained spectra. The standard deviation is mentioned to be sometimes less than the initial random standard deviation s_i. It is suggested that there are mainly two reasons for this: the spectra statistical averaging, and the yield to the standard deviation of the uncertainty linked with the surface heterogeneity in the real observational conditions, which could be less than the average estimation in Table 3.1.

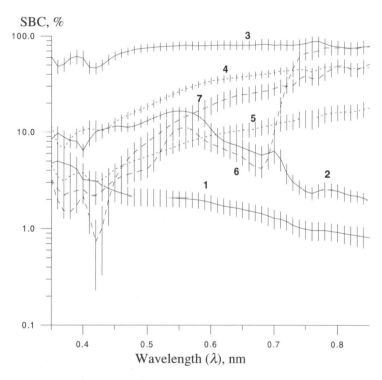

Fig. 3.17. Spectral brightness coefficients (SBC) of the typical natural surfaces. Average values of the SBC of the correspondent classes and the one standard deviation interval: 1 – pure lake water with low chlorophyll and mineral matter content; 2 – lake water with high chlorophyll and mineral matter content; 3 – snow; 4 – sand; 5 – black soil; 6 – green vegetation (grass); 7 – yellow vegetation (ripe grain crop)

The results of the accomplished classification are discussed in detail in the articles by Vasilyev A et al. (1997a, 1997b, 1997c). Figure 3.17 illustrates the typical spectra of the SBC for the natural surfaces. The SBC of three surface types (snow, sand, and pure lake water) are presented in Fig. 3.17. In addition, we should mention that the albedo coincides with the SBC for the orthotropic surface. However, we will analyze to what extent this approximation is correct for the considered surfaces. The analysis of the accomplished classification has shown that there is no dependence upon the viewing direction for the snow surface, i. e. the snow surface is close to the orthotropic. The uncertainty of the approximation is determined by the standard deviation for the class "snow" and is equal to 8% in average over the spectrum (Vasilyev A et al. 1997a, 1997b, 1997c). The description of the main classes of the SBC and their spectral values are presented in Tables A.4–A.7 of Appendix A.

The water surface is the most anisotropic among all natural surfaces as Fig. 3.18 demonstrates. The sharp maximum is a solar glare in the viewing direction to the Sun ($\varphi = 0°$) formed by the solar beams mirror-reflected from the waving water surface (curves 2 and 3 in Fig. 3.18). There is a weak minimum in the opposite direction (curve 4) that has been explained in the book by Mulamaa (1964). Regretfully there was no possibility to measure SBC dependence of the viewing angle and azimuth for the sand surface. The

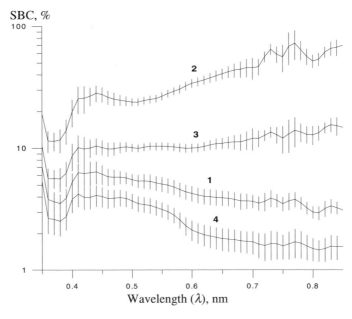

Fig. 3.18. Dependence between the spectral brightness coefficients of the water surface and the viewing direction. Average values of the SBC of the relevant classes and one standard deviation interval are shown: 1 – observation to nadir; 2 – the mirror reflection direction; 3 – the viewing angle corresponded to the mirror reflection at azimuth 45°; 4 – the same as 3 for azimuth 180°

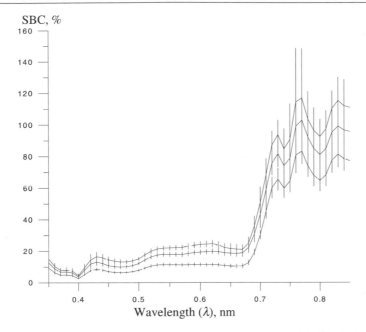

SBC, %

Fig. 3.19. Dependence of spectral brightness coefficients of moose marsh upon the viewing direction. Average values of SBC of the relevant classes and the 1 standard deviation interval are shown: *upper curve* – the direction of the back reflection: *middle curve* – the viewing angle, corresponding to back reflection for azimuth 135°; *lower* – observation to nadir and azimuth from 0 till 90°

sand has just been noted to be a back-reflecting surface (see Sect. 1.4). The similar property is also inherent in the moss marsh (qualitative pictures of the anisotropy for sand and marsh are close). The data in Fig. 3.19 obtained for the marsh surface allow the estimating of the analogous SBC dependence upon the viewing angle and azimuth for the sand surface. The SBC increase is apparent when approaching to the point opposite to the Sun ($\varphi = 180°$): their magnitudes exceed the magnitude to the nadir in 1.5–2 times.

For the analytical description of the anisotropy the following function is introduced:

$$R(\mu, \varphi) = \pi A \frac{1}{\mu} X_1(\mu - \mu') X_2(\varphi - \varphi') \,, \tag{3.14}$$

where A is the surface albedo, (μ, φ) is the definite certain selected direction of the reflection namely it is the direction of the mirror reflection or the back-reflection here, $X_1(\mu'')$ and $X_2(\varphi'')$ are the certain functions describing the anisotropy, when the viewing angle cosine μ and azimuth φ of the reflection are deviating from the selected direction: $\mu'' = \mu - \mu', \varphi'' = \varphi - \varphi'$. Factor π/μ is appearing to co-ordinate (3.14) with expression of the albedo through the SBC (3.11). From the same relation the normalizing conditions for the functions

$X_1(\mu'')$ and $X_2(\varphi'')$ are derived:

$$\int_{-\mu'}^{1-\mu'} X_1(\mu'')d\mu'' = 1 , \qquad \int_{-\varphi'}^{2\pi-\varphi'} X_2(\varphi'')d\varphi'' = 1 . \qquad (3.15)$$

Conditions (3.15) are similar thus, it is possible to construct the functions $X_1(\mu'')$ and $X_2(\varphi'')$ with the same analytical expression. These functions are to reach maximums for $\mu = \mu'$ and $\varphi = \varphi'$ i.e. $\mu'' = 0$ and $\varphi'' = 0$ and to describe the decreasing of the reflection for other quantities. The classical and well known Henyey-Greenstein function defined by (1.31) satisfies these conditions. It is rather appropriate for the application, and the integrals of types (3.15) are easy calculated for it. So the approximation on the base of the Henyey-Greenstein function is proposed:

$$X_1(\mu'') = \frac{g_1}{\frac{1}{((1-g_1)^2+2g_1\mu')^{1/2}} - \frac{1}{(1+g_1^2+2g_1\mu')^{1/2}}} \cdot \frac{1}{(1+g_1^2-2g_1\mu'')^{3/2}} ,$$

$$X_2(\varphi'') = \frac{g_2}{2\pi} \cdot \frac{\frac{1}{(1+g_2^2-g_2\varphi''/\pi)^{3/2}}}{\frac{1}{((1-g_2)^2+g_2\varphi'/\pi)^{1/2}} - \frac{1}{(1+g_2^2+g_2\varphi'/\pi)^{1/2}}} , \qquad (3.16)$$

where $0 < g_1 < 1$ and $0 < g_2 < 1$ are the approximation parameters.

Values A, g_1, and g_2 have been obtained by the simple counting of all possible SBC magnitudes over all measurements (over the initial data but not classification results) to estimate the accuracy of proposed approximations (3.14) and (3.16) and to evaluate the approximation parameters. The procedure has been conducted for the corresponding data of water and moss marsh surfaces with choosing the concrete magnitudes of the parameters providing the minimal standard deviation of the approximation. As the viewing angle did not exceed $45°$ (Sect. 3.1), the obtained values of albedo A have had no physical meaning (it has not been the albedo but a certain coefficient). Concerning the anisotropy parameters it has been obtained for water surface $g_1 = 0.7$, $g_2 = 0.7$ and for moss marsh: $g_1 = 0.2$, $g_2 = 0.5$.

Mention that the observational grid is not rather detailed over viewing angle and azimuth (Vasilyev A et al. 1997a, 1997b, 1997c), so the accuracy of the obtained coefficients is not rather high, namely they could be considered as a rough estimation with assuming their standard deviation is equal to 0.05. According to the same reason the spectral dependence of coefficients g_1 and g_2 (increasing from the UV to NIR spectral regions) could be ignored and the equal magnitudes could be attributed to all wavelengths. It should be emphasized, that the formally calculated standard deviation of approximations (3.14) and (3.16) turns out about 10%, which is close to the observational uncertainty. Hence, we can conclude that (3.14) and (3.16) are describing the anisotropy of the natural surfaces reflection exactly enough, though the small values of the standard deviation could only be the consequence of a small amount of the grid points.

3.5
The Problem of Excessive Absorption
of Solar Short-Wave Radiation in Clouds

Studies of the impact of aerosols and clouds on radiation balance and on radiative flux divergence in the atmosphere are of great importance for the analysis of various factors contributing to climate formation (Monin 1982; Hobbs 1993). The data of the surface pyranometric observations of the downward and upward short-wave radiation (SWR) fluxes permit us to calculate the *short-wave radiation budget* (SWRB) of the surface, whereas the satellite observational data characterize the *outgoing short-wave radiation* (OSWR). The difference between SWRB and OSWR for conditions of the cloudy and clear sky determines *short-wave radiative forcing* at the surface $C_s(S)$ and at the top of the atmosphere $C_s(TOA)$. The mean annual and mean global value of the short-wave cloud radiative forcing $C_s(TOA)$, at the top has been determined to vary from -45 to $-50\,\mathrm{mW\,cm}^{-2}$. The value $C_s(TOA) = C_s(S) + C_s(A)$ is evidently determined through the yields of the cloud radiative forcings of the surface $C_s(S)$ and atmosphere $C_s(A)$. The ratio that describes the cloud forcing of the system "atmosphere plus surface" has been introduced by the authors of the study by Cess et al. (1995)

$$f_s = \frac{C_s(S)}{C_s(TOA)} = \frac{(F_o - F_c)_{\text{bottom}}}{(F_o - F_c)_{\text{top}}} \ . \tag{3.17}$$

The analysis of the observational data of the cloud radiative forcing of the city Boulder (USA) has shown that $C_s(S) = -92.6\,\mathrm{mW\,cm}^{-2}$ and $C_s(TOA) = -63.2\,\mathrm{mW\,cm}^{-2}$ that leads to the value $f_s = 1.46$. The value calculated on the basis of numerical simulations should be close to unity: $f_s \sim 1$. Thus, it follows that the calculations are essentially underestimating the values of the absorbed SWR by the cloudy atmosphere by magnitude $C_s(TOA) - C_s(S) \sim 30\,\mathrm{mW\,cm}^{-2}$ (Hobbs 1993; Cess et al. 1995). It has been called *"excessive"* (or even *"anomalous"*) *cloud absorption of shortwave radiation* (Stephens and Tsay 1990; Cess et al. 1995; Pilewskie and Valero 1995, 1996; Ramanathan et al. 1995). This result has revealed a fundamental gap in present understanding of the cloud impact on SWRB. This obstacle has led to an emotional scientific discussion significantly changing the modern ideas about the role of cloudiness in climate and weather formation (Stephens and Tsay 1990; Cess et al. 1995; Charlock et al. 1995; King et al. 1995; Pilewskie and Valero 1995; Ramanathan et al. 1995; Stephens 1995,1996; Titov and Zhuravleva 1995; Yamanouchi and Charlock 1995; Cess and Zhang 1996; Valero et al. 1997; Zhang et al. 1998). The importance of the problem is seen even in the titles of articles ("An absorbing mystery", "Shortwave cloud forcing: a missing physics", "Anomalous absorption paradox" etc.).

3.5.1
Review of Conceptions for the "Excessive" Cloud Absorption of Shortwave Radiation

The explanations of the excessive absorption of SWR proposed presently can be divided into six main groups.

1. The excessive absorption is an artifact caused by observational errors and imperfectness of data processing (Stephens and Tsay 1990; Pilewskie and Valero 1995; Poetzsch-Heffter et al. 1995; Yamanouchi and Charlock 1995; Arking 1996; Taylor et al. 1996; Francis et al. 1997). Certain results of SWR observations under the conditions of cloudy atmosphere have provided the basis for this conclusion because of providing no significant values of the cloud radiative absorption. The optical and radiative properties of clouds are variable very much depending on the physical mechanism of their origin and in many cases they don't increase radiation absorption by the system "atmosphere plus surface" but on the contrary decrease it. It happens because the clouds are reflecting a significant part of incoming radiation preventing the absorption by the lower atmospheric layers and ground surface. It also should be mentioned that in many cases the observations don't provide a data array sufficient for the qualitative processing. Thus, observations in the cloudy atmosphere frequently haven't been accompanied with the corresponding observations in clear atmosphere at the same period, the ground albedo hasn't been measured every time and only reflected radiation has been registered. All these factors prevent adequate estimation of the radiative characteristics of the cloudy atmosphere.

2. The increased absorption in the cloudy atmosphere in comparison with the clear atmosphere could be explained with the radiation escaping through the cloud sides in the broken clouds, as it has not been registered during the observations at the cloud top and bottom. Either field (Hayasaka et al. 1994; Chou et al. 1995; Arking 1996) or simulated (Titov 1988, 1996a, 1996b; Romanova 1992) experiments could correspond to this group of studies. The methodology of estimating the radiation escaping through the cloud sides proposed in the study by Chou et al. (1995) a priori assumes the *absence* of true SWR absorption by clouds. The authors of another study (Hayasaka et al. 1994) have processed the observational data according to the method of study proposed by Chou et al. (1995). The result of this processing is naturally to provide the conclusion of SWR absorption absence by the cloud.

3. The excessive absorption is an apparent effect caused by the horizontal transport of radiation in the cloud layer due to the horizontal heterogeneity of the layer (stochastic layer structure). A detailed presentation of this approach is provided in the studies by Titov and Kasyanov (1997). In addition, it is necessary to distinguish the cases of the roughness of the top cloud surface (case 1) and of the heterogeneity of the inner cloud structure (extinction coefficient variations; case 2). The numerical analysis

has shown that the horizontal transport in the case of a stochastic cloud top structure is revealed as stronger than in the case of the cloud inner parameter variations. To estimate the absorption in the layer correctly, the scale of the reflected and transmitted irradiances averaging over the cloud horizontal extension should be 30 km for case 1 and 6 km for case 2 correspondingly. The case of the stochastic cloud top structure corresponds to real cumulus clouds and the case of the cloud inner parameter variations corresponds to real stratus clouds. Different combinations of the absorption and scattering coefficients in the cloud layer and different scales of the horizontal and vertical heterogeneity have been considered in the study by Hignett and Taylor (1996) and the authors has revealed that "the internal inhomogeneity in the cloud microphysics and in the macrophysical structure in terms of cloud thickness are both important in the determination of the cloud radiative properties".

4. In addition to other reasons the anomalous absorption in clouds is suggested to be explained with the water vapor absorption within the absorption bands in the NIR spectral region, which has not been accounted for before (Evans and Puckrin 1996; Crisp and Zuffada 1997; Nesmelova et al. 1997; O'Hirok and Gautier 1997; Savijarvi et al. 1997; Harshvardhan et al. 1998; Ramaswami and Freidenreih 1998). However, while computing, the detailed and careful accounting of the molecular absorption in the NIR region has not provided the observed magnitude of the cloud absorption (Kiel et al. 1995; Ramaswami and Freidenreih 1998). Besides, the results of spectral observations (Titov and Zhuravleva 1995) have demonstrated the strongest effect of the anomalous absorption in the visual spectral region, where the water vapor absorption is too weak. Thus, it is seen that the molecular absorption by water vapor in the NIR region is not enough for an explanation of anomalous absorption.

5. The microphysical properties of clouds have been implied as a reason of the excessive absorption in various studies (Ackerman and Cox 1981; Wiscombe et al. 1984; Hegg 1986; Ackerman and Stephens 1987). Very large drops of the cloud are considered in the studies by Ackerman and Stephens (1987) and Wiscombe et al. (1984); it is suggested the presence of them actually increases the radiation absorption within clouds, but it is too weak and insufficient to explain the anomalous absorption. The authors of another study (Hegg 1986) have calculated in detail the optical and radiative parameters of clouds containing two-layer particles with absorbing nuclei and a nonabsorbent shell and have not obtained high enough values of the absorption by clouds either. In all considered models, the noticeable absorption by clouds succeeds only when assuming a significant amount of the atmospheric aerosols (Wiscombe 1995; Bott 1997; Vasilyev A and Ivlev 1997).

6. The authors of three studies (Kiel et al. 1995; Hignett and Taylor 1996; Ramaswami and Freidenreich 1998) have considered the above-mentioned reasons in different combinations and they conclude that with certain

assumptions the calculated and observed values of the cloud radiation absorption turns out to be close to each other. Nevertheless, it is safe to say that there is no exhaustive explanation for the total set of observations. Thus, the problem has not been solved yet as the authors Wiscombe (1995), Lubin et al. (1996), Bott 1997, Ramanathan and Vogelman (1997), and Collins (1998) point out.

3.5.2
Comparison of the Observational Results of the Shortwave Radiation Absorption for Different Airborne Experiments

In the above-mentioned studies of radiation absorption by clouds (confirming or denying the excessive absorption), the satellite data and the data of the meteorological network have been mainly used. These observations were accomplished with different instruments during a long period that called for complicated statistical data processing. As a result, an averaging picture including different types of clouds has been obtained. The absence of either uniform data or a common methodology for data choice and processing is likely to lead to the contradictory conclusions in the studies hereinbefore described.

Let the airborne observations considered in the previous section be analyzed in terms of factor f_s. Absorption $R = (F^\downarrow - F^\uparrow)_{\text{top}} - (F^\downarrow - F^\uparrow)_{\text{base}}$ in the atmospheric layer with and without clouds is computed with the airborne measurements of SWR. Table 3.2 demonstrates the conditions and results of the airborne experiments and the values of factor f_s for the total (within spectral region 0.3–3.0 μm) and spectral (for wavelength 0.5 μm) radiation measurements as values of the total absorption in the layer of the clear or cloudy atmosphere. The results of the airborne observations are seen to allow fixing of the effect of the strong shortwave anomalous absorption ($f_s > 1$) in a set of cases. In other cases there is no influence of clouds on the radiation absorption ($f_s = 1$) and in some cases the strong reflection of solar radiation by clouds even prevents its absorption by the below cloud atmospheric layer and by the ground surface ($f_s < 1$).

3.5.3
Dependence of Shortwave Radiation Absorption upon Cloud Optical Thickness

In accordance with the results of the experiments either in pure and dust clear atmosphere or under overcast conditions the relative value of SWR absorption $b(\mu_0, \tau) = R/\pi S \mu_0$ is presented as a function of the optical thickness in the studies by Kondratyev et al. (1996, 1997a, 1997b) and Vasilyev A et al. (1994). The approximation of the experimental points has elucidated the linear dependence of function $b(\tau)$ that is confirming the analytical expression for SWR absorption presented in the book by Minin (1988). Table 3.2 demonstrates different magnitudes of factor f_s. It is close to unity for the thin clouds with optical thickness $\tau \leq 7$ especially in the pure atmosphere in the Arctic region. In cases with a high content of sand and black carbon aerosols it is valid $f_s \geq 2.5$ at wavelength 0.5 μm and $f_s \sim 1.5$ for total radiation over the shortwave spectral region (experiments 1, 2 and 4) that is pointing to the strong absorption of solar radiation in the atmosphere. Thus, the anomalous absorption obviously reveals itself under conditions of a high content of absorbing aerosols together with cloudiness of large optical thickness ($\tau > 15$) and for small solar zenith angles. Moreover, this effect is not displayed at all in the pure clouds of small optical thickness.

3.5.4
Dependence of Shortwave Radiation Absorption upon Geographical Latitude and Solar Zenith Angle

Presented in Table 3.2 are values of parameter f_s and absorption R, which demonstrate a decrease as they move from tropical to polar regions, which is in agreement with the analysis results in the studies by Kondratyev et al. (1996, 1997a, 1997b) and Vasilyev A et al. (1994). This tendency is broken for the industrial zones characterized with high pollution of the atmosphere (experiments 3–5) and in case 6 of two-layer cloudiness.

The detailed analysis of the mean monthly data sets of the total solar short-wave irradiance obtained from the ground and satellite observations during 46 months (from March 1985 till December 1988) has been accomplished in

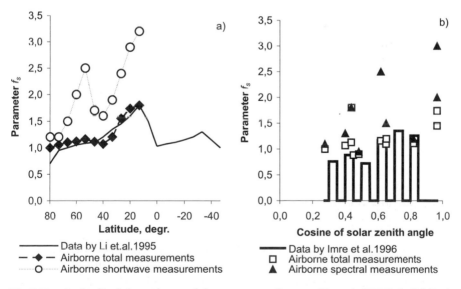

Fig. 3.20. a Latitudinal dependence of the parameter f_s as per Li et al. (1995) (*solid line*) and the values obtained from the airborne observations (*dashed and dotted lines*). *Squares* point to the values of f_s in total shortwave spectrum, *circles* point to the wavelength 0.5 μm; **b** Dependence of the parameter f_s of cosine of the solar incident angle as per Imre et al. (1996) (nomograph) and the values obtained from the airborne observation. *Squares* indicate the total spectrum data; *triangles* indicate the data at the wavelength 0.5 μm

the study by Li et al. (1995). The results of this study include the latitudinal dependence of parameter f_s cited in Fig. 3.20a as a solid line. The results of the airborne observations (Kondratyev 1972; Kondratyev et al. 1973a; Kondratyev and Ter-Markaryants 1976; Kondratyev and Binenko 1981; Kondratyev and Binenko 1984; Vasilyev O et al. 1987; Grishechkin et al. 1989; Vasilyev A et al. 1994) are presented in the same figure. Squares and dashed lines correspond to the total shortwave observations with the pyranometer, which almost coincide with the data of the study by Li et al. (1995). Circles and dotted lines correspond to the observations at a wavelength equal to 0.5 μm and they show crucially larger values than the results of the total observations while keeping the same latitudinal dependence. As hereinbefore described the values of parameter f_s exceeding 2.0 indicate the high content of the absorbing aerosols together with the large optical thickness of the cloud.

The variations of the anomalous absorption with solar zenith angle were studied in Imre et al. (1996) and Minnet (1999). The authors Imre et al. (1996) derived the relationship between parameter f_s and solar zenith angle, which we are citing in Fig. 3.20b (nomograph) together with our results of the airborne observations (Kondratyev 1972; Kondratyev et al. 1973a; Kondratyev and Ter-Markaryants 1976; Kondratyev and Binenko 1981, 1984; Vasilyev O et al. 1987; Grishechkin et al. 1989; Vasilyev A et al. 1994) (squares indicate total spectrum data, triangles indicate data at wavelength 0.5 μm). The solar angle dependence of the airborne data of the total irradiances is evidently coinciding with the data of Imre et al. (1996) while the dependence in question for wavelength 0.5 μm is significantly higher. It should be pointed out that the mentioned coincidence reflects the essence of the specific features of radiation absorption in cloudy atmosphere, though the results either by Imre et al. (1996) and Li et al. (1995) or by Kondratyev (1972), Kondratyev et al. (1973a), Kondratyev and Ter-Markaryants (1976), Kondratyev and Binenko (1981, 1984), Vasilyev O et al. (1987), Grishechkin et al. (1989), and Vasilyev A et al. (1994) were obtained with different instruments, methodologies of measurements and processing. Thus, the excessive (anomalous) absorption really exists and it is mostly evinced in the shortwave spectral region.

The main result of the study by Minnet (1999) is the following: "solar zenith angle is critical in determining whether clouds heat or cool the surface. For large zenith angles ($\mu_0 > 0.15$) the infrared heating of clouds is greater than the reduction in insolation caused by clouds, and the surface is heated by the presence of cloud. For smaller zenith angles, cloud cover cools the surface and for intermediate angles, the surface radiation budget is insensitive to the presence of or changes in, cloud cover." The linear dependence of the cloud radiative forcing upon the cosine of the solar zenith angle in the Arctic has been revealed in the study by Minnet (1999).

The impact of the thick cloudiness and black carbon aerosols on the solar radiation absorption has been revealed in the study by Liao and Seinfield (1998) to produce the forcing values three times higher than those under the cloud-free conditions. Moreover, it is increasing with the growth of cosine of the solar zenith angle. Thus, the absorbing aerosols within the clouds cause the cloud radiation absorption.

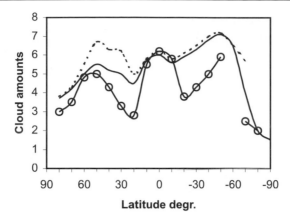

Fig. 3.21. The annual zonal cloud amount: (1) averaged over the latitude; (2) above the sea surface and (3) above the ground surface in 1971–1990 according to Matveev et al. (1986)

The common features of the considered relationship are clear because of the evident relation between the solar zenith angle and geographical latitude (keeping in mind that the radiative experiments are accomplished around midday). However, the original reason is not clear: whether it is the solar height or different cloud optical properties in different latitudinal zones.

It is obvious that for elucidation of the cloud absorption a sufficient amount of clouds is necessary. It is of special interest that the comparison of the latitudinal dependence of the cloud amount (Fig. 3.21) from the study by Matveev et al. (1986) and the dependence of parameter f_s characterizing the cloud radiative forcing as per Fig. 3.20b are seen to coincide qualitatively.

The airborne radiative experiments accomplished in the range of CAENEX, GAAREX, GARP and GATE programs have apparently demonstrated a significant absorption of SWR by clouds. In the remainder of this subsection the following thesis are given:

The excessive absorption of SWR is defined just by the optical properties of cloudiness and is not caused by the observational or processing uncertainties as some investigators have presented.

1. The relationship between the scattering and absorbing properties of stratus clouds and the geographical latitude, solar zenith angle, and type of the atmospheric aerosols within clouds is experimentally proved.

2. The increase in radiation absorption is stronger in thick cloud layers in a dusty atmosphere containing carbon or sand aerosols.

The effect of the excessive absorption is observed over the shortwave spectral region as a whole but it is especially high for the shorter wavelengths ($\lambda < 0.7\,\mu$). The existence of the anomalous absorption fundamentally changes the current understanding of the energetic budget of the atmosphere. In this connection, it is of great importance to account for the atmospheric heating caused by the cloud absorption of SWR for climate forecast simulations.

3.6
Ground and Satellite Solar Radiance Observation in an Overcast Sky

This section presents brief information about the experiments whose results have been used for the retrieval of the cloud optical parameters. There are ground observations with the spectral instruments described in various studies (Mikhailov and Voitov 1969; Kondratyev and Binenko 1981; Radionov et al. 1981; Gorodetskiy et al. 1995; Melnikova et al. 1997) and satellite observations with the POLDER instrument on board the ADEOS satellite (Deschamps et al. 1994; Breon et al. 1998).

3.6.1
Ground Observations

The ground observations have included the transmitted spectral radiance measurements for several viewing angles. The conditions of their accomplishment are listed in Table 3.3 (the numeration in the table continues Table 3.2). The first experiment was performed under overcast conditions at the drifting Arctic station SP-22 on the 13th August and on the 8th October 1979 (Radionov et al. 1981). The measurements had been carried out in the spectral interval 0.35–0.96 µm with resolution 0.001 µm, but the results were processed only at 11 spectral points in each spectrum. The error of the transmitted radiance measurements was evaluated within 3% (Mikhailov and Voitov 1969; Radionov et al. 1981). There were extended, horizontally homogeneous thick clouds during the experiment.

The second experiment was accomplished under the overcast condition in St. Petersburg's suburb on 12th April 1996 (Melnikova et al. 1997) with the spectral instrument, constructed by the authors of the study by Gorodetskiy et al. (1995) on the basis of the CCD matrix detector and with spectral resolution 0.002 µm and spectral range 0.35–0.76 µm (Gorodetskiy et al. 1995). Use of this spectrometer allowed registration of the signal within the spectral ranges 0.35–0.76 µm simultaneously in every spectral point. The instrument was characterized with small size and was PC or Notebook compatible thus, it was convenient for field observations, provided the diminishing of some observational uncertainties and allowed the initial data processing at once.

Table 3.3. Details of the ground radiative experiments

No.	Experiment	μ_0	φ, °N	Date	A_s	Other conditions
11	Arctic drifting station SP-22	0.500	85	13 August 1979	0.60	Surface is wet snow
12	Arctic drifting station SP-22	0.275	85	08 October 1979	0.90	Surface is fresh snow
13	Petrodvorets	0.620	60	12 April 1996	0.70	Surface is fresh snow

In all these cases, the data were obtained for 5 viewing angles (0°, 10°, 15°, 45°, 70°) and for 5 azimuth angles to control the cloudiness homogeneity. One set of measurements took about 10 minutes in the Arctic experiments. The measurements were accomplished at midday, when the solar zenith angle was changing weakly during the 10-minute period. The transmitted radiance for different azimuth angles and for the one viewing angle varying in the range of the measurement error was averaged in the data processing.

During the Arctic experiment the observations of the downwelling and upwelling irradiance were accomplished and ground albedo A was obtained in Radionov et al. (1981). Different types of snow cover were studied (fresh snow, wet snow and so on), and in all cases the spectral dependence of ground albedo A was weak. On the 13th August 1979, the ground surface was covered with wet snow and ground albedo A was about 0.6. On the 8th October 1979, the ground surface was covered with fresh snow and ground albedo A was about 0.9.

In addition, the observation of direct solar radiation was carried out in the clear sky during the Arctic experiment of 1979. It gave the opportunity of calibrating the instrument in units of solar incident flux πS at the top of the atmosphere necessary for the retrieval of optical thickness τ. The experiment on 12th April 1996 was accomplished in a similar manner excluding the measurement of direct solar radiation in the clear sky, hence the instrument was not calibrated and optical thickness τ could not have been obtained. Figure 3.22 illustrates the spectral irradiances for cosines 1.0, 0.985, 0.966, 0.707, 0.340.

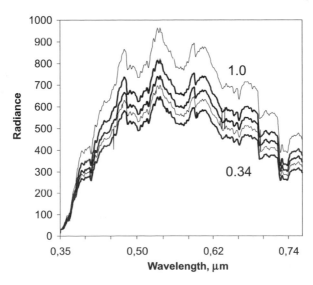

Fig. 3.22. Results of the transmitted radiance observation (relative units) for overcast sky on 12th April 1996

3.6.2
Satellite Observations

The POLDER radiometer consisted of three principal components: a CCD matrix detector, a rotating wheel carrying the polarizers and spectral filters, and wide field of view (FOV) telecentric optics as described in Deschamps et al. (1994). The optics had a focal length of 3.57 mm with a maximum FOV of 114°. POLDER acquired measurements in nine bands, three of which were polarized.

All POLDER measurements were sent to Centre National des Etudes Spatiales (CNES, France) where they were processed. One can find a detailed description in Breon et al. (1998). Processed data have 3 levels of products. Level-1 product consists of radiometric and geometric processing. It yields top-of-the-atmosphere geocoded radiances. Level-2 processing generates geophysical parameters from individual Level-1 products, which cover the fraction of the Earth observed during one ADEOS orbit with adequate illumination conditions. POLDER Level-2 product is taken here for interpreting.

Table 3.4. Details of the satellite experiments

No.	Experiment, geographic site	μ_0	φ, °N	Date	Image size (pixels)	τ_0	ω_0
14	The Southwest part of Europe, 1.65°E–32.04°E	0.7–0.9	43.7–47.8	24 June 1997	388	15	0.996
15	The Atlantic Ocean, and the South of France 24.80°W–3.24°E	0.7–0.9	43.7–47.8	24 June 1997	316	15	0.997
16	The North Sea and the West part of Scandinavia −0.48°E–17.22°E	0.6–0.8	57.7–60.8	24 June 1997	316	20	0.995
17	Scandinavia and the Baltic Sea 1.57°–38.88°E	0.6–0.8	57.7–60.8	24 June 1997	289	15	0.995
18	Baltic Sea and the Northwest part of Russia 27.65°–66.72°E	0.6–0.8	57.7–60.8	24 June 1997	316	7	0.995
19	Southeast Asia and the Pacific Ocean 121.63°–123.61°W	0.8–1.0	6.7–13.8	24 June 1997	585	40	0.995
20	The East part of Siberia, the Pacific Ocean, Sakhalin Island 127.60°–148.68°W	0.7–0.9	45.7–51.3	24 June 1997	585	30	0.997

The geometry for pixel was the following: the point remained within the POLDER field while the satellite passed over it. As the satellite passed over a target, from 6 up to 14 directional radiance measurements (for each spectral band) were performed aiming at the point. Therefore, POLDER successive observations allowed the measurement of the multidirectional reflectance properties of any target within the instrument swath.

Three wavelength channels with the centers at 443, 670 and 865 nm were available for our analysis. The radiance multidirectional data were given in units of the normalized radiance, i. e. the maximum spectral radiance divided by the solar spectral irradiance at nadir and multiplied by $\pi\mu_0$, where μ_0 was the cosine of the solar incident angle. The solar angle, azimuth angle, viewing directions and cloud amount were also included to the data array. The date of the observations under interpretation was 24 June 1997. Seven sites with extended cloud fields were chosen.

The information about the satellite images used for the optical parameters retrieval hereinafter are shown in Table 3.4. The values of the single scattering albedo and the optical thickness typical for most of the pixels of the image are presented in columns number eight and nine of the table. We should mention that images 14 and 15 demonstrate the same cloud field, as do images 16–18.

References

Ackerman SA, Cox SK (1981) Aircraft observations of the shortwave fractional absorption of non-homogeneous clouds. J Appl Meteor 20:1510–1515

Ackerman SA, Stephens GL (1987) The absorption of solar radiation by cloud droplets: an application of anomalous diffraction theory. J Atmos Sci 44:1574–1588

Anderson TW (1971) The Statistical Analysis of time series. Wiley, New York

Arking A (1996) Absorption of solar energy in the atmosphere: Discrepancy between a model and observations. Science 273:779–782

Berlyand ME, Kondratyev KYa, Vasilyev OB et al. (1974) Complex study of the specifics of the meteorological regime of the big city, case study Zaporoghye city (CAENEX-72). Meteorology and Hydrology (1):14–23 (in Russian)

Bott A (1997) A numerical model of cloud-topped planetary boundary layer: Impact of aerosol particles on the radiative forcing of stratiform clouds. QJR Meteorol Soc 123:631–656

Bréon F-M, CNES Project Team (1998) POLDER Level-2 Product Data Format and User Manual. PA.MA.O.1361.CEA Edn. 2 – Rev. 2, January 26th

Cess RD, Zhang MH (1996) How much solar radiation do clouds absorb? Response. Science 271:1133–1134

Cess RD, Zhang MH, Minnis P, Corsetti L, Dutton EG, Forgan BW, Garber DP, Gates WL, Morcrette JJ, Potter GL, Ramanathan V, Subasilar B, Whitlock CH, Yound DF, Zhou Y (1995) Absorption of solar radiation by clouds: Observation versus models. Science 267:496–499

Chapurskiy LI (1986) Reflection properties of natural objects within spectral ranges 400–2,500 nm. Part I. USSR Defense Ministry Press (in Russian)

Chapurskiy LI, Chernenko AP (1975) Spectral radiative fluxes and inflows in the clear atmosphere above the sea surface within the ranges 0.4–2.5 μ. Main Geophysical Observatory Studies 366, pp 23–35 (in Russian)

Chapurskiy LI, Chernenko AP, Andreeva NI (1975) Spectral radiative characteristics of the atmosphere during the sand storm. Main Geophysical Observatory Studies 366, pp 77–84 (in Russian)

Charlock TP, Alberta TL, Whitlock CH (1995) GEWEX data sets for assessing the budget for the absorption of solar energy by the atmosphere. GEWEX News WCRP 5:9–11

Chou M-D, Arking A, Otterman J, Ridgway WL (1995) The effect of clouds on atmospheric absorption of solar radiation. Geoph Res Lett 22:1885–1888

Collins W (1998) A global signature of enhanced shortwave absorption by clouds. J Geophys Res 103:31669–31679

Crisp D, Zuffada C (1997) Enhanced water vapor absorption within tropospheric clouds: a partial explanation for anomalous absorption. In: IRS'96 Current Problems in Atmospheric Radiation. Proceedings of the International Radiation Symposium, August 1996, Fairbanks, Alaska, USA. A. Deepak Publishing, pp 121–124

Danishevskiy YuD (1957) Actinometric instruments and methods of observations. Hydrometeorologic Press, Leningrad (in Russian)

Deschamps PY, Bréon FM, Leroy M, Podaire A, Bricaud A, Buriez JC, Sèze G (1994) The POLDER Mission: Instrument Characteristics and Scientific Objectives. IEEE Trans Geosc Rem Sens 32:598–615

Duran SB, Odell PL (1974) Cluster Analysis. A Survey. Springer, Berlin Heidelberg New York

Ermakov SM, Mikhailov GA (1976) Course of statistical modeling. Nauka, Moscow (in Russian)

Evans WFJ, Puckrin E (1996) Near-infrared spectral measurements of liquid water absorption by clouds. Geophys Res Lett 23:1941–1944

Francis PN, Taylor JP, Hignett P, Slingo A (1997) Measurements from the U.K. Meteorological office C-130 aircraft relating to the question of enhanced absorption of solar radiation by clouds. In: IRS'96 Current problems in Atmospheric Radiation. Proceedings of the International Radiation Symposium, August 1996, Fairbanks, Alaska, USA. A. Deepak Publishing, pp 117–120

Gorelik AL, Skripkin VA (1989) Methods of recognizing. High School, Moscow (in Russian)

Gorodetskiy VV, Maleshin MN, Petrov SYa, Sokolova EA, Pchelkin VI, Solovyev SP (1995) Small dimension multi-channel optical spectrometers. Optical J 7:3–9 (in Russian)

Grishechkin VS, Melnikova IN (1989) Investigations of radiative flux divergence in stratus clouds in Arctic. In: Rational Using of Natural Resources. Polytechnic University Press, Leningrad, pp 60–67 (in Russian)

Grishechkin VS, Melnikova IN, Shults EO (1989) Analysis of spectral radiative characteristics. LGU, Atmospheric Physics Problems 20. Leningrad University Press, Leningrad, pp 20–30 (in Russian)

Harshvardhan, Ridgway W, Ramaswamy V, Freidenreich SM, Batey MJ (1998) Spectral characteristics of solar near-infrared absorption in cloudy atmospheres. J Geophys Res 103(D22):28793–28799

Hayasaka T, Kikuchi N, Tanaka M (1994) Absorption of solar radiation by stratocumulus clouds: aircraft measurements and theoretical calculations. J Appl Meteor 1047–1055

Hegg D (1986) Comments on "The effect of very large drops on cloud absorption. Part I: Parcel models." J Atmos Sci 43:399–400

Hignett P, Taylor JP (1996) The radiative properties of inhomogeneous boundary layer cloud: Observations and modelling. QJR Meteorol Soc 122:1341–1364

Hobbs V (ed) (1993) Aerosol-Cloud-Climate Interaction. Academic Press, New York

Imre DG, Abramson EN, Daum PH (1996) Quantifying cloud-induced short-wave absorption: an examination of uncertainties and recent arguments for large excessive absorption. J Appl Met 35:1191–2010

Ivlev LS, Popova CI (1975) Optical constants of atmospheric aerosols substance. Izv. USSR High Schools, Physics 5:91–97 (in Russian)

Ivlev LS, Andreev SD (1986) Optical properties of atmospheric aerosols. Leningrad University Press, Leningrad (in Russian)

Kalinkin NN (1978) Numerical methods. Nauka, Moscow (in Russian)

Kiehl JT et al. (1995) Sensitivity of a GCM climate to enhanced shortwave cloud absorption. J Climate 8:2200–2212

King M D, Si-Chee Tsay, Platnick S (1995) In situ observations of the indirect effects of aerosols on clouds. In: Charlson RJ, Heitzenberg J (eds) Aerosol forcing of climate. Wiley, New York

Kolmogorov AN, Fomin SV (1999) Elements of the theory functions and the functional analysis. Dover Publications

Kondratyev KYa (1972) Complex Atmospheric Energetic Experiment. GARP Publ. Series. WMO, Geneva (12)

Kondratyev KYa, Ter-Markaryants NE (eds) (1976) Complex radiation experiment. Gydrometeoizdat, Leningrad (in Russian)

Kondratyev KYa, Binenko VI (eds) (1981) First global experiment FGGE, vol 2. Polar aerosols, extended cloudiness and radiation. Gydrometeoizdat, Leningrad, pp 89–91 (in Russian)

Kondratyev KYa, Binenko VI (1984) Impact of Clouds on Radiation and Climate. Gydrometeoizdat, Leningrad (in Russian)

Kondratyev KYa, Vasilyev OB, Grishechkin VS et al. (1972) Spectral shortwave radiative flux divergence in the troposphere. Doklady RAS, Iss Math Phys 207:334–336 (in Russian)

Kondratyev KYa, Vasilyev OB, Grishechkin VS et al. (1973a) Spectral radiative flux divergence of radiation energy in the troposphere within the spectral ranges 0.4–2.4 μ. I. Methodology of observations and data processing. Main Geophysical Observatory Studies. 322:12–23 (in Russian)

Kondratyev KYa, Vasilyev OB, Ivlev LS et al. (1973b) The aerosol influence on radiation transfer: possible climatic consequences. Leningrad University Press, Leningrad (in Russian)

Kondratyev KYa, Vasilyev OB, Grishechkin VS et al. (1974) Spectral shortwave radiative flux divergence in the troposphere and their variability. Izv. RAS, Atmosphere and Ocean Physics 10:453–503

Kondratyev KYa, Vasilyev OB, Ivlev LS et al. (1975) Complex observational studies above the Caspian Sea (CAENEX-73). Meteorol Hydrol 7:3–10 (in Russian)

Kondratyev KYa, Lominadse VP, Vasilyev OB et al. (1976) Complex study of radiation and meteorological regime of Rustavi city. (CAENEX-72). Metcorol Hydrol 3:3–14 (in Russian)

Kondratyev KYa, Binenko VI, Vasilyev OB, Grishechkin VS (1977) Spectral radiative characteristics of stratus clouds according CAENEX and GATE data. Proceedings of Symposium Radiation in Atmosphere, Garmisch-Partenkirchen 1976, Science Press, pp 572–577

Kondratyev KYa, Vasilyev OB, Fedchenko VP (1978) The attempt of the soils detection by their reflection spectra. Soil Sci 4:5–17 (in Russian)

Kondratyev KYa, Korotkevitch OE, Vasilyev OB et al. (1987) Color characteristics of Ladoga Lake waters. In: Complex remote lakes monitoring. Nauka, Leningrad, pp 55–60 (in Russian)

Kondratyev KYa, Vlasov VP, Vasilyev OB et al. (1987) Spectral optical characteristics of the melting snow cover (case studies the Onega Lake and the White Sea). In: Complex remote lakes monitoring. Nauka, Leningrad, pp 211–217 (in Russian)

Kondratyev KYa, Pozdnyakov DV, Isakov VYu (1990) Radiation – hydrooptics experiments in lakes. Nauka, Leningrad (in Russian)

Kondratyev KYa, Binenko VI, Melnikova IN (1996) Cloudiness absorption of solar radiation in visual spectral region. Meteorology and Hydrology 2:14–23 (in Russian)

Kondratyev KYa, Binenko VI, Melnikova IN (1997a) Absorption of solar radiation by clouds and aerosols in the visible wavelength region at different geographic zones. CAS/WMO working group on numerical experimentation, WMO, Geneva

Kondratyev KYa, Binenko VI, Melnikova IN (1997b) Absorption of solar radiation by clouds and aerosols in the visible wavelength region. Meteorology and Atmospheric Physics (0/319):1–10

Li Zhanging, Howard WB, Moreau L (1995) The variable effect of clouds on atmospheric absorption of solar radiation. Nature 376:486–490

Liao H, Seinfield JH (1998) Effect of clouds on direct aerosol radiative forcing of climate. J Geoph Res 103(D4):3781–3788

Lubin D, Chen J-P, Pilewskie P, Ramanathan V, Valero PJ (1996) Microphysical examination of excessive cloud absorption in the tropical atmosphere. J Geophys Res 101(D12):16,961–16,972

Makarova EA, Kharitonov AV, Kazachevskaya TV (1991) Solar irradiance. Nauka, Moscow (in Russian)

Marshak A, Davis A, Wiscombe W, Cahalan R (1995) Radiative smoothing in fractal clouds. J Geophys Res 100(D18):26247–26261

Matveev LT, Matveev YuL, Soldatenkov SA (1986) Global field of cloudiness. Gydrometeoizdat, Leningrad (in Russian)

Melnikova IN, Domnin PI, Varotsos C, Pivovarov SS (1997) Retrieval of optical properties of cloud layers from transmitted solar radiance data. Proceeding of SPIE, vol. 3237, 23-d European Meeting on Atmospheric Studies by Optical Methods, September 1996, Kiev, Ukraine, pp 77–80

Mikhailov VV, Voytov VP (1969) An improved model of universal spectrometer for investigation of short-wave radiation field in the atmosphere. In: Problems of Atmospheric Physics. 6:Leningrad University Press, Leningrad, pp 175–181 (in Russian)

Minin IN (1988) The theory of radiation transfer in the planets atmospheres. Nauka, Moscow (in Russian)

Minnet P (1999) The influence of solar zenith angle and cloud type on cloud radiative forcing at the surface in the Arctic. J Climate 12:147–158

Molchanov NI (ed) (1970) Set of computer codes of small electronic-digital computer "Mir". vol 1. Methods of calculations. Naukova dumka, Kiev (in Russian)

Monin AC (1982) Introduction to theory of climate. Gydrometeoizdat. Leningrad (in Russian)

Mulamaa YuAR (1964) Atlas of optical characteristics of waving sea surface. Estonian AS Press, Tartu (in Russian)

Nesmelova LI, Rodimova OB, Tvorogov SD (1997) Absorption by water vapor in the near infrared region and certain geophysical consequences. Atmosphere and Ocean Optics 10:131–135 (Bilingual)

O'Hirok, Gautier C (1997) Modelling enhanced atmospheric absorption by clouds. IRS'96. Current problems in Atmospheric Radiation. Proceedings of the International Radiation Symposium, August 1996, Fairbanks, Alaska, USA. A. Deepak Publishing, pp 132–134

Otnes RK, Enochson L (1978) Applied Time-Series Analysis. Toronto. Wiley, New York

Pilewskie P, Valero FPJ (1995) Direct observations of excessive solar absorption by clouds. Science 267:1626–1629

Pilewskie P, Valero FPJ (1996) How much solar radiation do clouds absorb? Response. Science 271:1134–1136

Poetzsch-Heffter C, Liu Q, Ruprecht E, Simmer C (1995) Effect of cloud types on the Earth radiation budget calculation with the ISCCP C1 dataset: Methodology and initial results. J Climate 8:829–843

Radionov VF, Sakunov GG, Grishechkin VS (1981) Spectral albedo of snow surface from measurements at drifting station SP-22. In: Kondratycv KYa, Binenko VI (eds) First global experiment FGGE, vol 2. Polar aerosols, extended cloudiness and radiation. Gidrometeoizdat, Leningrad, pp 89–91 (in Russian)

Ramanathan V, Subasilar B, Zhang GJ, Conant W, Cess RD, Kiehl JT, Grassl G, Shi L (1995) Warm Pool Heat Budget and Shortwave Cloud Forcing: a Missing Physics?. Science 267:500–503

Ramanathan V, Vogelman AM (1997) Greenhouse effect, atmospheric solar absorption and the Earth's radiation budget: From the Arrhenius–Langley era to the 1990s. Ambio 26:38–46

Ramaswamy V, Freidenreich SM (1998) A high-spectral resolution study of the near-infrared solar flux disposition in clear and overcast atmospheres. J Geophys Res 103(D18):23,255–23,273

Romanova LM (1992) Space variation of radiative characteristics of horizontally inhomogeneous clouds. Izv RAS Atmosphere and Ocean Physics 28:268–276 (Bilingual)

Savijärvi H, Arola A, Räisänen P (1997) Short-wave optical properties of precipitating water clouds. QJR Meteorol Soc 123:883–899

Shettle EP (1996) The data were tabulated in Naval Research Laboratory and were used to generated the aerosol models which are incorporated into the LOWTRAN, MODTRAN and FASCODE computer codes. Data form HITRAN-96 cd-rom media

Skuratov SN, Vinnichenko NK, Krasnova TM (1999) Observations of upwelling and downwelling solar shortwave irradiances with the stratospheric airplane "Geophysics" in the Tropics (Seishel Islands, February-March 1999). In: International Symposium of former USSR "Atmospheric radiation (ISAR-99)". St. Petersburg, NICHI, St.-Petersburg University, pp 58–59 (in Russian)

Sobolev VV (1972) The light scattering in the planet atmospheres. Nauka. Moscow (in Russian)

STANDARDS 4401–81. Standard atmosphere. USSR state standard (1981) Standard's Press, Moscow (in Russian)

Stephens G (1995) Anomalous shortwave absorption in clouds. GEWEX News, WCRP 5:5–6

Stephens G (1996) How much solar radiation do clouds absorb? Technical comments. Science 271:1131–1133

Stephens G, Tsay SC (1990) On the cloud absorption anomaly. Quart J Roy Meteorol Soc 116:671–704

Taylor JP, Edwards JM, Glew MD, Hignett P, Slingo A (1996) Studies with a flexible new radiation code. II. Comparison with aircraft short-wave observations. QJR Meteorol Soc 122:839–861

Titov GA (1988) Mathematic modeling of radiative characteristics of broken cloudiness. Atmosphere and Ocean Optics 1:3–18 (Bilingual)

Titov GA (1996) Radiation effects of inhomogeneous stratus-cumulus clouds: Horizontal transport. Atmosphere and Ocean Optics 9:1295–1307 (Bilingual)

Titov GA (1996) Radiation effects of inhomogeneous stratus-cumulus clouds: Absorption. Atmosphere and Ocean Optics 9(10):1308–1318 (Bilingual)

Titov GA, Ghuravleva TB (1995) Spectral and total absorption of solar radiation within broken cloudiness. Atmospheric and Ocean Optics 8:1419–1427

Titov GA, Zhuravleva TB (1995) Absorption of solar radiation in broken clouds. Proceedings of the Fifth ARM Science Team Meeting, San Diego, California, USA, 19–23 March, pp 397–340

Titov GA, Kasyanov EI (1997) Radiation properties of inhomogeneous stratus-cumulus clouds with the stochastic geometry of the top boundary. Atmosphere and Ocean Optics 10:843–855 (Bilingual)

Valero FPJ, Cess RD, Zhang M, Pope SK, Bucholtz A, Bush B, Vitko J,Jr (1997) Absorption of solar radiation by the cloudy atmosphere: interpretations of collocated aircraft measurements. J Geophys Res 102(D25):29,917–29,927

Vasilyev AV, Ivlev LS (1997) Empirical models and optical characteristics of aerosol ensembles of two-layer spherical particles. Atmosphere and Ocean Optics 10:856–868 (Bilingual)

Vasilyev AV, Melnikova IN, Mikhailov VV (1994) Vertical profile of spectral fluxes of scattered solar radiation within stratus clouds from airborne measurements. Izv. RAS, Atmosphere and Ocean Physics 30:630–635 (Bilingual)

Vasilyev AV, Melnikova IN, Poberovskaya LN, Tovstenko IA (1997a) Spectral brightness coefficients of natural ground surfaces in spectral ranges 0.35–0.85 μ on base of airborne measurements. I. Instruments and processing methodology. Earth Observations and Remote Sensing 3:25–31 (Bilingual)

Vasilyev AV, Melnikova IN, Poberovskaya LN, Tovstenko IA (1997b) Spectral brightness coefficients of natural ground surfaces in spectral ranges 0.35–0.85 μ on base of airborne measurements. II. Water surface. Earth Observations and Remote Sensing 4:43–51 (Bilingual)

Vasilyev AV, Melnikova IN, Poberovskaya LN, Tovstenko IA (1997c) Spectral brightness coefficients of natural ground surfaces in spectral ranges 0.35–0.85 μ on base of airborne measurements. III. Ground surface. Earth Observations and Remote Sensing 5:25–32 (Bilingual)

Vasilyev OB (1986) To the methodology of spectral brightness coefficients and spectral albedo of natural objects. In: Zanadvorov PN (ed) Possibility of studies of natural resources with remote methods. Leningrad University Press, Leningrad, pp 95–105 (in Russian)

Vasilyev OB, Grishechkin VS, Kashin FV et al. (1982) Studies of the spectral transmissivity of the atmosphere, spectral phase functions and determination of the aerosol parameters. In: Atmospheric physics problems. Iss 17, Leningrad University Press, Leningrad, pp 230–246 (in Russian)

Vasilyev OB, Grishechkin VS, Kondratyev KYa (1987) Spectral radiative characteristics of the free atmosphere above the Ladoga Lake. In: Complex remote lakes monitoring. Nauka, Leningrad, pp 187–207 (in Russian)

Vasilyev OB, Grishechkin VS, Kovalenko AP et al. (1987) Spectral information – measuring system for airborne and ground study of the shortwave radiation field in the atmosphere. In: Complex remote lakes monitoring. Nauka, Leningrad, pp 225–228 (in Russian)

Vasilyev OB, Contreras AL, Velazques AM et al. (1995) Spectral optical properties of the polluted atmosphere of Mexico City (spring–summer 1992). J Geophys Res 100(D12):26027–26044

Wiscombe WJ (1995) An absorbing mystery. Nature 376:466–467

Wiscombe WJ, Welch RM, Hall WD (1984) The effect of very large drops on cloud absorption. Part I:Parcel models. J Atmos Sci 41:1336–1355

Yamanouchi T, Charlock TP (1995) Comparison of radiation budget at the TOA and surface in the Antarctic from ERBE and ground surface measurements. J Climate 8:3109–3120

Zhang MH, Lin WY, Kiehl JT (1998) Bias of atmospheric shortwave absorption in the NCAR Community Climate Models 2 and 3:Comparison with monthly ERBE/GEBA measurements. J Geophys Res 103:8919–8925

Zuev VE, Krekov GM (1986) Optical models of the atmosphere.(Recent problems of the atmospheric optics, vol. 2). Gydrometeoizdat, Leningrad (in Russian)

CHAPTER 4

The Problem
of Retrieving Atmospheric Parameters
from Radiative Observations

This chapter presents a general statement of the problem of determination of atmospheric and surface parameters from observational results of radiative characteristics. The methods of determining the parameters of the radiative transfer theoretical model providing the minimal standard deviation (SD) between the numerical and measured results for the correspondent characteristics are considered below in detail. The choice of a concrete set of the parameters, the influence of systematic uncertainties of the numerical simulations and the technical realization of the considered methods are discussed further.

4.1
Direct and Inverse Problems of Atmospheric Optics

Hereinbefore described in Chaps. 1 and 2 we have demonstrated the possibilities for solving the problem of calculating the solar radiance and irradiance after setting the parameters of the atmosphere and ground surface (volume coefficients of absorption and scattering, phase function, and surface albedo). Furthermore, the results of the characteristic radiative observations have been presented in Chap. 3. Therefore, this gives us the possibility of relating the problem of selecting the atmospheric parameters, which allowed computing values to be equal to the measured characteristics. The problems considered in Chap. 2, i.e. calculations of the observational characteristics with the chosen parameters of the atmosphere and surface, are specified as *direct problems of atmospheric optics*. Contrary to this, the problems considered below, i.e. determination of the atmospheric and surface parameters from observational results of the radiative characteristics, are specified as *inverse problems of atmospheric optics*.

The solution of the direct problem implies the creation of *the mathematical model of observations*, on the basis of which one can relate the physical notions concerning the interaction of radiation with atmosphere and surface (see Chap. 1). We should point out two important obstacles for further consideration.

Firstly, the choice of the physical and consequently mathematical models of the mentioned processes is ambiguous. Actually, while creating the mathematical descriptions, different idealizations of the concrete physical processes together with simplifications and approximations are inevitable, so any model is simpler than the reality is, so it is inadequate when compared to the reality to a certain degree. Hence, the choice of the concrete model together with its parameters is always ambiguous and it is defined either with the physical processes put to the model, or with the degree of approximation of the description of these processes. For example, if we are considering only the radiative transfer, the parameters of the model will be the following: optical thickness, single scattering albedo, and phase function (see Sect. 1.3). Then we could account for the processes of the radiation-media interaction defining the mentioned values (see Sect. 1.2), and the parameters of the model will be: vertical profiles of the pressure, temperature, concentrations of the atmospheric gases, and volume coefficients of the aerosol scattering and absorption.

Secondly, the number of parameters describing the mentioned processes is always *finite* in *the range of the chosen model*. It is readily seen from the technical point of view and needs no comment. However, from the other side the number of the measured characteristics is finite too. Actually, if even the continuous spectrum of the irradiance or radiance is registered, really it is representing as a finite array of the measured characteristics (see Sect. 3.1). The opposite case is impossible because of digitations of the output signal.

Thus, it is safe to say without the generality loss that while solving the direct problem we realize an algorithm allowing the calculation of a strictly limited set of values through a strictly limited set of parameters. This statement is expressed with the mathematically formal relation:

$$\tilde{\mathbf{Y}} = \mathbf{G}(\mathbf{U}) \,, \tag{4.1}$$

where $\tilde{\mathbf{Y}} \equiv (\tilde{y}_i)$, $i = 1,\ldots,N$ is the set, i.e. the vector, of the calculated values, corresponding to real N measurements; $\mathbf{G}$ is the operator of the direct problem solving, i.e. the realization of a certain (concretely chosen as has been pointed out above) mathematical model of the observational process; $\mathbf{U} = (u_j)$, $j = 1,\ldots,M$ is the vector of parameters of the model in question. In general the components of vectors $\tilde{\mathbf{Y}}$ and $\mathbf{U}$ could be inhomogeneous, i.e. could have different meaning and different units (it is always so for vector $\mathbf{U}$). We should mention that vector $\mathbf{U}$ includes *all* necessary parameters for solving the direct problem (not only parameters characterizing the atmosphere and surface but also the solar zenith angle, value of the incident flux at the top of the atmosphere, spectroscopic parameters for computing the volume coefficient of the molecular absorption – Sect. 1.2 etc.), and vector $\tilde{\mathbf{Y}}$ contains *only* the observational results.

The formal statement of the inverse problem is determination (in atmospheric optics it is accepted to say *retrieval*) of the components of parameter vector $\mathbf{U}$ with the specified concrete values of observational result vector $\mathbf{Y}$. However, there is no sense in retrieving all parameters included in vector $\mathbf{U}$. Actually, some parameters of vector $\mathbf{U}$, for example the solar zenith angle, are known (exacter: are supposed to be known). Therefore, from the components of vector $\mathbf{U}$ let us select vector $\mathbf{X} \equiv (x_k)$, $k = 1\ldots,K$, which has to be retrieved. The concrete variants of this selection are considered in the study by Timofeyev

(1998) where it is proposed to classify the inverse problems coming from the type of known and desired parameters. We will return to the topic of choice in Sect. 4.3, and now let us assume that the concrete parameters contained in vector X are specified. Equation (4.1) could now be rewritten as:

$$\tilde{Y}(X) = G(X, U \setminus X) , \qquad (4.2)$$

where $U \setminus X$ is the set of vector U components not included in vector X, i.e. the known parameters of the direct problem. Thus: $G(U) = G(X, U \setminus X)$, i.e. solution of the direct problem is not to depend on which parameters are to be retrieved.

The inverse problem could be formulated as a determination of vector X from the equation:

$$G(X, U \setminus X) = Y . \qquad (4.3)$$

However, in a general case system (4.3) may have no solution. Indeed, as has been shown above, the operator of the direct problem G is just an approximation of reality. Hence, even if we supposed that it reflected reality exactly, vector Y would not be adequate to reality because of the systematic and random observational uncertainties. Thus, a set of possible solutions of the direct problem $\tilde{Y}(X)$ could disagree with a set of possible values of the observational results Y. In addition, the case of the nonexistence of the solution for (4.3) is quite a likely one, even in the simplest variant of the linear operator G. It is connected with the general properties of the abstract linear operators (Tikhonov and Aresnin 1986; Kolmogorov and Fomin 1989). However in our version of the inverse problem statement it is evident: if observations $\{y_i\}$ are linearly independent and their quantity exceeds the quantity of the parameters under retrieval ($M > K$), the system of the linear equations will be unsolved. Therefore, generally the inverse problem of atmospheric optics can be formulated as follows: to find a set of parameters of the direct problem so that its solution would be as close as possible to the observational results. In mathematical wording given in the book by Tikhonov and Aresnin (1986), it means to find value X, for which the minimum is reached:

$$\min_{X \in T} \varrho(Y, \tilde{Y}(X)) = \min_{X \in T} \varrho(Y, G(X, U \setminus X)) , \qquad (4.4)$$

where T is the set of possible solutions, $\varrho(\ldots)$ is the certain measure in space of the observational vectors, i.e. the metrics (more details are in the book by Kolmogorov and Fomin 1989). Note that in particular cases the minimum in question could be equal to zero, i.e. the equality in relation $Y = G(X, U \setminus X)$ is possible.

The essential factor, which is to be accounted for while solving the inverse problem, is the observational uncertainty. These questions will be considered in further detail and here we only mention that unknown parameters X are determined with the uncertainty as well. Hence, accounting for the uncertainty is an alienable and important stage of the inverse problem solving in atmospheric optics. Besides, as the base of the inverse problem solving consists of the comparison between the observational results and solution of the direct problem, the inverse problems are solved with the accuracy defined with the uncertainty of the selected model parameters choice, i. e. with the concrete choice of operator **G**. Hence, the stage of the choice of method for the direct problem solving is the most important part of solving the inverse problem. Besides, as has been mentioned above, the operator of the direct problem solving is inevitably approximated in any case; hence, the account of the approximation influence on the results is necessary as well.

In conclusion, the following general scheme for numerically solving the inverse problems in atmospheric optics could be proposed:

1. Studying the contemporary theory of the physical processes forming the measured characteristics.

2. Choosing a concrete mathematical model of the observations together with its parameters, realization of this model on computer.

3. The error analysis of the direct problem.

4. Dividing the parameters of the mathematical model to the known ones and to the subjects of the retrieval.

5. Choosing the method for solving the inverse problem. Estimating its accuracy.

6. Realization of the solving algorithm on computer.

7. Observational data processing, the analysis and interpretation.

Excluding the first one, which has been considered in Chap. 1 we will discuss all listed stages further, applying them to concrete inverse problems. However, the survey is more appropriate in a different order from that listed above. We should mention that firstly the described scheme has been proposed according to the results of the accomplished observations, so the actual problem of the optimal experiment planning will not be touched upon. Secondly, the presented algorithm has a more complicated logic in practice; in particular, returning to previous stages with the purpose of verifying the model and modernization of the numerical methods are possible. Thus, the numerous consequent versions of the processed results presented in the studies by Chu et al. (1989, 1993) and Steele and Turko (1997) are the standard situation while processing the observational data of atmospheric optics. In fact, it is well known to specialists: the results of the field observations in majority is impossible to process once and for all, there is always something to improve.

We will not review the huge volume and variety of recent inverse problems of atmospheric optics and methods of their solution. As has been mentioned hereinbefore a certain classification of these problems was presented in the study by Timofeyev (1998), and concerning the solution methods there has been no classification for them yet. Here we will confine ourselves only to the concrete inverse problems of retrieval of the atmospheric and surface parameters from the results of the airborne and satellite observations of the solar spectral radiance and irradiance in the atmosphere considered in Chap. 3.

It is possible to distinguish two essentially different cases: clear and overcast sky.

In case of the overcast sky, we succeeded in obtaining the explicit analytical solution, i. e., to write the components of vector X through the results of observations Y as explicit analytical expressions. Moreover, these expressions are not the approximations or empirical formulas, which are often used, but the consequences of the rigorous relations of the radiative transfer theory. We should point out that deriving similar relations for the inverse problem of atmospheric optics is a rather rare case against the backcloth of the recent mass enthusiasm for the numerical solving of the inverse problems on PC. Actually, it corresponds to the philosophical traditions of physics according to which the analytical methods of description of the natural phenomena are preferable.

As follows from the results of the well-known study by Tikhonov (1943) concerning the mathematical aspects of the inverse problem theory: if the inverse problem solution is the limited set of continues functions[1] (the analytical solution is the limited set), this solution will be stable. It has been shown in the book by Prasolov (1995) that the analysis of the stability of the inverse problem solution (robustness) in the limited class of functions is reduced to the statement of the intervals of the continuity of the functions describing the solution. It follows from Chebyshev theorems about the solution stability in the polynomials basis and from the Weierstrass theorem about the existence of the uniform limit (converging to the solution) in the continuous function space. In the case of the analytical solution, its analysis for the continuity is not complicated. Further, the corresponding results will be presented while in detail considering the possibilities of the analytical approaches. The derivations of the pointed analytical relations will be shown in Chap. 6, and the analysis of the results of the observational data processing for the cloudy atmosphere will be considered in Chap. 7.

Regretfully, a similar analytical solution for the clear atmosphere has not succeeded. It is easy to understand it basis on general principles. The variant of the overcast sky, when only the diffused radiation is measured, and the variant of the pure clear atmosphere, when only the direct radiation is accounted for (the optical thickness is easily obtained from Beer's law) are the limit cases of very strong diffusion or its absence. The real clear atmosphere is an intermediate case from the point of view of the diffuse strength and the intermediate cases are usually more complicated than the limit ones. So, while processing the vertical profiles of the spectral irradiances, (Chap. 3) the inverse problem has been put as a problem of numerical choice of the parameters satisfying the above-formulated demand of the minimum $\min_{X \in T} \varrho(Y, G(X, U \setminus X))$. The search for the minimum (4.4) is not physical but a mathematical problem. Thus, in this chapter this solution will be considered from the mathematical side while accounting for the physical conditions and observational errors.

---

[1] In the original wording by Andrey Tikhonov, the term "continues mapping in the compact space" is used. It is more general than that which we are using but these terms coincide in the case of finite dimensioned space, which we are considering.

The solution of the inverse problem for the irradiance observations in the atmosphere and its results will be described in Chap. 5.

Before we present the concrete formulas and algorithms of the search for minimum (4.4), we will mark that the mathematical aspects of the mentioned problem solving are presented often in a rather abstract manner (Kondratyev and Timofeyev 1970) (coming from the approaches of variation calculus and the theory of self conjugate operators in Gilbert space, Elsgolts 1969; Kolmogorov and Fomin 1989). Sometimes it is complicated in practical applications of the abstract expressions and they are perceived as formal receipts for the problem solving without the real physical meaning. Besides, the important questions of the choice of the mathematical model for the direct problem solving, the choice of its concrete parameters and their influence are out of the scope of such a presentation. Our experience of solving the inverse problems of atmospheric optics demonstrates that the understanding of the physical meaning of the relations in use plays an important role together with the formal mathematical approaches. Thus, we will try to present the indicated mathematical approaches not from the abstract positions but from the applied ones in the simplest manner not ignoring even the technical aspects of the realization. To understand such a presentation knowledge of linear algebra (Ilyin and Pozdnyak 1978) and mathematical statistics (Cramer 1946) is enough. We should mention that it is very convenient for comprehension and analysis of the described approaches to consider them applying to the problems of the minimal dimensions (one-dimensional and two-dimensional).

The methodology presented below is not the only approach to the search for minimum (4.4). In fact, the stated problem relates to the class of mathematical extreme problems, whose solutions are well known nowadays (Vasilyev F 1988). For example, in practice such elementary manner as a sorting of a limited quantity of the vector $\mathbf{X}$ variants (Kaufman and Tanre 1998) is often used for the solution search. However, the methodology described below is the mathematically faultless one and allows for the correct account of the observational uncertainties that is particularly important. Its application becomes increasingly popular with the development of the possibilities of computer techniques.

We will begin the presentation from the definition of the distance between the vectors. Let us use the standard Euclid metrics (Kolmogorov and Fomin 1989) i. e. assume the following:

$$\varrho(\mathbf{Y}^{(1)}, \mathbf{Y}^{(2)}) = \frac{1}{N}\sqrt{\sum_{i=1}^{N}(y_i^{(1)} - y_i^{(2)})^2} \,. \tag{4.5}$$

The matter of Euclid metrics (4.5) is the SD of two vectors, i. e. from the physical point of view we are interested in the closeness between the observational results and the direct problem solution in average over the entire observational data set $i = 1, \ldots, N$. The choice of this metric is predetermined because only it succeeds the construction of the real algorithms for the search of the metrics minimum. For example, if we take not an average difference between the observational and calculation results but the one, maximum over all points $i = 1,\ldots,N$, the path described below will become impassable.

The distance between the observational and calculated values of $R \equiv \varrho(Y, G(X, U \setminus X))$ is called *a discrepancy*. Thus, finally it is possible to define the formulated problem as the revealing of the values of the vector X components through the known observational vector Y corresponding to the minimum of discrepancy:

$$R = \frac{1}{N}\sqrt{\sum_{i=1}^{N}(y_i - \tilde{y}_i)^2} \,, \quad \tilde{Y} = G(X, U \setminus X) \,. \tag{4.6}$$

The problem formulated in this manner constitutes the matter of *a least-squares technique (LST)*, proposed by CF Gauss. The following section contains the consequent elucidating of the LST, its specifics and modification.

4.2
The Least-Square Technique for Inverse Problem Solution

Write the solution of the direct problem explicitly through the vector components of the observations and initial parameters:

$$\tilde{y}_i = g_i(x_1,\ldots,x_K) \,, \quad i = 1,\ldots,N \,, \tag{4.7}$$

where $g_i(\ldots)$ are certain functions where the components of vector $U \setminus X$ are included, however we will not write them further in the explicit relations. Substituting (4.7) to the expression for discrepancy (4.6) and considering the square of discrepancy R^2 as a function of variables x_k, $k = 1,\ldots,K$ to obtain its extremums we derive the following equation system:

$$\frac{\partial R^2}{\partial x_k} = 0 \,,$$

i. e. the same in detail:

$$\sum_{i=1}^{N}(y_i - g_i(x_1,\ldots,x_K))\frac{\partial g_i(x_1,\ldots,x_K)}{\partial x_k} = 0 \,, \quad k = 1,\ldots,K \,. \tag{4.8}$$

In the common case of nonlinear functions g_i the direct obtaining of the solutions of system (4.8) and their analysis for the minimum of discrepancy are rather complicated. Thus, to begin with, consider the case of linear functions g_i, which could be further generalized to the nonlinear dependence. Besides the problems of obtaining the parameters of the linear dependence with LST often appear, for example these very problems have been solved during the secondary processing of the airborne irradiance data (Sect. 3.2).

Equation (4.7) in the case of the linear dependence is written as follows:

$$\tilde{y}_i = g_{i0} + \sum_{k=1}^{K} g_{ik} x_k \,. \tag{4.9}$$

Coefficients g_{i0}, g_{ik} are not to be the identical constants at all. They can be rather complicated functions of vector $\mathbf{U}\backslash\mathbf{X}$. It should only be noted that the coefficients are constants from the sense of the considered relationship between the observations and desired parameters because all parameters of vector $\mathbf{U}\backslash\mathbf{X}$ are known and fixed within the range of the concrete inverse problem. The substitution of (4.9) to equation system (4.8) leads to the system of K linear algebraic equations with K unknowns:

$$\sum_{j=1}^{K} x_j \left(\sum_{i=1}^{N} g_{ij} g_{ik} \right) = \sum_{i=1}^{N} (y_i - g_{i0}) g_{ik} \,, \quad k = 1, \ldots, K \,. \tag{4.10}$$

Rewrite (4.10) in the matrix form using above-defined vectors $\mathbf{X} \equiv (x_k)$, $\mathbf{Y} \equiv (y_i)$ and introducing vector $\mathbf{G}_0 \equiv (g_{i0})$ together with matrix $\mathbf{G} \equiv (g_{ik})$, $i = 1, \ldots, N, k = 1, \ldots, K$:

$$(\mathbf{G}^+\mathbf{G})\mathbf{X} = \mathbf{G}^+(\mathbf{Y} - \mathbf{G}_0) \tag{4.11}$$

where the sign "+" specifies the matrix transposition. The vectors are assumed as columns; the first indices of the matrix are assumed as indices of a line while writing system (4.11), and we will stick to this order. Multiplying both parts of (4.11) from the left-hand side to combination $(\mathbf{G}^+\mathbf{G})^{-1}$ the desired solution is obtained:

$$\mathbf{X} = (\mathbf{G}^+\mathbf{G})^{-1}\mathbf{G}^+(\mathbf{Y} - \mathbf{G}_0) \,, \tag{4.12}$$

We should mention that matrix $(\mathbf{G}^+\mathbf{G})$ of equation system (4.11) is symmetric $(\sum_{i=1}^{N} g_{ij} g_{ik} = \sum_{i=1}^{N} g_{ik} g_{ij})$ and positive defined (as per Sylvester criterion (Ilyin and Pozdnyak 1978)). Hence, solution (4.12) exists, it is unique (because the determinant of the positive defined matrix exceeds zero) and corresponds to the discrepancy minimum (because the positive defined matrix $(\mathbf{G}^+\mathbf{G})$ is its second-order derivative). Equation (4.12) is called *a solution of the system of linear equations* $\mathbf{G}_0 + \mathbf{G}\mathbf{X} = \mathbf{Y}$ with LST. Further, we will use this terminology.

The following standard normalizing approach (Box and Jenkins 1970) is recommended here and further to diminish the possible uncertainty connecting with accumulation of the computer errors of the rounding-off during the practical calculations with (4.12). Specify system (4.11) as $\mathbf{AX} = \mathbf{B}$ for a brevity and introduce operator $d_k = \sqrt{a_{kk}}$, $k = 1, \ldots, K$. Pass to system $\mathbf{A}'\mathbf{X}' = \mathbf{B}'$, where $a'_{jk} = a_{jk}/(d_j d_k)$, $b'_k = b_k/d_k$ and after its solution $\mathbf{X}' = (\mathbf{A}')^{-1}\mathbf{B}$ obtain final results $x_k = x'_k/d_k$. The effective square root technique (Kalinkin 1978) is appropriate for the matrix $\mathbf{A}'$ inversion owing to its symmetry and positive definiteness. The computing of the factors in (4.12) is to be accomplished from the right-hand side to the left-hand side; hence, all operations will be reduced to the multiplying of the vector by the matrix.

Hereinbefore we have assumed that the yield to the discrepancy of all squares of the differences between the observational and calculation results is the same. However, it is often desirable to account for the individual specific of these yields. In this case, we use the generalization of *the least-squares technique* – *the least-squares technique "with weights"* (Kalinkin 1978). Write the equation for the discrepancy (4.6) as:

$$R^2 = \sum_{i=1}^{N} w_i(y_i - \tilde{y}_i)^2 \Big/ \sum_{i=1}^{N} w_i , \tag{4.13}$$

where $w_i > 0$ is a certain "weight", attributed to point i. Then for linear dependence (4.9) system (4.10) transforms to:

$$\sum_{j=1}^{K} x_j \left(\sum_{i=1}^{N} w_i g_{ij} g_{ik} \right) = \sum_{i=1}^{N} (y_i - g_{i0}) w_i g_{ik} , \quad k = 1, \ldots, K . \tag{4.14}$$

Not a vector but the diagonal weight matrix $\mathbf{W} \equiv (w_{ij})$, $w_{ii} = w_i$, $w_{i,j \neq i} = 0$, $i = 1, \ldots, N, j = 1, \ldots, N$, is necessary to introduce for writing equation system (4.14) and for solving it in the matrix form. Then the matrix of system (4.14) is written as $(\mathbf{G}^+\mathbf{W}\mathbf{G})$, the free term is written as $\mathbf{G}^+\mathbf{W}(\mathbf{Y} - \mathbf{G}_0)$ and the solution is written as:

$$\mathbf{X} = (\mathbf{G}^+\mathbf{W}\mathbf{G})^{-1}\mathbf{G}^+\mathbf{W}(\mathbf{Y} - \mathbf{G}_0) . \tag{4.15}$$

It is important to mention that explicit expressions (4.14) are more convenient to use during the practical calculations of the matrix and free term. The meaning of the introduced weight matrix $\mathbf{W}$ will become clear in the following section. Mention here, that the solution of the problem with LST does not depend on the absolute magnitudes of the weights, i.e. the multiplying of all values w_i by the constant does not change the values of desired parameters $\mathbf{X}$. In particular, if all w_i are equal, then solution (4.15) will coincide with the case of the solution "without weights" (4.12).

In principle, weights w_i could be chosen from different views. The situation when the inverse square of the mean square uncertainty of the observations is taken as a weight is rather usual, i.e. $w_i = 1/s_i^2$, where s_i is the SD of the y_i observation. The theoretical reasons for this choice will be presented in the following section. Now we should mention its obvious meaning: the greater uncertainty the less its yield to the discrepancy and the demand to the closeness of corresponding values y_i and $\tilde{y}_i$ is weaker. The other important case of using the weights is passing to the relative value of the discrepancy, i.e. summarizing of the squares of not absolute but relative deviations y_i from $\tilde{y}_i$ in (4.13). Equality $w_i = 1/y_i^2$ is evidently valid in this case. If the relative value of the discrepancy is calculated and the relative SD of series points δ_i is fixed then the following will be inferred: $w_i = 1/(\delta_i^2 y_i^2) = 1/\sigma_i^2$. That is to say, that the case of the relative discrepancy minimization together with the specifying of the relative SD is almost equivalent to the case of the absolute discrepancy minimization with the specifying of the absolute value of the SD for every point. The calculation scheme with "weight" has been used for accounting the observational uncertainties in Sect. 3.2.

Parameters x_k desired with LST are to be linearly independent, otherwise the matrix of equation system (4.11) would be degenerate and the inverse matrix would not exist. However, there are the cases, when the linear constraints between the desired parameters are to be accounted for, this very situation has been described in Sect. 3.2 during the secondary processing of the sounding results. We can write the mentioned constraints in a general form as:

$$c_{j0} + \sum_{k=1}^{K} c_{jk}x_k = 0 , \quad j = 1,\ldots,J . \tag{4.16}$$

Obviously, conditions (4.16) are to be linearly independent and $J < K$ (otherwise, the linear dependent lines should be excluded from system (4.16) by decreasing value J). As per *the theorem of basis minor* (Ilyin and Pozdnyak 1978) there are J independent columns in the conditions (4.16). We will assume they are the first ones from the left-hand side (otherwise, components x_k should be renumbered). Divide vector $\mathbf{X}$ into two parts: $\mathbf{X}^{(J)} \equiv (x_k), k = 1,\ldots,J$ and: $\mathbf{X}^{(K-J)} \equiv (x_k), k = J + 1,\ldots, K$. Then conditions (4.16) are written in the matrix form as:

$$\mathbf{C}_0 + \mathbf{C}^{(J)}\mathbf{X}^{(J)} + \mathbf{C}^{(K-J)}\mathbf{X}^{(K-J)} = 0 , \tag{4.17}$$

where $\mathbf{C}_0 \equiv (c_{j0})$, $\mathbf{C}^{(J)} \equiv (c_{jk})$, $k = 1,\ldots,J$, $\mathbf{C}^{(K-J)} \equiv (c_{jk})$, $k = J + 1,\ldots,K$, $j = 1,\ldots,J$. Matrix $\mathbf{C}^{(J)}$ is non-degenerate, hence system (4.17) is soluble relating to $\mathbf{X}^{(J)}$:

$$\mathbf{X}^{(J)} = (\mathbf{C}^{(J)})^{-1}(-\mathbf{C}_0 - \mathbf{C}^{(K-J)}\mathbf{X}^{(K-J)}) . \tag{4.18}$$

On the basis of (4.18) the expression of vector $\mathbf{X}$ as a whole through its independent part $\mathbf{X}^{(K-J)}$ is inferred:

$$\mathbf{X} = \mathbf{B}_0 + \mathbf{B}\mathbf{X}^{(K-J)} , \tag{4.19}$$

where vector $\mathbf{B}_0 \equiv b_{k0}$ and matrix $\mathbf{B} \equiv b_{kl}$ possess a similar structure:

$$b_{k0} = ((\mathbf{C}^{(J)})^{-1}(-\mathbf{C}_0))_k \quad \text{and} \quad b_{kl} = ((\mathbf{C}^{(J)})^{-1}(-\mathbf{C}^{(K-J)}))_{kl} ,$$
$$\text{for} \quad k = 1,\ldots,J , \quad l = 1,\ldots,J ;$$
$$b_{k0} = 0 , \quad b_{kk} = 1 \quad \text{and} \quad b_{kl \neq k} = 0 ,$$
$$\text{for} \quad k = J + 1,\ldots, K ; \quad l = J + 1,\ldots, K .$$

Substituting (4.19) to initial equation system $\mathbf{GX} = \mathbf{Y}$, writing its solution with LST for independent variables $\mathbf{X}^{(K-J)}$ and passing again to the whole vector $\mathbf{X}$

with (4.19), the following is obtained while taking into account matrix general property $(GB)^+ = B^+G^+$ (and with adding the weights):

$$X = B_0 + B(B^+G^+WGB)^{-1}B^+G^+W(Y - G_0 - GB_0) \,. \tag{4.20}$$

Equations system (3.7) has been solved in Sect. 3.2 just using relation (4.20) with account of restrictions to the parameters (3.8). We should mention, that the matrix of system (4.20) being a subject to inversion is still symmetric and positive defined, with concern to all similar matrices, which will be presented below. To compute the product of several matrices effectively is to use the above-described approach of multiplying a matrix from the right-hand side by a vector with consequent choosing of the last matrix columns of the product as such vectors.

Now consider the general case of nonlinear relationship (4.7) between observations Y and parameters X. Take certain initial values of parameters $X_0 \equiv \{x_{0,k}\}$ and expand (4.7) into a Taylor series accounting for only the linear item:

$$\tilde{y}_i = g_i(x_{0,1}, \ldots, x_{0,K}) + \sum_{k=1}^{K} \frac{\partial g_i(x_1, \ldots, x_K)}{\partial x_k}(x_k - x_{0,k}) \,, \quad i = 1, \ldots, N \,. \tag{4.21}$$

Difference $\tilde{y}_i(x_1, \ldots, x_K) - g_i(x_{0,1}, \ldots, x_{0,K}) = \tilde{y}_i(x_1, \ldots, x_K) - \tilde{y}_i(x_{0,1}, \ldots, x_{0,K})$ is a linear function of parameter difference $x_k - x_{0k}$ (in the considered approximation). It allows the constructing of the iteration algorithm for the nonlinear dependence using the above-obtained solution for the case of the linear one. This standard approach of reducing the nonlinear problems to the linear ones is known as *a linearization*. System (4.21) is converted to the matrix form as:

$$\tilde{Y}(X) = G_0 \cdot (X - X_0) + \tilde{Y}(X_0) \,, \tag{4.22}$$

where G_0 is the matrix of partial derivative $(\partial g_i(x_1, \ldots, x_K)/\partial x_k), i = 1, \ldots, N, k = 1, \ldots, K$, calculated in point X_0. This specification of the matrix of derivatives is convenient because in the linear case of (4.9) the matrix G evidently has the same meaning; hence, the successiveness of the specifications is kept. The operator of the direct problem solution also keeps initial specification $G(X, U \setminus X)$, but for a brevity we will further write just $G(X)$. Iterationally applying the above-considered solution with LST as per (4.15) to (4.22) we obtain:

$$X_{n+1} = X_n + (G_n^+WG_n)^{-1}G_n^+W(Y - G(X_n)) \,, \tag{4.23}$$

where $n = 0, 1, 2, \ldots$ is a number of the iteration.

There are three difficulties of the practical application of (4.23):

- indeterminacy of the zeroth approximation choice;

- necessity of elaborating the criteria for the iteration interruption;

- possible large spread in the desired values during the consequent iterations.

In concrete problems, the choice of zeroth approximation X_0 is usually accomplished from the physical reasons. This choice is bound up with the "guessing" of the solution. Indeed, the closer the zeroth approximation is to a final solution the less number of the iterations is necessary and the better their convergence is. Usually, certain a priori mean values are taken as a zeroth approximation. It could be mean-climatic data for the problems of atmospheric optics. Sometimes there is a possibility to obtain anywise the approximate solution rough as it is, and this solution is to take as X_0. Usually such a choice of zeroth approximation essentially increases the effectiveness of iteration process (4.23). Mention that owing to the problem of the nonlinearity, the solution could be not unique, i. e. to depend on the concrete choice of X_0. These questions we will discuss in Sect. 4.4 in detail.

Standard condition $\varrho(X_{n+1}, X_n) < \varepsilon$, where $\varrho(\ldots)$ is a certain metric, and ε is a parameter describing the solution accuracy, is used theoretically as a criteria for breaking off the iteration. Usually the Euclid metric is used as $\varrho(\ldots)$ (4.10), because it is coordinated with the metric of the observations. Nevertheless, the other variants are possible, for example, the rigorous condition: $\max_{k=1,\ldots,K} |x_{n+1,k} - x_{n,k}| < \varepsilon$ (Box and Jenkins 1970). However, everything is much more complicated during the practical calculations. The accumulation of the errors of the computer calculation together with possible special features of the discrepancy behavior around the minimum point leading to value $\varrho(X_{n+1}, X_n)$ is finishing to diminish with n increasing, hence, the condition for the breaking of the iterations could be not valid for too small ε. Thus, to provide the solution independency of the concrete choice of value ε, the other conditions are often used for breaking off the iterations. Thus, the effective way is analyzing value $\varrho(X_{n+1}, X_n)$ as a function of n and breaking off the iteration when its stable decreasing changes to the oscillations around a certain magnitude (Vasilyev O and Vasilyev A 1994). In the simplest variant, the decision about the breaking off is assumed in the interactive regime. Another simple way is a choice of the solution corresponding to the minimum of the discrepancy for the fixed iteration number. Note that the peculiarities of the iteration convergence are caused by the conditions of the concrete problem and need the special study within the range of the preliminary numerical experiments (Vasilyev O and Vasilyev A 1994).

It is easy to understand the reason for the appearance of the strong spread of values $\varrho(X_{n+1}, X_n)$ (i. e. the large difference of the desired values of two neighbor iterations) from the physical meaning. Indeed, the matrix of the partial derivatives G_n depends on current magnitude X_n, and vector X_n is quite possibly falling within the area where the measured values will extremely weakly depend on some component of vector $x_{n,j}$. However, it means that magnitude $x_{n,j}$ could vary strongly without essential influence to the measured values. Algorithm (4.23) operates in this very way. There are diverse approaches to remove this difficulty. They all are based on the paradoxical idea of *convergence retarding*, which does not allow the vector X_n values to distinguish too strong at the neighbor iterations (the more haste the less speed). We will return to this question repeatedly, and now consider one of the simplest possibilities.

Write initial equation (4.22) for iteration number n and add product $\mathbf{G}_n(\mathbf{X}_n - \mathbf{X}_0)$ to both parts. After accomplishing elementary manipulations and solving the equation with LST, as has been shown by Timofeyev et al., (1986), the solution could be expressed in the form:

$$\mathbf{X}_{n+1} = \mathbf{X}_0 + (\mathbf{G}_n^+ \mathbf{W} \mathbf{G}_n)^{-1} \mathbf{G}_n^+ \mathbf{W}(\mathbf{Y} - \mathbf{G}(\mathbf{X}_n) + \mathbf{G}_n(\mathbf{X}_n - \mathbf{X}_0)) . \tag{4.24}$$

In algorithm (4.24), all iterations are counted off $\mathbf{X}_0$ that turns out a certain obstacle for too strong spread of the values. At any rate, according to the practical application, (4.24) demonstrates a higher effectiveness than (4.23) does, in spite of the increased calculation volume. Furthermore, we will use only algorithm (4.24) for solving the problem with LST.

Quite often the additional conditions (constraints and restrictions) are imposed to desired parameters $x_1, \ldots, x_k$ based on the physical reasons, i.e. the problem of searching not absolute but conditional extremums appears from the mathematical point of view. This problem is more complicated and common methods of its solution, e.g., the classical method of Lagrange indeterminate multipliers (Vasilyev F 1988), do not always blend with the LST ideology.

In some separate cases, we succeeded in accounting for the constraints and restrictions to the desired parameters using special approaches. For example, in the above-considered case of the linear constraints between parameters $x_1, \ldots, x_k$ expressed through vector $\mathbf{B}_0$ and matrix $\mathbf{B}$ the following is elementarily obtained:

$$\mathbf{X}_{n+1} = \mathbf{X}_0 + \mathbf{B}(\mathbf{B}^+ \mathbf{G}_n^+ \mathbf{W} \mathbf{G}_n \mathbf{B})^{-1} \mathbf{B}^+ \mathbf{G}_n^+ \mathbf{W}(\mathbf{Y} - \mathbf{G}(\mathbf{X}_n) + \mathbf{G}_n(\mathbf{X}_n - \mathbf{X}_0)) . \tag{4.25}$$

Algorithm (4.25) is a generalization of algorithm (4.20) for the case of nonlinear problems.

Some special difficulties will arise if the restrictions to the possible values of the parameters are written as inequalities. For example, practically all parameters (gases and aerosols contents, surface albedo, etc.) are to be non-negative proceeding from their physical meaning. In this case, the rather evident way of the removal of restrictions is the conversion of the values to their logarithms (Virolainen 2000; Potapova 2001). However, strictly speaking, in this case the values of the logarithms providing the minimum of the discrepancy mustn't correspond to the values of the parameters providing the same minimum. That is to say, that taking logs brings an additional uncertainty to the solution obtained with LST. Hence, in spite of its simplicity and attractiveness it is necessary to use this approach carefully, studying its "pluses and minuses" when applying it to concrete problem conditions.

At the same time there is a general method allowing the *approximate* accounting of any complicated constraints and restrictions to the retrieved parameters – *the method of penalty functions* (Vasilyev F 1988).

Let J conditions of the constraints be imposed on the desired parameters, which could be written without breaking off the generality as:

$$c_j(x_1, \ldots, x_K) = 0 , \quad j = 1, \ldots, J . \tag{4.26}$$

Functions c_j are supposed to be continuous and differentiable within the whole region of the argument values. Note that in the method of penalty functions unlike the linear case (4.16) under conditions (4.26) the relations between the number of constraints J and number of parameters K could be arbitrary, particularly, $J \geq K$ is admissible (and certainly concrete values c_j could be independent of all arguments at once). The search of the minimum of the following value instead of discrepancy R minimum (4.13) is a matter of the method of penalty functions:

$$R_C^2 = R^2 + R_H^2 = \sum_{i=1}^{N} w_i(y_i - \tilde{y}_i)^2 \Big/ \sum_{i=1}^{N} w_i + \sum_{j=1}^{J} h_j^2 c_j^2(x_1, \ldots, x_K) \Big/ \sum_{i=1}^{N} w_i ,$$

(4.27)

where h_j is a certain constant. The idea of the method is elementary. Indeed, additional sum R_H^2 in (4.27) with functions c_j yields nothing to discrepancy R^2 if conditions (4.26) are severely satisfied. The less constraint conditions (4.26) are satisfied (i.e. the farther values of c_j from zero), the greater the yield of the additional sum is to total value R_C^2. This yield is like a penalty for the violation of the constraint conditions, hence the name of the method appears (penalty functions are expressions $h_j c_j(x_1, \ldots, x_K)$). During the search of minimum R_C^2 the solution tends to the parameter values, when the additional yield of the conditions is minimal, i.e. to the most exact satisfying of constraint conditions (4.26). The choice of constants $h_j, j = 1, \ldots, J$ in (4.27) are arbitrary enough. It is clear that the greater they are, the more exact constraint conditions (4.26) are satisfied by the solution. Theoretically, constants h_j have to tend to the infinity (Vasilyev F 1988), but practically the greater h_j are, the more nonlinear the problem is and the more difficult the calculation algorithm is adapted to the problem. Thus, it is necessary to select constants h_j carefully in practice. Usually all constants h_j are selected equal to each other, i.e. the algorithm is managed by one parameter of penalty functions $h = h_1 = \ldots = h_j$.

To solve the problem of the search for the value minimum (4.27) we are applying linearization: at first, the solution of linear functions g_i and c_j is obtained and then the nonlinear case is reduced to the linear one. From equation system $\partial R_C^2 / \partial x_k = 0$ the following is obtained with linear dependences (4.9) and (4.16) instead of (4.14):

$$\sum_{j=1}^{K} x_j \left(\sum_{i=1}^{N} w_i g_{ij} g_{ik} + \sum_{l=1}^{J} h_l^2 c_{lj} c_{lk} \right) = \sum_{i=1}^{N} (y_i - g_{i0}) w_i g_{ik} - \sum_{l=1}^{J} h_l^2 c_{l0} c_{lk} ,$$

(4.28)

$$k = 1, \ldots, K .$$

Introducing the vector and matrix of the constraints $\mathbf{C_0} \equiv (c_{j0})$, $\mathbf{C} \equiv (c_{jk})$, $j = 1, \ldots, J, k = 1, \ldots, K$ and also diagonal matrix $\mathbf{H} \equiv (h_{jl})$, $h_{jj} = h_j^2$, $h_{j,l \neq j} = 0$ analogous to matrix $\mathbf{W}$ the solution of system (4.28) is obtained:

$$\mathbf{X} = (\mathbf{G}^+\mathbf{W}\mathbf{G} + \mathbf{C}^+\mathbf{H}\mathbf{C})^{-1}(\mathbf{G}^+\mathbf{W}(\mathbf{Y} - \mathbf{G_0}) - \mathbf{C}^+\mathbf{H}\mathbf{C_0}) .$$

(4.29)

Note that in practice especially with the equality of all h_j, explicit expressions (4.28) are to be used for the calculations. In the nonlinear case functions g_i and c_j are expanded into Taylor series and with considering only the linear term the equation for the iteration is obtained:

$$(G_n^+ W G_n + C_n^+ H C_n)(X_{n+1} - X_n) = G_n^+ W(Y - G(X_n)) - C_n^+ H C(X_n), \quad (4.30)$$

where G_n and C_n are the matrices of partial derivatives $(\partial g_i / \partial x_k)$ and $(\partial c_j / \partial x_k)$ for $i = 1, \dots, N$ and $j = 1, \dots, J$, correspondingly, and $C(X_n)$ is the vector of function c_j (4.26). All vectors and matrices are calculated for argument X_n. Applying to (4.30) the above-described approach of improving the iterations convergence, namely adding to both parts combination $(G_n^+ W G_n + C_n^+ H C_n)(X_n - X_0)$ the iteration algorithm of LST is obtained with taking into account conditions (4.26) according to the penalty functions method:

$$X_{n+1} = X_0 + (G_n^+ W G_n + C_n^+ H C_n)^{-1} \quad (4.31)$$
$$[G_n^+ W(Y - G(X_n) + G_n(X_n - X_0)) + C_n^+ H(-C(X_n) + C_n(X_n - X_0))] .$$

An important point of general expression (4.31) is that the parameter values of the previous step of the iterations are defined in the range of the current iteration; hence, they can be used as constants in the penalty functions. For example, demand that the desired parameters of the current iteration don't differ too much from their values at the previous iteration, i. e. we use the above-discussed approach of the convergence retarding. The constraint conditions evidently correspond to it:

$$X_{n+1} - X_n = 0 , \quad (4.32)$$

and X_n here is not a variable but the constant. In this case, matrix $C_n^+ H C_n$ coincides with matrix H, as C_n is the identity matrix. Also equality $C(X_n) = 0$ is correct as per to conditions (4.32) and (4.31) converts to the algorithm with improved convergence, proposed in the study by Polyakov (1996):

$$X_{n+1} = X_0 + (G_n^+ W G_n + H)^{-1} \quad (4.33)$$
$$\times [G_n^+ W(Y - G(X_n) + G_n(X_n - X_0)) + H(X_n - X_0)] .$$

In algorithm (4.33) the greater the weight magnitudes are, the closer the values of the previous and following iterations, thus the smooth (without spread) convergence of the iterations with correct selections of h_j could be provided.

We should mention one other particular case of applying the penalty function method (Gorelik and Skripkin 1989), which could be used for the inverse problems of atmospheric optics. The situation frequently met in practice, is the case, where the part of the desired parameters (or even all) could be equal to an integer only. For example, the problem of accounting for a certain factor influencing the radiative transfer, which could be described by introducing to vector X a certain parameter that can be equal to zero or unity, i. e. "to turn on" or "turn off" this factor. These problems can be solved with the method of the reselection of all possible variants of the parameter values if their quantity is small, however with a large quantity of parameter values it is unreal. Then the penalty function method is appropriate, as the constraint equations have the form $x_k(x_k - 1) = 0$.

Lastly, the case where we impose inequality on the desired parameters is considered. They can be written without the generality loss as:

$$\psi_j(x_1,\ldots,x_K) \geq 0 , \quad j = 1,\ldots,J . \tag{4.34}$$

But conditions (4.34) could be *approximately* reduced to the above-considered constraint conditions (4.26) if the differentiable functions are used instead ψ_j. These functions are close to zero, when the inequality is valid and they are big enough when the inequality is not valid. The following condition could be used as the simplest constraint satisfying the desired properties:

$$c_j = \exp(-h_j\psi_j(x_1,\ldots,x_k)) , \quad j = 1,\ldots,J . \tag{4.35}$$

Constants h_j have the same meaning as the above-considered ones in relations (4.26) because the matrix of the partial derivatives from condition (4.35) is equal to:

$$\frac{\partial c_j}{\partial x_k} = -h_j c_j \psi_j \frac{\partial \psi_j}{\partial x_k} .$$

Thus, formally, for coordination with the above-derived equations, the following should be set everywhere:

$$\mathbf{H} \equiv (h_{jl}) , \quad h_{jj} = h_j^2 , \quad h_{j,l \neq j} = 0 , \quad \text{and} \quad (\mathbf{C}_n)_{jk} = -c_j\psi_j \frac{\partial \psi_j}{\partial x_k} .$$

The selection of values h_j is arbitrary to a certain degree, the larger they are the more exact the inequalities are, and the stronger the exponent nonlinearity is. The method of penalty functions is obviously converted to the case when constraints (4.26) and conditions (4.34) are imposed on the parameters.

4.3
Accounting for Measurement Uncertainties and Regularization of the Solution

We will begin to consider the impact of the observational uncertainties on the inverse problem solution from an elementary but rather important relation. Let some parameters **X** be expressed linearly through observational results **Y**:

$$\mathbf{X} = \mathbf{A}\mathbf{Y} + \mathbf{A}_0 , \tag{4.36}$$

where **A** and $\mathbf{A}_0$ are the specified matrix and vector. Note that all relations obtained in the previous section finally have just a similar view, though vector **X** is treated here in a wider sense: as some value linearly dependent on **Y**.

Observational data $\mathbf{Y}$ contain the random errors characterized with the SD of components y_i, $i = 1,\ldots,N$. In general, the errors could correlate, i.e. they are interconnected (although everybody aims to avoid this correlation with all possible means in practice). Thus, the observational errors are described with symmetric covariance matrix $\mathbf{S}_Y$ of dimension $N \times N$, which can be obtained conveniently by writing schematically according to Anderson (1971) as:

$$\mathbf{S}_Y = \sum (\mathbf{Y} - \tilde{\mathbf{Y}})(\mathbf{Y} - \tilde{\mathbf{Y}})^+ , \tag{4.37}$$

where $\tilde{\mathbf{Y}}$ is the exact (unknown) value of the measured vector, $\mathbf{Y}$ is the observed value of the vector (distinguishing from the exact value owing to the observational errors), the summation is understood as an averaging over all statistical realizations of the observations of the random vector (over the general set).

The relation for covariance matrix of the errors $\mathbf{S}_X$ of parameters $\mathbf{X}$, of dimension $K \times K$ written in the same way as (4.37). Then, substituting relation (4.36) to it, the following is obtained:

$$\mathbf{S}_X = \sum (\mathbf{AY} - \mathbf{A}\tilde{\mathbf{Y}})(\mathbf{AY} - \mathbf{A}\tilde{\mathbf{Y}})^+ = \mathbf{A}\left(\sum (\mathbf{Y} - \tilde{\mathbf{Y}})(\mathbf{Y} - \tilde{\mathbf{Y}})^+\right)\mathbf{A}^+ ,$$
$$\mathbf{S}_X = \mathbf{A}\mathbf{S}_Y\mathbf{A}^+ . \tag{4.38}$$

A set of important consequences directly follows from (4.38)

Consequence 1. Equation (4.38) expresses the relationship between the covariance matrices of observational errors $\mathbf{Y}$ and parameters $\mathbf{X}$ linearly linked with them through (4.36), i.e. allows the finding of errors of the calculated parameters from the known observational errors. Namely, values $\sqrt{(\mathbf{S}_X)_{kk}}$ are the SD of parameters x_k, values $(\mathbf{S}_X)_{kj}/\sqrt{(\mathbf{S}_X)_{kk}(\mathbf{S}_X)_{jj}}$ are the coefficients of the correlation between the uncertainties of parameters x_k and x_j. In the particular case of non-correlated observational errors that is often met in practice, (4.38) converts to the explicit formula convenient for calculations:

$$(\mathbf{S}_X)_{kj} = \sum_{i=1}^{N} a_{ki}a_{ji}s_i^2 , \quad k = 1,\ldots,K , \quad j = 1,\ldots,K , \tag{4.39}$$

where a_{ki} are the elements of matrix $\mathbf{A}$, s_i is the SD of parameter y_i. In the case of the equally accurate measurements, i.e. $s = s_1 = \ldots = s_N$, the direct proportionality of the SD of the observations and parameters follows from (4.39):

$$(\mathbf{S}_X)_{kj} = s^2 \sum_{i=1}^{N} a_{ki}a_{ji} .$$

Consequence 2. From the derivation of (4.38) the general set could be evidently replaced with a finite sample from M measurements $\mathbf{Y}^{(m)}$, $m = 1,\ldots,M$,

i. e. S_Y in (4.37) is obtained as an estimation of the covariance matrix using the known formulas:

$$(S_Y)_{ij} = \frac{1}{M-1} \sum_{m=1}^{M} (y_i^{(m)} - \bar{y}_i)(y_j^{(m)} - \bar{y}_j), \quad \bar{y}_i = \frac{1}{M} \sum_{m=1}^{M} y_i^{(m)},$$

$$i = 1,\ldots,N, \quad j = 1,\ldots,N.$$

Then the analogous estimations are inferred for matrix S_X with (4.38). On the one hand, if just random observational errors are implied, then all M measurements will relate to one real magnitude of the measured value. But on the other hand the elements of matrix S_Y could be treated more widely, as characteristics of variations of the vector Y components caused not by the random errors only but by any changes of the measured value. In this case, (4.38) is the estimation of the variations of parameters X by the known variations of values Y

Consequence 3. Consider the simplest case of the relations similar to (4.36) – the calculation of the mean value over all components of vector Y i. e. $x = \frac{1}{N} \sum_{i=1}^{N} y_i$ (here $K = 1$, so value X is specified as a scalar). Then $a_{ki} = 1/N$ for all numbers i and the following is derived from (4.38) for the SD of value x:

$$s(x) = \frac{1}{N} \sqrt{\sum_{i=1}^{N} \sum_{j=1}^{N} (S_Y)_{ij}}. \tag{4.40}$$

For the non-correlated observational errors in sum (4.40) only the diagonal terms of the matrix remain and it transforms to the well-known *errors summation rule*:

$$s(x) = \frac{1}{N} \sqrt{\sum_{i=1}^{N} (S_Y)_{ii}}. \tag{4.41}$$

SD of the mean value decreases with the increasing of the quantity of the averaged values as $\sqrt{N}$ (for the equally accurate measurements $s(x) = s(y)/\sqrt{N}$), as per (4.41). As not only the uncertainties of the direct measurements could be implied under S_Y, the properties of (4.40) and (4.41) are often used during the interpretation of inverse problem solutions of atmospheric optics. For example, after solving the inverse problem the passage from the optical characteristics of thin layers to the optical characteristics of rather thick layers or of the whole atmospheric column essentially diminishes the uncertainty of the obtained results (Romanov et al. 1989). Note also that we have used the relations similar to (4.41) in Sect. 2.1 while deriving the expressions for the irradiances dispersion (2.17) in the Monte-Carlo method.

Consequence 4. Analyzing (4.41) it is necessary to mention one other obstacle. It is written for the real numbers, but any presentation of the observational results has a discrete character in reality, i.e. it corresponds finally to integers. The discreteness becomes apparent in an uncertainty of the process of the instrument reading. Hence, real dispersion $s(x)$ could not be diminished infinitely, even if $N \to \infty$ [indeed the length value measured by the ruler with the millimeter scale evidently can't be obtained with the accuracy 1 μm even after a million measurements, although it does follow from (4.41)]. Regretfully, not enough attention is granted to the question of influence of the measurement discreteness on the result processing in the literature. The book by Otnes and Enochson (1978) could be mentioned as an exception. However, this phenomenon is well known in practice of computer calculations where the word length is finite too. It leads to an accumulation of computer uncertainties of calculations, and special algorithms are to be used for diminishing this influence even during the simplest calculation of the arithmetic mean value (!) (Otnes and Enochson 1978). As per this brief analysis, the discreteness causes the underestimation of the real uncertainties of the averaged values.

Consequence 5. In addition to the considered averaging, the interpolation, numerical differentiation, and integration are the often-met operations similar to (4.36). Actually, they are all reduced to certain linear transformations of value y_i and could be easily written in the matrix form (4.36). Thus, (4.38) is a solution of the problem of uncertainty finding during the operations of interpolation, numerical differentiation, and integration of the results. Note that in the general case the mentioned uncertainties will correlate even if the initial observational uncertainties are independent.

Consequence 6. Matrix S_X does not depend on vector A_0 in (4.36). Assuming $A_0 = AY_0$, where Y_0 is the certain vector consisting of the constants, (4.38) turns out valid not for the initial vector only but for any $Y + Y_0$ vector, i.e. the covariance error matrix of parameters vector X does not depend on the addition of any constant to observation vector Y.

Consequence 7. Consider nonlinear dependence $X = A(Y)$. It could be reduced to the above-described linear relationship (4.36) using linearization, i.e. expanding $A(Y)$ into Taylor series around a concrete value of Y and accounting only for the linear terms as shown in the previous section. Then the elements of matrix A will be partial derivatives $a_{ki} = \partial(A(Y))_k/\partial y_i$, all constant terms as per consequence 6 will not influence the uncertainty estimations and the same formula as (4.38) will be obtained. For example, the uncertainties of the surface albedo have been calculated in this way with the covariance matrix of the irradiance uncertainties obtained at the second stage of the processing of the sounding results in Sect. 3.3. The uncertainties of the retrieved parameters, while solving the inverse problem in the case of the overcast sky have been calculated in this way, as will be considered in Chap. 6. Note, that relation (4.38) is *an approximate estimation* of the parameters of uncertainty in the nonlinear case because for exact estimation all terms of Taylor series are to be accounted. The accuracy of this estimation is higher if the observational uncertainties (i.e. the matrix S_X elements are less).

Return to the inverse problem solution and to begin with again consider the case of the linear relationship of observational results Y and desired parameters X (4.9): $\tilde{Y} = G_0 + GX$. Let the observational errors obey the law of normal distribution, in which probability density depends only on the above-defined $\bar{Y}, S_Y$ and is equal to:

$$\varrho(Y) = \frac{1}{(2\pi)^{N/2}|S_Y|^{1/2}} \exp\left(-\frac{1}{2}(Y - \bar{Y})^+ S_Y^{-1}(Y - \bar{Y})\right).$$

Abstract from the above-discussed non-adequacy of the operator of the inverse problem solution and assume that the difference of real observational results Y and calculated values $\tilde{Y}$ is caused only by the random error. Then vector X, which true value $\bar{Y}$ corresponds to (i.e. $\tilde{Y} = \bar{Y}$), is to be selected as an inverse problem solution. Substituting this condition to the formula for the probability density, we obtain it as a function of both the observational and desired parameters: $\varrho(Y, X)$. Then use the known *Fisher's scoring method in the maximum likelihood estimation* according to which the maximum of the combined probability density is to correspond to the desired parameters. Writing explicitly the argument of the exponent through parameter x_k the maximum is found from equation $\partial \varrho(Y, X)/\partial x_k = 0$ that gives the system of the linear equations:

$$\sum_{j=1}^{K} x_j \left(\sum_{i=1}^{N} \sum_{l=1}^{N} g_{ij}(S_Y^{-1})_{il} g_{lk} \right) = \sum_{i=1}^{N} \sum_{l=1}^{N} (y_i - g_{i0})(S_Y^{-1})_{il} g_{lk} \quad k = 1, \ldots, K.$$

(4.42)

The problem solution is obtained after writing (4.42) in matrix form:

$$X = (G^+ S_Y^{-1} G)^{-1} G^+ S_Y^{-1} (Y - G_0).$$

(4.43)

It is to be pointed out that if equality $W = S_Y^{-1}$ is assumed then (4.43) will almost coincide with solution (4.15) for LST with weights. In particular, for the case of non-correlated observational random uncertainties obeying Gauss distribution, matrix S_Y is the diagonal one and solution with LST (4.15) is an estimation of maximal likelihood (4.43). This statement is a kernel of the known Gauss-Markov theorem (see for example Anderson 1971) – a severe ground of selecting the inverse squares of the observational SD as weights of the LST. It is evident that relation $W = S_Y^{-1}$ is directly applied to all further algorithms of LST described by (4.20), (4.23)–(4.25), (4.28), (4.30) and (4.32).

As (4.43) has linear constraint form (4.36) between Y and X, the covariance matrix of the uncertainties of the retrieval parameters S_X is obtained with (4.36). Substituting the expression $A = (G^+ S_Y^{-1} G)^{-1} G^+ S_Y^{-1}$ from (4.43) to (4.38) and accounting the symmetry of matrix $(G^+ S_Y^{-1} G)^{-1}$ the following relation is inferred:

$$S_X = (G^+ S_Y^{-1} G)^{-1}.$$

(4.44)

Equation (4.44) allows finding estimations of the uncertainty of the retrieved parameters through the known observational uncertainty, i.e. it almost solves the problem of their accounting. Equation (4.44) evidently keeps its form for nonlinear algorithms, if matrix $\mathbf{G}$ is to be taken at the last iteration. Note that (4.44) relates also to the penalty functions method (4.30) and (4.31). As the additional yield to discrepancy at the last iteration is zeroth for this method (at least, theoretically), hence the matrix of the system (4.36) is similar to above matrix $\mathbf{A}$.

The main stage of the inverse problem solving with LST and of the method of the maximal likelihood (4.43) is solving a linear equation system, i.e. the inversion of its matrix. However, in the general case the mentioned matrix could be very close to a degenerate one. Then, with real computer calculations, matrix $(\mathbf{G}^+\mathbf{S}_Y^{-1}\mathbf{G})^{-1}$ is unable to inverse or the operation of the inversion is accompanied with a significant calculation error. The reason of this phenomenon is connected with the *incorrectness* of the majority of the inverse problems of atmospheric optics (that is a general property of inverse problems). The detailed theoretical analysis of the incorrectness of the inverse problem together with the numerous examples of the similar problems is presented in the book by Tikhonov and Aresnin (1986). The simple enough interpretation was performed in the previous section while discussing the phenomenon of the strong spread of the desired values during the consequent iterations. Technically, the incorrectness appears as mentioned difficulties of matrix $(\mathbf{G}^+\mathbf{S}_Y^{-1}\mathbf{G})^{-1}$ inversion, i.e. its determinant closeness to zero. Note that not all concrete inverse problems are incorrect, however, the solving methods of the incorrect inverse problems should always be applied if the correctness does not follow from the theory. It is necessary because the analysis of the incorrectness is technically inconvenient, as it needs a large volume of calculations (Tikhonov and Aresnin 1986). Thus, further we will consider the problem of the parameters $\mathbf{X}$ retrieval from observations $\mathbf{Y}$ as an incorrect one. Assume for brevity the linear case of the formulas and then automatically apply the obtained results to the algorithm recommended for the nonlinear inverse problems.

The method of the incorrect inverse problems solving is their *regularization* – the approach (in our concrete case of the linear equation system) of replacing the initial system with another one close to it in a certain meaning and for which the matrix is always non-degenerate (Tikhonov and Aresnin 1986). Further, we consider two methods of regularization usually applied for the inverse problems solving in atmospheric optics.

The simplest approach of regularization is adding a certain a priori non-degenerate matrix to the matrix of the initial system. Instead of solution (4.43), consider the following:

$$\mathbf{X} = (\mathbf{G}^+\mathbf{S}_Y^{-1}\mathbf{G} + h^2\mathbf{I})^{-1}\mathbf{G}^+\mathbf{S}_Y^{-1}(\mathbf{Y} - \mathbf{G}_0), \qquad (4.45)$$

where $\mathbf{I}$ is the unit matrix, h is a quantity parameter. It is evident that solution (4.45) tends to "the real" one (4.43) with $h \to 0$. Thus, the simple algorithm follows: the consequence of solutions (4.45) is obtained while parameter h decreases and value $\mathbf{X}$ with the minimum discrepancy is assumed as a solution. This approach is called "*the regularization by Tikhonov*" (although it had been known for a long time as an empiric method, Andrey Tikhonov gave the rigorous proof of it (Tikhonov and Aresnin 1986)).

The regularization by Tikhonov is easy to link with the considered in the previous section method of penalty functions. Indeed, if there are conditions $x_k = 0$ then the solution with the penalty functions method (4.28) converts directly to (4.45). As the rigorous equality $x_k = 0$ is not succeeded, the factor h is selected as small as possible. Thus, the regularization by Tikhonov corresponds with imposing the definite constraint on the solution, namely the requirement of the minimal distance between zero and the solution, i. e. the reduction of the set of the possible solutions of the inverse problem. Theoretically, all regularization approaches are reduced to imposing the definite constraint on the solution. Requirement $x_k = 0$ means that the components of vector $\mathbf{X}$ should not differ greatly from each other, i. e. it aborts the possibility of strongly oscillating solutions. However in fact, it is the way to diminish the strong spread of solutions during the iterations of nonlinear problems. Actually, nowadays the regularization by Tikhonov is applied to all standard algorithms of nonlinear LST (see for example Box and Jenkins 1970).

All desired parameters $\mathbf{X}$ in the considered statement of the atmospheric optics inverse problems have physical meaning. Hence, definite information about them is known *before the accomplishment of observations* $\mathbf{Y}$, and it is called an *a priori information*. Assuming that parameters $\mathbf{X}$ are characterized by a priori mean value $\bar{\mathbf{X}}$ and by a priori covariance matrix $\mathbf{D}$. Suppose that the parameters uncertainties obey Gauss distribution, i. e.:

$$\varrho(\mathbf{X}) = \frac{1}{(2\pi)^{N/2}|\mathbf{D}|^{1/2}} \exp\left(-\frac{1}{2}(\mathbf{X} - \bar{\mathbf{X}})^+\mathbf{D}^{-1}(\mathbf{X} - \bar{\mathbf{X}})\right) .$$

We should point out that mentioned a priori characteristics $\bar{\mathbf{X}}$ and $\mathbf{D}$ are the information about the parameters known in advance without considering the observations, in particular, it relates also to an a priori SD of parameters $\mathbf{X}$. Accounting for the above-obtained probability density of the observational uncertainties $\varrho(\mathbf{Y}, \mathbf{X})$, and supposing the absence of correlation between the uncertainties of the observations and desired parameters, the criterion of the maximal likelihood is required for their joint density $\varrho(\mathbf{Y}, \mathbf{X})\varrho(\mathbf{X})$. For convenience difference $\mathbf{X} - \bar{\mathbf{X}}$ is considered as an independent variable. The following can be inferred after the manipulations analogous to the derivation of (4.43):

$$\mathbf{X} = \bar{\mathbf{X}} + (\mathbf{G}^+\mathbf{S}_Y^{-1}\mathbf{G} + \mathbf{D}^{-1})^{-1}\mathbf{G}^+\mathbf{S}_Y^{-1}(\mathbf{Y} - \mathbf{G}_0 - \mathbf{G}\bar{\mathbf{X}}) . \qquad (4.46)$$

Solution (4.46) is known as *a statistical regularization method* (Westwater and Strand 1968; Rodgers 1976; Kozlov 2000). The regularization is reached here by adding inverse covariance a priori matrix $\mathbf{D}^{-1}$ to the matrix of the equation system. Indeed, it is easy to test that solution (4.46) exists even in the worst case $\mathbf{G}^+\mathbf{S}_Y^{-1}\mathbf{G} = 0$. On the other hand the larger the a priori SD of parameters, the less the yield of matrix $\mathbf{D}^{-1}$ to (4.46) and in the limit, when $\mathbf{D}^{-1} = 0$, solution (4.46) converts to solution without regularization (4.43). Statistical regularization (4.46) is much more convenient than (4.45), which is because it requires no iteration selection of parameter h (though it requires a priori information), thus it is mostly used for the inverse problems of atmospheric optics. Note that the solution dependence of $\bar{X}$ disappears for the nonlinear problems, where just the difference between the parameters is considered during the expansions into Taylor series, i. e. the statistical regularization is equivalent to the adding of D^{-1} to the matrix subject to inversion. Parameters $\bar{X}$ are usually chosen as a zeroth approximation. Using the following identity:

$$(G^+ S_Y^{-1} G + D^{-1})^{-1} G^+ S_Y^{-1} = DG^+ (GDG^+ + S_y)^{-1} , \tag{4.47}$$

which is elementarily tested by multiplying both parts from the left-hand side by combination $G^+ S_Y^{-1} G + D^{-1}$ and from the right-hand side by combination $GDG^+ + S_y$. For some types of problems, it is more appropriate to rewrite solution (4.46) in the equivalent form not requiring the covariance matrix inversion:

$$X = \bar{X} + DG^+ (GDG^+ + S_Y)^{-1} (Y - G_0 - G\bar{X}) . \tag{4.48}$$

Compute the uncertainties of obtained parameters X using observational uncertainties S_Y, i. e. the *posterior* covariance matrix of the parameters X uncertainties. According to the definition, the following is correct: $S_X = \sum (X - X)(\tilde{X} - X)^+$, where X is solution (4.48), and $\tilde{X}$ is the random deviation from it caused by the observational uncertainties. Substituting (4.48) to matrix S_X definition, accounting $\tilde{Y} = G_0 + G\tilde{X}$, after the elementary manipulations we are inferring $S_X = D - DG^+ (GDG^+ + S_Y)^{-1} GD$. Note that a certain positively defined matrix is subtracted from the a priori covariance matrix in this expression, thus the observations cause the decreasing of the a priori SD of the parameters, which has a clear physical meaning: the observations cause precision of the a priori known values of the desired parameters. For the further transformation of matrix S_X, the following relation is to be proved:

$$(D^{-1})^{-1} - (G^+ S_Y^{-1} G + D^{-1})^{-1} = DG^+ (GDG^+ + S_Y)^{-1} GD .$$

Use for that the identity $A^{-1} - B^{-1} = B^{-1}(B - A)A^{-1}$ with accounting (4.47). Finally, the following is obtained:

$$S_X = (G^+ S_Y^{-1} G + D^{-1})^{-1} . \tag{4.49}$$

It should be emphasized that (4.49) has the same form as (4.44) in spite of the complicated method of deriving it, namely: the covariance matrix of the uncertainties of the desired parameters is just the inverse matrix of the algebraic equation system subject to solving, i. e. it is directly obtained in the process of calculation.

As has been mentioned hereinbefore, posterior SD $\sqrt{(S_X)_{kk}}$ obtained with (4.49) are always not exceeded by a priori values $\sqrt{(D)_{kk}}$. The ratio of these SD characterizes the information content of the accomplished observations relative to the parameter in question. The lower this ratio the more information about the parameter is contained in the observational data. It is curious that proper observational results are not needed for the calculation of posterior SD (4.49) in the linear case; it is enough to know the algorithm of the only solution of the direct problem (matrix $\mathbf{G}$). Thus, calculating the possible accuracy of the parameters retrieval and the information content estimation could be done even at the initial stage of the solving process before the accomplishment of the observations. Strictly speaking, this confirmation is not correct for the nonlinear case, when the matrix of the derivatives $\mathbf{G}$ depends on solution $\mathbf{X}$; nevertheless, even in this case (4.49) is often used for analyzing the information content of the problem before the observations.

The choice of a priori covariance matrix $\mathbf{D}$ causes some difficulties while using the statistical regularization method. If there are sufficient statistics of the direct observations of the desired parameters then matrix $\mathbf{D}$ will be easily calculated. Otherwise, we need to use different physical and empirical estimations and models. The a priori models will be discussed in Chap. 5 for the concrete problem of the processing of sounding results. Note that in the case of the necessity of matrix $\mathbf{D}$ interpolation it is elementarily recalculated with (4.38) as per consequence 5. It should be mentioned that the results of the covariance matrix calculation have to be presented with a rather high accuracy without rounding off the correlation coefficients. Otherwise, the errors of rounding cause the distortions of the matrix structure (according to consequence 4), those, in turn, lead to difficulties in the use of the matrix. In particular, all reference data about the correlation coefficients of the atmospheric parameters are presented with accuracy up to 2–3 signs, hence, these matrices are not to inverse while using them. However, the difficulties with matrix $\mathbf{D}$ inversion could be principal, as this matrix would be degenerate if the desired parameters strongly correlate to each other.

To overcome the mentioned difficulties and to optimize the algorithm it is necessary to transform the desired parameters to independent ones for those there are no correlations for and the matrix $\mathbf{D}$ is diagonal. This transformation is provided by matrix $\mathbf{P}$ consisting of the eigenvectors of matrix $\mathbf{D}$, incidentally matrix $\mathbf{D}$ converts to diagonal matrix $\mathbf{L}$ with the known formulas of the coordinates conversion $\mathbf{L} = \mathbf{PDP}^{-1}$ (Ilyin and Pozdnyak 1978). The inverse transformation to the desired parameters $\mathbf{P}^{-1}\mathbf{S}_X\mathbf{P}$ is to be realized after the calculation of the posterior covariance matrix and we infer the following solution of (4.46) with accounting for eigenvectors orthogonality ($\mathbf{P}^{-1} = \mathbf{P}^{+}$):

$$\mathbf{X} = \bar{\mathbf{X}} + \mathbf{P}^{+}(\mathbf{PG}^{+}\mathbf{S}_Y^{-1}\mathbf{GP}^{+} + \mathbf{L}^{-1})^{-1}\mathbf{PG}^{+}\mathbf{S}_Y^{-1}(\mathbf{Y} - \mathbf{G}_0 - \mathbf{G}\bar{\mathbf{X}}) . \qquad (4.50)$$

The method of the revolution (Ilyin and Pozdnyak 1978) should be used for calculating the eigenvectors and eigenvalues of matrix $\mathbf{D}$. Although it is slow, it works successfully for the close (multiple) eigenvalues. To prevent the accuracy lost during the eigenvalue calculations the following approach of normalizing is recommended. The a priori SD of parameter x_k is assumed as a unit of measurement, i. e. introduce vector $d_k = \sqrt{(\mathbf{D})_{kk}}$ and pass to the values:

$$x_k' = x_k/d_k , \quad \bar{x}_k' = \bar{x}_k/d_k , \quad g_{0k}' = g_{0k} , \quad g_{ik}' = g_{ik}d , \quad (\mathbf{D}')_{ik} = (\mathbf{D})_{ik}/(d_i d_k) ,$$

where matrix $\mathbf{D}'$ is the correlation one. After solving the inverse problem with the primed variables pass to the initial units of measurements $x_k = x'_k d_k$, $(S_K)_{ik} = (S'_X)_{ik} d_i d_k$. In addition, note that the eigenvalues of the covariance matrix could become negative owing to the above-mentioned distortions while rounding. The regularization by Tikhonov is recommended against this phenomenon when matrix $\mathbf{D}' + h^2 \mathbf{I}$ is used instead of matrix $\mathbf{D}'$ with the consequent increasing of value h up to the negative eigenvalues disappearing.

Only several maximal eigenvalues of matrix $\mathbf{D}$ differ from zero in the strong correlation between the desired parameters often met in practice. Specify their number as m. Then all calculations would be accelerated if only m pointed eigenvalues remain in matrix $\mathbf{L}$ (it becomes of the dimension $m \times m$) and matrix $\mathbf{P}$ contains only m corresponding columns (dimension is $m \times K$). This approach is the kernel of the known *method of the main components*. Specifying the obtained matrices as $\mathbf{L}_m$ and $\mathbf{P}_m$ the following is obtained from (4.50):

$$\mathbf{X} = \bar{\mathbf{X}} + \mathbf{P}_m^+ (\mathbf{P}_m \mathbf{G}^+ \mathbf{S}_Y^{-1} \mathbf{G} \mathbf{P}_m^+ + \mathbf{L}_m^{-1})^{-1} \mathbf{P}_m \mathbf{G}^+ \mathbf{S}_Y^{-1} (\mathbf{Y} - \mathbf{G}_0 - \mathbf{G}\bar{\mathbf{X}}) . \qquad (4.51)$$

Sometimes we can succeed in reducing the volume of calculations by an order of magnitude and more using (4.51) instead of (4.50).

The criteria of selection of value m in (4.51) could be different. The mathematical criteria are based on the comparison of initial matrix $\mathbf{D}$ and matrix $\mathbf{P}_m^+ \mathbf{L}_m \mathbf{P}_m$, which have to coincide for $m = K$ in theory. Correspondingly, value m is selected proceeding from the permitted value of their noncoincidence. The comparison of every element of the mentioned matrices is needless. Usually the comparison of the diagonal elements (dispersions) or of the sums of these elements (the invariant under the coordinates conversion (Ilyin and Pozdnyak 1978)) is enough. The objective physical selection of value m is proposed in the informatic approach by Vladimir Kozlov (Kozlov 2000), though it is not convenient for all types of inverse problems because of very awkward calculations. According to Consequence 2 from (4.38), the variation of the observations caused by the a priori variations of the parameters is $\mathbf{GDG}^+$. We will use the eigenbasis of this matrix, i. e. the independent variations of the observations. Then eigenvalues of matrix $\mathbf{GDG}^+$ are the "valid signal" that is to be compared with the noise, i. e. with the SD of the observations. If the observations are of equal accuracy and don't correlate with SD equal to s then number m is a number of the eigenvalues exceeding s^2. The case of non-equal accuracy and correlated observations (just that is realized in the sounding data processing) is more complicated. In this case the observations are preliminary to reduce to the independency and to the unified accuracy $s = 1$. This transformation is based on the theorem about *the simultaneous reducing of two quadratic forms to the diagonal form* (Ilyin and Pozdnyak 1978) and is provided with matrix $\mathbf{P}_Y \mathbf{L}_Y^{-1/2}$, where $\mathbf{P}_Y$ is the matrix of eigenvectors $\mathbf{S}_Y$, and $\mathbf{L}_Y$ is the diagonal matrix from eigenvalues $\mathbf{S}_Y$ corresponded to them. Thus, according to (4.38) the selection of number m is determined by the number of the eigenvalues of matrix $\mathbf{P}_Y \mathbf{L}_Y^{-1/2} \mathbf{GDG}^+ \mathbf{L}_Y^{-1/2} \mathbf{P}_Y^+$, which exceed unity. Note that matrix $\mathbf{G}$ varies from iteration to iteration in the nonlinear case, but such awkward calculations are unreal to be accomplished. That's why it is preliminarily calculated using matrix G_0 with a strengthening of the selection conditions for the guarantee, i. e. comparing the eigenvalues not with unity but with the less magnitude.

Finally, we present the concrete calculation algorithms of the nonlinear inverse problems. The general algorithm of the penalty functions method (4.30) is converted to the form:

$$X_{n+1} = X_0 + P_m^+ (P_m G_n^+ S_Y^{-1} G_n P_m^+ + L_m^{-1} + P_m C_n^+ HC_n P_m^+)^{-1} P_m \qquad (4.52)$$
$$[G_n^+ S_Y^{-1}(Y - G(X_n) + G_n(X_n - X_0)) + C_n^+ H(-C(X_n) + C_n(X_n - X_0))] \; .$$

The algorithm with improved convergence (4.32), which has been used in the sounding data processing, transforms to:

$$X_{n+1} = X_0 + P_m^+ (P_m G_n^+ S_Y^{-1} G_n P_m^+ + L_m^{-1} + P_m HP_m^+)^{-1} P_m$$
$$\times \; [G_n^+ S_Y^{-1}(Y - G(X_n)G_n(X_n - X_0)) + H(X_n - X_0)] \; . \qquad (4.53)$$

In both cases the posterior covariance matrix is calculated with the following formula:

$$S_X = P_m^+ (P_m G_n^+ S_Y^{-1} G_n P_m^+ + L_m^{-1})^{-1} P_m \; .$$

4.4
Selection of Retrieved Parameters in Short-Wave Spectral Ranges

Hereinbefore the mathematical aspects of the inverse problems have been mainly considered. In addition to the availability of the formal-mathematical algorithms, the analysis of the physical meaning of the obtained results is of great importance. In particular, for the inverse problems of atmospheric optics it is important to answer the question: to what extent the retrieved parameters correspond to their real values in the atmosphere at the moment of the observation. The comparison of the results of the inverse problem solution with the data of direct measurements of the retrieved parameters answers this question sufficiently clearly and unambiguously. However, in the general case, the possibility of parallel direct measurements is limited. For example, during the airborne observations the vertical profiles of the temperature, contents of absorbing gases and parameters of the aerosols would have been measured simultaneously with the radiances and irradiances, if there had been an opportunity. The situation with the satellite observations is even worse; because the simultaneous airborne observations of the mentioned parameters are necessary, that needs developing and financing the scientific programs at the state level. Thus, the simultaneous direct measurements to test the retrieved parameters are too expensive. In this connection, the way proposed by the authors of the book by Gorelik and Skripkin (1989) has to be mentioned, where the expenditures for the technical solution of the problem (costs of the instruments, experiments, data processing etc.) are included in the total value, which is assumed as the minimum for the inverse problem solution. In that statement, the optimal ones will be the observations, where the demanded compromise between the exactness of the parameter retrieval and needed expenditures for obtaining them is reached, contrary to the observations providing the maximal exactness. Note that testing the solution of the inverse problem by a comparison with the independent measurements strictly speaking is reasonable for the direct measurements only. If the parameters for the comparison have been also obtained from the solution of another inverse problem, it is possible to discuss the comparing of the instruments and methodics only.

Accounting for the above-mentioned difficulties together with the fact that there has been no direct simultaneous observations for the considered soundings hereinafter consider the problem of the analysis of the adequacy of the inverse problem solution with the theoretical means.

Either the observation or the direct problem solution contains systematic uncertainties. These uncertainties evidently cause the minimum of discrepancy $\varrho(Y, \tilde{Y}(X))$ reached while the inverse problem solving will not correspond to the minimum of the discrepancy of true values of the observational data and direct problem solution. Take into account that the desired parameters are linearly expressed through the difference of the observations and direct problem solution in the formulas of Sects. 4.2 and 4.3, i.e. $X = A(Y - \tilde{Y})$, where A is a certain linear "solving" operator. Then writing $Y = Y' + \Delta Y$, $\tilde{Y} = \tilde{Y}' + \Delta \tilde{Y}$, where Y' is the true mean value of the measured characteristic, $\tilde{Y}'$ is the absolutely exact solution of the direct problem, ΔY, $\Delta \tilde{Y}$ are the corresponding systematic uncertainties of the observations and calculations, we are obtaining $X = A(Y' - \tilde{Y}') + A(\Delta Y - \Delta \tilde{Y})$. The first item is the desired adequate value X, but the second item means its distortion by a random shift. As the random observational uncertainty causes the obtaining of the vector of parameters X either with the random uncertainty, the mentioned systematic shift is to be estimated from its comparison with the random uncertainty of vector X. If the systematic shift is not less than the random uncertainty is then the result ignoring this shift will be evidently inauthentic. In practice, it is more convenient to compare not the retrieval errors but the errors of the observation and direct problem solution (Zuev and Naats 1990).

The systematic uncertainties of the observations are always much more than the random ones, so value $\Delta \tilde{Y}$ is of main interest. The simple receipt of its accounting is presented in the book by Zuev and Naats (1990); if it is essentially less than the random uncertainty is, then a subject to $\Delta \tilde{Y}$ will not be needed, otherwise it should be added to the random uncertainty. With this adding, the observations become less accurate and it causes the corresponding increase of the random uncertainty, i.e. SD of the retrieved parameters, and the systematic shift does not cause the escape of parameters vector X out of the admissible range of the confidence interval. Thus, the reliability of the result is reached by increasing the SD. Quite often this fact is difficult to be accepted psychologically, particularly, in limits of the "fight for accuracy" traditional in the observational technology. However, it is obvious: in general form while solving the inverse problems the measurements provide not only the instrument readings but the results of their numerical modeling as well, so both processes influence the accuracy. On the basis of the above arguments, the authors of another study (Zuev and Naats 1990) have inferred the existence of a certain limit to the observational accuracy, conditioned by possibilities of the contemporary methods of the direct problem solutions of atmospheric optics. Beyond this limit, the further increasing of the accuracy becomes useless (but accentuate, that it is valid only in ranges of the considered approach of the inverse problem solving).

The algorithm of the direct problem has to account for all factors influencing the radiation transfer maximally accurate and full for decreasing uncertainty $\Delta \tilde{\mathbf{Y}}$. However, the similar algorithm could turn out rather complicated and awkward for the practical application. Besides, the operational speed and memory limits of computers demands the appropriate algorithms and computer codes for the inverse problem solving. Therefore, different simplifications and approximations are inevitable in the radiative transfer description. It leads to the necessity of elaboration and realization of two algorithms while solving the inverse problems of atmospheric optics. The first algorithm is *an etalon* one that solves the direct problem in detail with sufficient accuracy; and the second algorithm is *an applied* one proceeding from the concrete technical demands and possibilities (in the limit the applied algorithm might coincide with the etalon one, but in reality it is almost impossible). The accuracy estimation of the simplifications and approximations of the applied algorithm obtained by the comparison of the corresponded results of two (applied and etalon) algorithms is to be used as an uncertainty of direct problem solution $\Delta \tilde{\mathbf{Y}}$.

In the aspect of the accuracy of the direct problem solution, a quite important question is the selection of the set of parameters $\mathbf{X}$ subject to retrieval. In practice, the total selection of the retrieved parameters is always evident and is defined by the problems, which the experiment has been planned for. Particularly the inverse problem of atmospheric optics formulated concerning the atmospheric parameters (Timofeyev 1998) is to obtain the vertical profiles of the temperature, contents of the gases absorbing radiation, aerosol characteristics, and ground surface parameters. However, as has been mentioned in Sect. 4.1 the direct problem algorithm depends on a wider set of parameters in reality. For example, the parameters of the separate lines of the atmospheric gases absorption (see Sect. 1.2) are needed for the volume coefficient of the molecular absorption. However, all parameters of the direct problem solution (all components of vector $\mathbf{U}$) without excluding are known without absolute exactness, but with a certain error. Thus, the problem of general selection of retrieved parameters $\mathbf{X}$ could be formulated as follows: it is not only to select vector $\mathbf{X}$ but to take into account the influence of the uncertainty of the initial parameters, whose magnitudes are assumed to be known, i.e. $\mathbf{U} \setminus \mathbf{X}$.

The above-formulated problem of taking into account the uncertainty of components $\mathbf{U} \setminus \mathbf{X}$ is solved elementarily. Indeed, let us set $\mathbf{X} = \mathbf{U}$, i.e. will assume all parameters of the direct problem to be unknown. Then using the method of statistical regularization and setting the a priori mean values and covariance matrix for $\mathbf{X} = \mathbf{U}$ we obtain the analogous posterior parameters after the inverse problem solving, with the solution depending on the a priori covariance matrix, in particular, the posterior SD depends on the a priori one. Thus, we will take into account the influence of the a priori indetermination of all parameters of the direct problem to the solution of the inverse problem. Further we can divide vector $\mathbf{X} = \mathbf{U}$ into two parts: $\mathbf{X}^{(1)}$ are the retrieved parameters (their analysis has the meaning) and $\mathbf{X}^{(2)}$ are the parameters for which the uncertainty of the initial values setting is taken into account correctly.

However, in practice this path is unrealizable, it is enough to weigh up the number of the parameters describing the molecular absorption lines. Thus, only the set of parameters, whose magnitudes are not initially defined, are included to vector $\mathbf{X}$, and other parameters $\mathbf{U} \setminus \mathbf{X}$ are assumed as the exactly known ones. The influence of the uncertainty of the $\mathbf{U} \setminus \mathbf{X}$ assignment is estimated from the dependence of the exactness of the direct problem solution upon this uncertainty, and it is to be considered as a part of systematic uncertainty $\Delta \tilde{\mathbf{Y}}$. This estimation is usually accomplished either from the physical reasons (in this case there is a possibility to neglect the inaccurate assignment of the parameters) or from the results of the numerical experiments, i. e. the direct problem solving with varying values $\mathbf{U} \setminus \mathbf{X}$ in limits of the fixed accuracy (Mironenkov et al. 1996). Note, that the possibilities of the modern computers open large perspectives for the pointed numerical experiments. For example, it is possible to obtain the reliable assessment of the complex effect of the indeterminacy of the assignment of all vector $\mathbf{U} \setminus \mathbf{X}$ components to the direct problem solution, after varying all components of vector $\mathbf{U} \setminus \mathbf{X}$ at once with the method of statistical modeling and accumulating the representative sample.

Concerning the dividing of the retrieved parameters $\mathbf{X} = \mathbf{U}$ to the analyzed $\mathbf{X}^{(1)}$ and non-analyzed $\mathbf{X}^{(2)}$ ones, it should be noted that this dividing is to be accomplished based on the reasons of the retrieval accuracy only. Namely, the retrieved parameters $\mathbf{X}^{(2)}$ could be meaningless if their posterior dispersion is close to the a priori one. However, the latter recommendation is rather relative either, because even small preciseness of some physical parameters might be the rather actual one. Quite often the vector $\mathbf{X}^{(1)}$ components are selected based on the problem stated while accomplishing the observations, and as a result the precise data are thrown out to "a tray" – to vector $\mathbf{X}^{(2)}$. Therefore, for example in the study by Mironenkov et al. (1996), only the possibility and accuracy of the total content of the gases absorbing radiation is analyzed while processing the data of the ground observations of atmospheric transparence within IR spectral region. At the same time the product of the solar constant, instrument sensitivity, and aerosol extinction is accepted as a retrieved parameter in this method, that could give useful information about the aerosol extinction spectrum within the IR range while taking into account the smooth spectral dependence of the two first factors.

According to the physical meaning, the part of the retrieved parameters presents the vertical profiles (of the temperature or gases content). The problem arises of describing these profiles with the finite set of parameters. Then two approaches are used: the approximation of the profile by the discrete altitude grid and approximation of the profile by a certain function. In fact, both approaches are equivalent, because any discrete grid supposes the interpolation to the intermediate altitudes that the definite function accomplishes. However, it is desirable to distinguish these approaches in the aspect of the application.

While approximating the profile by the altitude grid, it is evident that the lower the altitude step the more accurate the approximation. There is no problem of selecting the grid in the range of the etalon algorithms. The grid provided by the algorithm should be as detailed as possible. However, during the construction of the applied algorithm, the less number of points that are in the grid the less the number of the retrieved parameters that is available, hence, the shorter computing time is used. Therefore, the problem of the optimal altitude grid selection providing the maximal accuracy with the minimal points quantity arises. Regretfully, this problem has not often been studied in the theoretical aspect. Thus, different empirical approaches have to be used for the optimal grid selection. In particular, we have used the path described below.

Write the variations of the calculated values through the variations of the retrieved components using the linear item of the Taylor series:

$$\Delta y_i = \sum_{j=1}^{N} \left(\frac{\partial y_i}{\partial x_k} \right) \Delta x_k ,$$

where x_k is the profile of the retrieved parameter, variation Δx_k corresponds to the a priori SD. The corresponding term $(\partial y_i/\partial x_k)\Delta x_k$ is calculated for every altitude level k of the initial maximally detailed grid. The excluding of the level corresponds to the replacement of its derivative with the arithmetic mean value over two neighbor levels and it is replaced with zero at the last level (the top of the atmosphere). The increasing of derivatives $(\partial y_i/\partial x_k)\Delta x_k$ regulates and consequently excludes the levels until variation Δy_i maximal over all numbers i remains less than the fixed magnitude is. The parameter for the break of the excluding is obviously linked with observation uncertainty y_i. We have used the value equal to one third of the SD. We should mention that the obtained grids (and the altitudes of the top of the atmosphere) essentially differ for the vertical profiles of different parameters, but the grid over them all will be the suitable one. Quite often, the vertical grid is selected similar to the standard models, radiosounding data, etc. without the above-described details, i. e. without the accuracy estimation that is not methodically correct on our opinion.

The second approximation of the profile with a certain function is used in the algorithms of the operative data processing because it allows for a decrease of the quantity of the retrieved parameters by many times. Usually the function is constructed using the mean standard profiles cited in the references. However, it is necessary to accomplish the analysis of its accuracy with the etalon algorithm and a maximally detailed grid (Mironenkov et al. 1996).

The essential feature of inverse problems in the shortwave spectral range is the necessity of aerosol optical parameters retrieval. The volume coefficients of the aerosol scattering and absorption depend not only on altitude but on wavelength as well. Thus, parameterization of both the altitudinal and spectral dependence is necessary. In some particular problems, we succeed in describing the spectral dependence with a function of small quantity of the parameters (Polyakov et al. 2001). However, the specification of the spectral dependence as a grid over wavelengths is to be considered as a general case. In fact, there is no problem with this grid selection: the wavelength, which the processed characteristics are presented for, is to be used. The etalon algorithms should be elaborated in this way only. Nevertheless, the above-mentioned problem of the grid optimization over wavelengths arises again in the applied algorithm. The derivatives with respect of the volume coefficients of the aerosol extinction and scattering at the excluded wavelengths are replaced with the interpolated values (at all altitudes) for this grid selection. The point of the spectral grid will be excluded if the maximal variation of the measured characteristics during this replacement does not exceed the fixed uncertainty. At first, the spectral grid should be defined and then the altitudinal one is defined for every remained wavelength points. The spectral grid for the surface albedo retrieval is selected almost the same way.

Parameterization of the phase function of the atmospheric aerosols is the especially complicated problem of selecting the concrete set of parameters in the short wavelength range. The necessity of the solution of this problem is connected with minimization of the quantity of parameters in the applied algorithm. Indeed, the phase function is technically impossible to retrieve as a table over scattering angle in addition to the tables of dependences upon the altitude and wavelength. Thus, it should be described with a small quantity of parameters. The Henyey-Greenstein function (1.31) could be an example of such a parameterization. However, as it has been mentioned in Sect. 1.2 this function describes the real phase functions with a low accuracy. Regretfully, the attempts of finding a similar function with a small quantity of parameters and describing any aerosol phase function with sufficient accuracy have not been successful yet. Hence, the uncertainty of the aerosol phase function parameterization has still been one of the strongest and irremovable sources of the systematic errors while elaborating the applied algorithms of the inverse problems solving. The concrete choice of parameterization for the sounding data processing we will discuss in Sect. 5.1. Note, that radiative characteristics measured by different ways respond differently to the parameterization accuracy. For example, the irradiance being the integral over the hemisphere is essentially more weak connected with the shape of the phase function than the radiance is; the latter is almost directly proportional to the phase function (for example the single scattering approximation). Thus, the inadequacy of the phase function statement is the most serious obstacle in the interpretation of the satellite observations of the diffused solar radiance.

In addition to the listed problems, there is a general difficulty for the inverse problems solving – the probable ambiguity of the obtained results. Actually, the desired minimum of the discrepancy might not be single in the nonlinear case. The numerical experiments allow conclusion of the uniqueness of the solution after keeping the definite statistics.

The relationship between the inverse problem solution and observational variations within the range of the random SD is studied in *the numerical experiments of the first kind*. For this purpose, the direct problem is solved with the definite magnitudes of the parameters, and then the obtained solution is distorted by the random uncertainty using the method of statistical modeling on the basis of the known SD of the observations. After that, the inverse problem is solved for these data and its result is compared with the initial parameters. If the inverse problem solution coincides with the initially stated parameters after a sufficient quantity of this statistical testing, it should be concluded that the random observational error does not cause the solution ambiguity (and the confident probability could be accessed). It is especially appropriate to solve the direct problem with the Monte-Carlo method as it allows for easy simulation of the results just as random values.

As the random observational error is not usually large, the indeterminacy of the choice of zeroth approximation could significantly affect the solution ambiguity while solving the nonlinear inverse problems. Thus, *the numerical experiments of the second kind* are necessary, where the dependence of the solution upon zeroth approximation choice is studied, while allowing the variations of this approximation to be as large as possible (Zuev and Naats 1990). To reduce the computing time it is appropriate to combine the numerical experiments of the first and second kinds and to model both the random error and indeterminacy of the zeroth approximation. Just this approach has been applied in the study by Vasilyev O and Vasilyev A (1994) to this class of problems and to the concrete problem of the sounding data processing during the procedure of testing the computer codes. The solution uniqueness has remained with the variation of the zeroth approximation within three a priori SD of parameters. Note that this complex approach to the implementation of the numerical experiment opens wide perspectives when taking into account the possibilities provided by modern computers (Mironenkov et al. 1996). In particular, it is possible to vary statistically the totality of the direct problem parameters together with the zeroth approximation, a priori covariance matrix etc. It should be emphasized that with the accumulation of sufficient statistics of such complex numerical experiments, it is possible to estimate the accuracy of the inverse problems solution without simplification formulas similar to (4.49).

References

Anderson TW (1971) The Statistical Analysis of Time Series. Wiley, New York
Box GEP, Jenkins GM (1970) Time series analysis. Forecasting and control. Holden-day, San Francisco
Chu WP, Chiou EW, Larsen JC et al. (1993) Algorithms and sensitivity analyses for Stratospheric Aerosol and Gas Experiment II water vapor retrieval. J Geoph Res 98(D3):4857–4866
Chu WP, McCormick MP, Lenoble J et al. (1989) SAGE II Inversion algorithm. J Geoph Res 94(D6):8339–8351
Cramer H (1946) Mathematical Methods of Statistics. Stockholm
Elsgolts LE (1969) Differential equation and variation calculus. Nauka, Moscow (in Russian)
Gorelik AL, Skripkin VA (1989) Methods of recognition. High School, Moscow (in Russian)
Ilyin VA, Pozdnyak EG (1978) Linear algebra. Nauka, Moscow (in Russian)
Kalinkin NN (1978) Numerical methods. Nauka, Moscow (in Russian)
Kaufman YJ, Tanre D (1998) Algorithm for remote sensing of tropospheric aerosol form MODIS. Product ID: MOD04, (report in electronic form)

Kolmogorov AN, Fomin SV (1999) Elements of the theory function and the functional analysis. Dover Publications

Kondratyev KYa, Timofeyev YuM (1970) Thermal sounding of the atmosphere from satellites. Gydrometeoizdat, Leningrad (in Russian)

Kozlov VP (2000) Selected works on the theory of the experiment planning and inverse problems of the optical sounding. St. Petersburg University Press, St. Petersburg (in Russian)

Mironenkov AV, Poberovskiy AV, Timofeyev YuM (1996) The methodics of the interpretation of infrared spectra of the direct solar radiation for definition of the total content of atmospheric gases. Izv. RAS, Atmospheric and Ocean Physics 32:207–215 (Bilingual)

Otnes RK, Enochson L (1978) Applied Time-Series Analysis. Toronto. Wiley, New York

Polyakov AV (1996) To the question of using a priori statistical information for the solution of nonlinear inverse problems of the atmospheric optics. Earth Observations and Remote Sensing 3:11–17 (Bilingual)

Polyakov AV, Timofeyev YuM, Poberovskiy AV, Vasilyev AV (2001) Retrieval of vertical profiles of coefficients of aerosol extinction in the stratosphere from results of Ozon-Mir instruments measurements. Izv. RAS, Atmospheric and Ocean Physics 37:213–222 (Bilingual)

Potapova IA (2001) The method for the interpretation of data of Lidar sounding of aerosols. Thesis on III Int Conf "Natural and anthropogenic aerosols", 2001, St. Petersburg. Chemistry Institute, St. Petersburg University Press, pp 57 (in Russian)

Prasolov AV (1995) Analytical and numerical methods of dynamic processes studying. St. Petersburg University Press, St. Petersburg (in Russian)

Rodgers CD (1976) Some theoretical aspects of remote sounding in the Earth's atmosphere. J Quant Spectroscopy Radiative Transfer 11:767–777

Romanov PYu, Rozanov VV, Timofeyev YuM (1989) On accuracy of functionals retrieval from microphysical characteristics of stratospheric aerosols by cosmic observations of the atmospheric transmission and solar areole. Earth Observations and Remote Sensing, pp 35–42 (Bilingual)

Steele HM, Turco RP (1997) Separation of aerosol and gas components in the Halogen Occultation Experiment and the Stratospheric Aerosol and Gas Experiment (SAGE) II extinction measurements: Implication for SAGE II ozone concentrations and trends. J Geoph Res 102(D16):19665–19681

Tikhonov AN (1943) On stability of inverse problems. Dokl. Acad. Sci. USSR 39:195–198 (in Russian)

Tikhonov AN, Aresnin VYa (1986) Methods of solution no correct problem. Nauka, Moscow (in Russian)

Timofeyev YuM (1998) On inverse problems of the atmospheric optics. Izv. RAS Atmosphere and Ocean Physics 34:793–798 (Bilingual)

Timofeyev YuM, Rozanov VV, Poberovskiy AV, Polyakov AV (1986) Multispectral method of definition of vertical profiles of O_3, NO_2 content and aerosol extinction of the radiation in the atmosphere. Meteorology and Hydrology, pp 66–73 (in Russian)

Vasilyev FP (1988) Numerical methods of extremal problems solution. Nauka, Moscow (in Russian)

Vasilyev OB, Vasilyev AV (1994) The informatic providing of the retrieval of optical parameters for atmospheric layers from spectral irradiance observations at different levels in the atmosphere. III. The obtaining of optical parameters in the inhomogeneous multilayer atmosphere (numerical experiment). Atmospheric and Ocean Optics 7:625–632 (Bilingual)

Virolainen YaA (2000) Ground observations of heat IR radiation as an informational source about gas composition of the atmosphere. Candidate Thesis, St. Petersburg (in Russian)

Westwater ER, Strand ON (1968) Statistical information content of radiation measurements used in indirect sensing. J Atm Sci 25:750–758

Zuev VE, Naats IE (1990) Inverse problems of the atmospheric optics. (Recent problems of the atmospheric optics, vol 7). Gydrometeoizdat, Leningrad (in Russian)

Determination of Parameters of the Atmosphere and the Surface in a Clear Atmosphere

This chapter provides a concrete statement of the complex inverse problem of the retrieval of atmospheric and surface parameters from the observational data considered in Chap. 3. The problem is solved applying the methods discussed in Chap. 4. This chapter concerns the following: the etalon algorithm and accuracy estimation of different simplifications and approximations while computing irradiances, the approach and formulas for calculating the derivatives of the irradiances with the Monte-Carlo method, and inverse problem solving with analysis of the obtained results.

5.1
Problem statement. Standard calculations of Solar Irradiance

The results of soundings (the airborne observations of solar spectral irradiances) in the clear atmosphere have been presented in Chap. 3. These observations were intended for the calculations of spectral radiative flux in the atmospheric layers; this analysis is also presented in Chap. 3. However, the contemporary algorithms of the inverse problems solving described in the previous chapter give the possibility of reprocessing the mentioned experimental data aimed at a more complete and correct extraction of the information concerning the aerosol and gaseous composition of the atmosphere, and to the approbation of the operative approaches of similar observational results. Incidentally, the processed results have lost their actuality from the point of real-time monitoring, but they have not become outdated as a series of unique experimental observations, useful for adequate modeling of the optical properties of atmospheric aerosols and for correct comparison between the results of model calculations and experimental data. At the same time, the existing data set allows elaboration of the approaches for real-time monitoring of the composition and structure of atmosphere and elucidating the technical and methodological shortcomings of the accomplished experiments for the purpose of its removal during further experiments.

To solve the stated problem we will follow the scheme presented in Sect. 4.1. Its first stage is the model selection for the direct problem solving and the estimation of the uncertainties of the obtained results. As per Sect. 4.4 the etalon algorithm for modeling the observational values while taking into account the processes of the radiation interaction in the atmosphere with maximal accuracy is needed for the direct problem solving. The applied algorithm containing the definite simplifications and approximations proceeding from the technical demands to the calculations is used in the process of the iteration solving of the inverse problem. The accuracy of these approximations is defined by the comparison of the results of the etalon and applied algorithms. To simplify the presentation we will estimate the accuracy of the corresponding approximations while describing certain elements of the etalon algorithm, i. e. present the etalon and applied algorithms simultaneously. For all the calculations considered here, the authors have used the aerosol model of the atmosphere described in the study by Krekov and Rakhimov (1986), and the model of the profiles of the temperature, pressure, and absorbing gases (Anderson et al. 1996) with adding the profiles from the program code Gometran (Rozanov et al. 1995). To select the data from the models the computer tools presented in the study by Vasilyev A (1996) have been applied.

For the direct problem solving, i. e. for the model calculations of the measured solar irradiances, the Monte-Carlo method has been chosen; the expediency of this choice has been described in Sect. 2.5. Here we just emphasize the simplicity and flexibility of this method, which allows "turning on" and "turning off" different concrete variants of the description of the processes of radiative transfer, i. e. to transform the etalon algorithm to the applied one without difficulty. The following model is considered as an atmospheric model both for the etalon and applied algorithms: the reflecting characteristics of the surface and the optical characteristics of the aerosols are specified directly and the volume coefficients of the molecular scattering and absorption are calculated with the formulas of Sect. 1.2. Thus, in a general problem statement the vertical profiles of the temperature and absorbing gases are used as parameters, which the measured values depend on, and hence, are the subjects for retrieval together with the above-described parameters during the inverse problem solving.

The atmospheric pressure is accepted as a vertical coordinate in the observations of solar radiation (Chap. 3). Hence, it is necessary to pass from the altitude scale to the pressure one during the mathematical modeling of the observational process. For this transformation it is enough to take into consideration that the optical thickness has no dimensions, then, $\tau = \int \alpha_z(z)dz = \int \alpha_P(P)dP$, where $\alpha_z(z)$ is the volume extinction coefficient connected with altitude z (for example, in km^{-1}), $\alpha_P(P)$ is the volume extinction coefficient connected with pressure p (for example, in $mbar^{-1}$). The following is obtained using hydrostatic equation $\frac{dP}{P} = -\frac{g(z)\mu(z)}{RT(z)}dz$, where $g(z)$ is the free fall acceleration, $\mu(z)$ is the air molecular mass, $T(z)$ is the air temperature and R is the gas constant:

$$\alpha_P(P(z)) = \alpha_z(z)\frac{RT(z)}{g(z)\mu(z)P(z)} . \tag{5.1}$$

Recalculation of the volume coefficients is carried out with (5.1) where the subscript at extinction coefficient α_P is omitted and the pressure is used as a vertical axis and considered as an independent variable. Incidentally, all formulas of Sect. 2.1 keep their form with accounting the optical thickness definition as:

$$\tau(P) = \int\limits_0^P \alpha(P)dP .$$

In this book we consider the unpolarized radiative transfer, however, discussing the accuracy estimation of the calculation of the radiative characteristic, it is desirable to determine the uncertainty of this approximation. In the rigorous problem statement, the equation of radiative transfer while accounting for polarization is rather complicated. However, in the etalon algorithm polarization is accounted for approximately because only a crude estimation of the accuracy is necessary. It should be implemented by dividing radiation (photon) into two components and considering the transformation of the phase function that depends on the relationship between the mentioned components. In this approximation, we assume that all scattering events happen at one scattering azimuth corresponding to the maximal influence of polarization. Polarization is not accounted for while passing to the applied algorithm. The comparison of the calculation results with and without polarization has demonstrated a decrease of the influence of polarization from the UV to NIR spectral regions, which is evident, because the yield of scattered radiation to the solar irradiance changes in the same way. The uncertainty caused by ignoring the polarization could be estimated as 1.5% on the average. It is close to the maximal uncertainty of the irradiance observations in the UV and VD spectral regions and it is essentially lower than the observational uncertainty in the NIR region.

The model of the ground surface (3.16) described in Sect. 3.4 together with the concrete parameters values is used for accounting for the anisotropy. The following scheme of modeling the photon reflecting from the surface equivalent to the scheme expressed by (3.14)–(3.16) is applied for simplicity. The modeling of the photon interaction with the surface is carried out as has been described in Sect. 2.1 by recalculation of the photon weight with the specified surface albedo. While modeling, the Henyey-Greenstein function will be used further in (3.16) without the normalizing factor, i.e. in initial form (1.31). For that function, the equation of modeling (2.27) is solved explicitly, which yields the following formulas for the desired declination (η'', φ'') from the initial direction:

$$\eta'' = \frac{(1 + g_1^2)(2g_1 b_1^2 + 2b_1(1 - g_1)) - (1 - g_1)^2}{(2g_1 b_1 + 1 - g_1)^2} , \quad b_1 = \beta ,$$

$$\varphi'' = \pi \frac{(1 + g_2^2)(2g_2 b_2^2 + 2b_2(1 - g_2)) - (1 - g_2)^2}{(2g_2 b_2 + 1 - g_2)^2} , \quad b_2 = \beta ,$$

$$(5.2)$$

where β is a random value. After the modeling of the declination of the reflection from the main direction (η'', φ'') with (5.2), the new direction of the photon is (η, φ) and it is recalculated through the initial direction (η, φ) while considering the sense of the main direction (Sect. 3.4):

$$\eta := -\eta\eta'' + c\sqrt{(1 - \eta^2)(1 - (\eta'')^2)} \cos(\Delta\varphi)$$

$$\Delta\varphi = k\pi + \varphi''\,.$$

(5.3)

The formulas of the recalculation are analogous to (2.21) for the scattering angle but in (5.3) azimuth $\Delta\varphi$ is computed in the same coordinate system as φ. Parameter k in (5.3) is equal to zero for the case of the mirror reflection (water surface) and to unity for the backward reflection (sand surface). Parameter c could be equal to two discrete magnitudes $+1$ and -1. The concrete magnitude is selected from condition $\eta < 0$, i.e. the photon moves up after the reflection (if this condition fulfills for both alternatives of value c, one of them will be selected randomly).

The comparison of the calculation results with a subject to reflection anisotropy and without it has been accomplished for the cases of the water and sand surfaces, while the snow surface has been assumed as an orthotropic one (Sect. 3.4). Note that these anisotropic models were constructed just for the surfaces above which the sounding had been carried out. The results have demonstrated that the influence of anisotropy on the observation uncertainty is negligible in the UV region, and it is about the SD of the upwelling irradiance (1–2%) in the VD and NIR regions. Thus, in the applied algorithm we ignore anisotropy, however we are accounting for its influence on the accuracy. It should be emphasized that the influence of anisotropy on the irradiance for the highly anisotropic surface (water) turns out to be much weaker than for the slightly anisotropic surface (sand). This phenomenon could be easily explained with the following. The albedo of the water surface is small and decreases, while passing from the UV to the VD region, hence the influence of the surface properties on the upwelling irradiance is also small. The albedo of the sand surface is rather significant and increases from the UV to VD region, thus its reflecting properties greatly affect the upwelling irradiance especially in the VD and NIR regions.

Simulation of monochromatic radiative transfer has been considered in Sect. 2.1. However, according to (1.23) solar irradiance for the real observations is the integral with instrumental function (3.1):

$$F(\lambda_i) = \int_{\lambda_i - \Delta\lambda}^{\lambda_i + \Delta\lambda} F(\lambda')f_\lambda(\lambda' - \lambda_i)d\lambda'\,.$$

(5.4)

The problem of its calculation with (5.4) connected with the complicated spectral behavior of the volume coefficient of molecular absorption κ_m expressed by (1.29) leads to the corresponding spectral behavior of monochromatic irradiance $F(\lambda')$, so the direct integration of (5.4) needs a lot of computing time.

The general scheme that allows us to avoid the calculations of the problems of multiple light scattering with (5.4) is presented in several studies (Lenoble 1985; Minin 1988; Tvorogov 1994). This approach is based on passing from the solving of the transfer equation at a fixed wavelength to its solving for the atmosphere with the specified parameters monotonically dependent on wavelength (on integration variable). This problem has not been solved completely yet and the existing algorithms are based on certain approximations. Thus, expanding the transmission function into a sum of exponents (as has been proposed in the study by Minin (1988)) or taking into account the photon free path (Lenoble 1985) is provided by assumption about the atmosphere homogeneity. While using the Monte-Carlo method, the most adequate approach is the passing to probability density of appearing the definite magnitude of volume molecular absorption coefficient κ_m. However, the modern algorithms of this passage (e. g. in the study by Tvorogov 1994) demand very awkward preliminary calculations, which are ill adapted to the computing of the derivatives with respect to κ_m, which is necessary for the inverse problems solving.

Nevertheless, taking into account the demands to the etalon algorithm we are assuming (5.4) as an initial one while noting the following. Based on the general formal scheme of the Monte-Carlo method wavelength λ' could be simulated for computing integral (5.4) by probability density $f_\lambda(\lambda')$. Then the method of double randomization is applied as per the book by Marchuk et al. (1980), whose kernel consists of the inessentiality of integration order for the Monte-Carlo method. Hence, it is enough to simulate only one photon trajectory for every random wavelength. As a result, we will estimate the values of the desired integral (5.4) from the counters magnitudes. After the modeling of random wavelength while accounting for the triangle instrumental function of the K-3 spectrometer and (2.6) the following is obtained:

$$
\lambda = \lambda_i - \Delta\lambda(1 - \sqrt{2\beta}) , \quad \beta \leq 1/2 ,
$$
$$
\lambda = \lambda_i - \Delta\lambda(1 - \sqrt{2 - 2\beta}) , \quad \beta \geq 1/2 .
$$
$$(5.5)$$

The instrumental function of the K-3 spectrometer is known with in an error of 1% (Sect. 3.1) that is comparable with the observational uncertainty. However, this uncertainty is significant only in the spectral intervals with the molecular absorption bands. Thus, its account as an additional yield to the uncertainty of the direct problem solution should be accomplished only within the following spectral intervals selected from the analysis of the radiative flux divergence (Sect. 3.3): 330–360 nm (O_3), 676–730 nm (H_2O), 756–780 nm (O_2), 804–850 nm (H_2O), 910–978 nm (H_2O). In addition, the solar constant should be assumed invariant within a narrow spectral interval $[\lambda_i - \Delta\lambda, \lambda_i + \Delta\lambda]$ to avoid the difference between the observational data and the calculation results, because the solar irradiances have been corrected with the incident solar spectrum taking into account the instrumental function, while processing the sounding data.

Thus, computing with (5.4) is reduced, in fact, to constructing the maximally fast and accurate algorithm of the profile κ_m calculation with (2.18) for the wavelength randomly selected within the interval $[\lambda_i - \Delta\lambda, \lambda_i + \Delta\lambda]$. From that point of view, we are using *the algorithm of the simplifying account of the yield of the spectral lines wings to the absorption* elaborated by Virolainen and Polyakov

(1999). According to this algorithm wavelength interval $[\lambda_i - \Delta\lambda, \lambda_i + \Delta\lambda]$ is divided into narrow intervals of $1\,\mathrm{cm}^{-1}$ length. When wavelength λ' gets into the definite narrow interval the spectral lines within the distance $\pm 2\,\mathrm{cm}^{-1}$ from it only are considered with rigorous (2.18)–(2.19) and the yield of the farthest lines to coefficient κ_m is calculated with the approximated formula:

$$\kappa_m := \kappa_m + \frac{P}{P*} \sum_{i=1}^{M} n_i \left(\frac{T*}{T}\right)^{l(i)} \sum_{k=1}^{4} R_{ik}(\nu - \nu')^{4-k}$$

$$R_{ik} = \frac{b_k}{\pi} \sum_{j=1}^{K(i)} S_{ij} \frac{W_{ij}(T)}{W_{ij}(T*)} \left(\frac{T*}{T}\right)^{m_{ij}} \frac{d_{ij}}{(\nu - \nu_i)^{6-k}} ,$$

(5.6)

where $\nu = 1/\lambda'$, ν' is the right boundary of the interval of the rigorous accounting of the spectral lines; the rest of the specifications are similar to (2.18)–(2.19). According to Virolainen and Polyakov (1999) the magnitudes of the empirical coefficients are: $b_1 = -1.40276$, $b_2 = 2.35451$, $b_3 = -1.93698$, $b_4 = 0.99854$, for $\nu < \nu'$; $b_1 = 10.6522$, $b_2 = 1.2675$, $b_3 = 2.14156$, $b_4 = 0.99750$, for $\nu > \nu'$. As coefficients R_{ik} for every concrete gas depend only on temperature, they are preliminarily tabulated and interpolated over the look-up tables during the calculation process. This procedure decreases the computing time by an order of magnitude. The uncertainty of approximation (5.6) is equal to about the decimals of a percent as per the analysis of the study by Virolainen and Polyakov (1999) and can be neglected in most cases.

All gases, about which the spectroscopic information (Sect. 1.2) is available, should be included in the computing code of the etalon algorithm. Nevertheless, only the gases, which the processing results are sensitive to, are to be taken into account by the applied algorithm. Thus, we are estimating the ratio of the maximal irradiance variation caused by the variation of the concrete gas content to the SD of the observations, i.e. the value called usually "*signal-to-noise ratio (SNR)*", and the necessity of the gas account in the applied algorithm is considered according to this ratio. During this calculation, the contents of the considered gases should be tripled to estimate the limit variation (excluding O_2). The condition of the relative humidity less than 95% should be tested while tripling the H_2O content otherwise, the content corresponding to 95% relative humidity is used.

The results of the calculations are presented in Fig. 5.1. Only five absorbing gases have been accounted for in the applied algorithm: H_2O, O_3, O_2, NO_2 and NO_3. Concerning the absence of gas SO_2 and the presence of gas NO_3 in the final list, the following should be pointed out. The SO_2 absorption band is rather strong in the UV region but it almost coincides with the strongest ozone band. Spectral interval 340–380 nm, where these bands could be divided, provides too weak SO_2 absorption to be registered by the K-3 instrument (Rozanov et al. 1995). As for NO_3, it is characterized with a very strong absorption band in the VD spectral region, but it is traditionally assumed to be decomposed by solar light so its content is negligibly low in the atmosphere. Nevertheless, according to the modern data some troposphere photochemical reactions (in

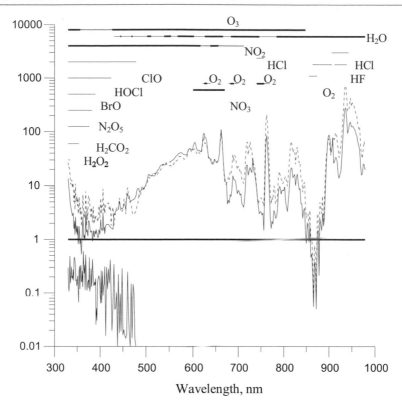

Fig. 5.1. The ratio of the irradiance variation to the SD ("signal-to-noise") when comparing the calculations with different degree of the accounting of the molecular absorption. *Lower curve:* five gases (O_3, H_2O, NO_2, O_2, NO_3) are taken into account comparing with all atmospheric gases; *upper curve* – accounting of all atmospheric gases comparing with the case of the neglecting of molecular absorption; *solid line* corresponds to the minimal SD of the measured irradiance, *dashed line* corresponds to the SD equal to 1%. *Thick lines* mark the complete intervals of the molecular absorption

particular, heterogeneous) cause not only decomposition of NO_3, but also its generation in daytime (Rudich et al. 1998). The existence of NO_3 in daytime atmosphere is confirmed with the observations of its content with *the method of the spectral transparency* (Weaver et al. 1996). At last, the NO_3 absorption band has been directly revealed in the spectral behavior of transmittance in the clear sky (Vasilyev O et al. 1995).

The same estimation of the signal/noise ratio (SNR) should be used for selection of the concrete spectral intervals. The selected intervals are presented in Table 5.1. It should be noted that in spite of the traditional conceptions there is no spectral interval within the VD region, where neglecting the molecular absorption is possible a priori, if the accuracy of the observations or calculations is about 1% (Ivlev and Vasilyev A 1998). Remember that the mentioned selection of the gases and wavelengths intervals has been accomplished after carrying

Table 5.1. Wavelength intervals for accounting for the molecular absorption of the atmospheric gases

Gas	Wavelength interval (nm)
H_2O	444–446, 468–470, 502–510, 538–552, 566–602, 626–666, 684–746, 784–978
O_3	330–356, 426–848
O_2	626–632, 686–694, 758–774
NO_2	330–616, 638–656
NO_3	598–672

out the concrete observations with the concrete K-3 instrument. In principle, application of these data to other experiments and instruments is admissible, but the analogous estimations and calculations are desirable to implement.

In spite of using the fast algorithm for the calculation of the volume coefficient of the molecular absorption as per (5.6), its application takes time unacceptable for the computer codes of the operative data processing. Therefore, the passage, which is based on using the look-up tables of the preliminary computed absorption cross-sections $C_a(\lambda, P, T)$, could essentially simplify the scheme of calculations in the applied algorithm.

The spectroscopic data of O_3, NO_2 and NO_3 gases have been initially presented as look-up tables (Sect. 1.2). The temperature dependence of the absorption cross-section of O_3 in 330–356 nm is described according to Bass and Paur (1984) as follows:

$$C_a(\lambda_j, T) = C_0(\lambda_j) + C_1(\lambda_j)T + C_2(\lambda_j)T^2 . \tag{5.7}$$

This very interval should be used for the interpolation over wavelengths and temperature dependence of the absorption cross-section (1.28). The absorption cross-sections $C_a(\lambda, P, T)$ of H_2O and O_3 are also computed preliminarily as look-up tables. The grid over wavelength is selected as an irregular one to describe all specific features of the spectrum on the one hand and to minimize the wavelength quantity in the grid on the other hand. Some ideas of the algorithm presented in the study by Pokrovsky (1967) constitute the basis of this grid selection. The centers of the spectral lines are selected as basic points λ_j because they correspond to maximal values $C_a(\lambda, P, T)$. Other points are defined by the consequent dividing of the basic grid proceeding from the comparison of exact and interpolated values $C_a(\lambda, P, T)$. Then the following parameterization of dependence $C_a(\lambda, P, T)$ is used:

$$C_a(\lambda_j, P, T) = R(\lambda_i)C_0(\lambda_j) \left(\frac{P}{P*} \right)^{C_1(\lambda_j)} \left(\frac{T*}{T} \right)^{C_2(\lambda_j)} , \tag{5.8}$$

where $R(\lambda_j)$ is the correcting factor. To minimize the systematic uncertainty of (5.8) this factor is selected after the preliminary computing of every indication of the instrument corresponding to wavelength λ_j. Equation (5.8) has been written on the basis of the known approximation of the coefficient of gas molecular absorption with the power function of temperature and pressure (Kondratyev and Timofeyev 1970). Coefficients $C_0(\lambda_j)$, $C_1(\lambda_j)$, $C_2(\lambda_j)$ are selected with LST. The influence of the uncertainty of (5.8) parameterization on the calculations of solar irradiances does not exceed their random SD excluding several points of spectrum.

The necessity of the parameterization of the dependence of the aerosol phase function upon scattering angle, while solving the inverse problems of atmospheric optics has been described in Sect. 4.4. To select the appropriate parameterization, the uncertainties of the calculation of irradiances, while using the approximations of the tabulated phase functions in the etalon algorithm have been analyzed by Krekov and Rakhimov (1986). In other words, while using the SNR, the signal is assumed as a difference between the irradiance calculations for the table and analytical approximation of the phase function. To estimate the maximal influence of the aerosols on radiative transfer, the ground values of the volume coefficients of the aerosols scattering and absorption (Krekov and Rakhimov 1986) have been increased by five times. Unexpectedly the good susceptibility of the solar semispherical irradiances (especially of the upwelling irradiance) to the shape of the aerosol phase function has been revealed. This susceptibility is likely to be caused by essentially different yields of the radiation scattered to different directions of the phase function to the upwelling irradiance. The yield of the single scattered light caused by the radiation scattering under the small angles exceeds the yield of the highest orders of the scattering for the aerosol extended phase functions.

The SNR has turned out to exceed unity for the Henyey-Greenstein function (1.31) and even for its two-parametric modification (Minin 1988), which is expressed by two Henyey-Greenstein functions: one is extended forward and the other one is extended backward. The situation has improved for the results of using the two-parametric tabulated model described in the study by Vasilyev O and Vasilyev A (1994) based on classifying the experimentally observed phase functions, but the uncertainty influence is also too strong in this case. The attempts to construct the analogous two-parametric model have provided no success on the basis of the initially tabulated phase functions (Krekov and Rakhimov 1986). However, higher accuracy has been reached in the calculation with the analytical presentation of the phase function not at all angles but at several fixed ones. Based on the strong correlation between the cross-sections of scattering and directional scattering revealed by some authors (Gorchakov and Isakov 1974; Gorchakov et al. 1976) and testing the set of parameterizations, it is possible to obtain the best results with the following expressions:

$$x_a(\chi,\lambda) = \exp(a_i(\chi,\lambda) + b_i(\chi,\lambda) \ln \sigma_a(\lambda) + c_i(\chi,\lambda) \ln^2 \sigma_a(\lambda))/C$$

$$C = \frac{1}{2} \int_{-1}^{1} \exp(a_i(\chi,\lambda) + b_i(\chi,\lambda) \ln \sigma_a(\lambda) + c_i(\chi,\lambda) \ln^2 \sigma_a(\lambda))d\chi , \qquad (5.9)$$

where $\sigma_a(\lambda)$ is the volume coefficient of the aerosol scattering, $a_i(\chi,\lambda)$, $b_i(\chi,\lambda)$, $c_i(\chi,\lambda)$ are the tables of the coefficients obtained for a certain set of wavelengths λ and cosines of scattering angles χ. Coefficients $a_i(\chi,\lambda)$, $b_i(\chi,\lambda)$, $c_i(\chi,\lambda)$ for intermediate λ values are derived using linear interpolation. Subscript i means the dependence of the model upon altitudinal zone. The atmosphere has been divided into three zones: 0–4 km (the surface layer as per Krekov and Rakhimov 1986), 4–11 km (the troposphere) and higher than 11 km (the stratosphere and upper atmosphere). Factor C is introduced to the denominator of (5.9) to satisfy normalizing condition (1.18).

The SNR for model (5.9) is minimal over all considered approximation although it exceeds unity. Thus, (5.9) will be used further for the phase function parameterization in the applied algorithm. The evident shortcoming of this parameterization (5.9) is its explicit referencing to the a priori aerosol model [through coefficients $a_i(\chi,\lambda)$, $b_i(\chi,\lambda)$, $c_i(\chi,\lambda)$], however, the problem of the adequacy of our a priori notions about the reality is the general difficulty of all inverse problems of atmospheric optics, especially of the retrieval of aerosol parameters (Zuev and Naats 1990). Although, dividing the atmosphere looks like an artificial adjustment to the initial model structure by Krekov and Rakhimov (1986), it could be grounded. Actually, according to studies by Vasilyev A and Ivlev (1995, 1996) in the general case coefficients $a_i(\chi,\lambda)$, $b_i(\chi,\lambda)$, $c_i(\chi,\lambda)$ depend on the type of the aerosol substance only, and it is accounted for during the selection of the altitude zones in (5.9).

Using model (5.9), where the phase function is uniquely defined with the aerosol scattering coefficient means the excluding of the parameters describing the phase function shape and, hence, the rejection of the retrieval of all phase function characteristics. The only parameters to obtain are the volume coefficients of scattering and absorption. It seems to contradict the above confirmation about the strong relationship between the irradiances and the phase function shape. However, we are using not real phase functions but the analytical approximations similar to the real ones in shape. For the real phase function, a strong correlation either with the scattering coefficient or with the aerosol substance is observed (Barteneva et al. 1967, 1978; Gorchakov and Isakov 1974; Gorchakov et al. 1976; Vasilyev A and Ivlev 1995, 1996). Thus, the parameters of the phase function shape are hardly retrieved in the limits of the concrete a priori model (Vasilyev O and Vasilyev A 1989a, 1989b, 1994a).

The weak point of parameterization (5.9) is a tabulated relationship between phase function and scattering angle. Certainly, the analytical parameterization is preferable. Nevertheless, all our attempts to find the phase function analytical presentation even for the calculations of the semispherical irradiances have failed. The rude numerical estimation indicates the needed accuracy of the phase function approximation about 5–10% for computing the upwelling irradiance above the dark (water) surface with the accuracy about 1%. It is a very rigorous demand because the accuracy of the field observations of the phase function is the same.

Note lastly, that all described approaches of the estimation of uncertainty have been implemented while computing the irradiances above the surfaces with small albedo (water surfaces) because the scattered radiation yields maximal solar upwelling irradiance. The link between the upwelling irradiance calculation and phase function parameterization is essentially weaker for the bright surfaces (sand, snow) and the problem of the aerosol phase function approximation is not so important for the present instance.

As the width of the instrumental function of the K-3 spectrometer is 6 nm, in addition to the evident spectral dependence of the molecular absorption within this interval there is the spectral dependence of the volume coefficients of the molecular scattering, and of the aerosol scattering and absorption. For example, the difference between the values of the volume coefficient and molecular scattering in interval 6 nm reaches 10% in the UV region. The estimation of the influence of ignoring this dependence on the uncertainty of the solar irradiance calculation shows that the spectral dependence of the molecular scattering needs to be accounted for and the aerosol dependence could be neglected. The latter provides essentially for faster work of the algorithm.

The general estimation of the simplifications and approximations introduced to the applied algorithm is demonstrated in Table 5.2. Note that the uncertainties caused by the molecular absorption and aerosol scattering have been calculated for the corresponding extremal models, and hence, could be considered as the limit ones. The uncertainties presented in Table 5.2 are the systematic ones, so their mutual compensation could occur. Thus, the comparison between the results of the calculations with "turning on" (applied algorithm) and "turning off" (etalon algorithm) all possible simplifications has been accomplished to estimate the total accuracy of the direct problem

Table 5.2. Estimation of the uncertainty of the irradiance calculations

Source of the uncertainty	Uncertainty (%)	Comments
Considering the absorption by five gases only	0.1–0.5	At separate wavelengths λ_i
Considering the molecular absorption in the selected spectral intervals only	0.3	At separate wavelengths λ_i
Tabulating the cross-sections of the molecular absorption	0.7–2.0	At separate wavelengths λ_i
Ignoring the polarization	1.5	Decreasing from the UV to the VD and NIR spectral regions
Parameterization of the aerosol phase function	1.0–2.0	Maximal for $F^{\uparrow}$ above the water
Ignoring the anisotropy of the reflection from the surface	1.0–2.0	Maximal for the sand surface
Ignoring the spectral dependence of the aerosol characteristics within the interval of the instrumental function width	0.5	

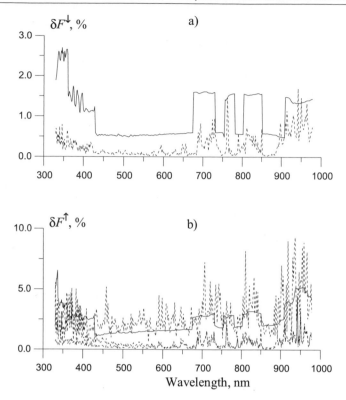

Fig. 5.2a,b. Comparison of the limited accuracy of the observations – *solid lines* and estimation of the calculation uncertainty – *dashed lines*: **a** for downwelling irradiance; **b** for upwelling irradiance: *upper dotted line* – observation above the water surface, *lower lines* – above the sand and snow surfaces

solution. Figure 5.2 illustrates the results of the comparison together with the limit accuracies of the observations including the uncertainty of the instrumental function equal to 1%. The uncertainty of the downwelling irradiance calculation is seen significantly less than the uncertainty of its observations (excludes the molecular absorption bands in the NIR spectral region). The analogous picture is demonstrated for the upwelling irradiances above the sand and snow. Regretfully, the uncertainty of the calculations of the upwelling irradiances above the water surface exceeds the observational uncertainty. The physical reason of this obstacle is the following. The upwelling irradiance measured above the dark water surface has formed owing to the radiation scattering in the atmosphere, and therefore, it needs higher accuracy of the numerical simulation. The obtained uncertainties of the calculation should be added to the observational uncertainties during the inverse problem solving as per Sect. 4.4.

Scrutinize the technical testing of the computer codes used for the calculations. The correspondence of the computer code to the initial mathematical algorithm or, in other words, the analysis of the coincidence of the calculation results with the certain expected values is interpreted as code testing (Borovin et al. 1987). The considered code is related to the class, where the analytical ("hand") testing is principally impossible (Borovin et al. 1987), so the main way of testing is the careful individual verification of the separate blocks during the program debugging stage. Besides, the comparison of the results of our calculations with the data presented in the book by Lenoble (1985, Tables 9–11) has been carried out. The deviation of the results of the calculation with the tested code from the "exact" data (Lenoble 1985) has turned out to be less than 1%. Taking into account that these "exact" data (Lenoble 1985) have been averaged over independent calculations implemented with seven different methods with the accuracy within 1%, the mentioned coincidence of the results of the tested computer code and the "exact" data could be accepted as the complete one.

The comparison of the calculation results with the observational data is shown in Fig. 5.3 as another technical test of the calculation algorithm. These calculations are accomplished for the mean model of the atmosphere (Anderson et al. 1996) together with the ground aerosol model (Krekov and Rakhimov 1986). We have not naturally expected the coincidence but the qualitative similarity of the spectra shapes, the widths and positions of the molecular

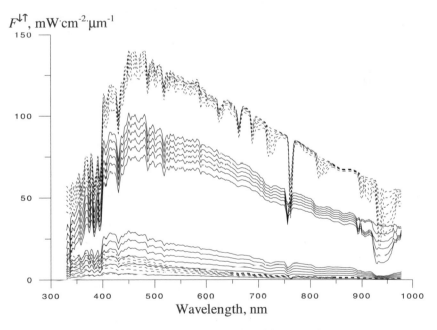

Fig. 5.3. Comparison of the calculation results (*dotted lines*) with the experimental data (*solid lines*). The airborne sounding 16th May 1984, water surface, incident solar angle 43°. Vertical profiles of the downwelling (*upper group of the curves*) and upwelling (*lower group of the curves*) irradiances. Each group consists of 6 curves from 500 mbar (*uppers*) to 1000 mbar (*lowers*) at every 100 mbar

absorption bands indicates the correctness of the computer code. On the other hand, the essential numerical difference between the observations and calculations shows the susceptibility of the observed irradiances to the atmospheric parameters (which have differed from the model values during the observations).

5.2
Calculation of Derivative from Values of Solar Irradiance

In addition to the irradiance calculations, the derivatives of the irradiances with respect to all retrieved parameters are necessary for the inverse problem solving using the methods presented in Chap. 4.

The derivatives computing in the Monte-Carlo method is based on the differentiation of formal Neumann series (2.22) (Marchuk 1988), i. e. of the general form of the solution of the equation of radiative transfer (Lenoble 1985; Marchuk 1988). We will keep the same specifications as in Sect. 2.1. Let the derivative of direct problem (2.22) solution to be inferred:

$$\Psi B = \Psi q + \Psi K q + \Psi K^2 q + \Psi K^3 \, q^+ \ldots$$

with respect to a certain parameter a, i. e. value $\partial(\Psi_a \, B_a)/\partial a$ where the subscript of the integral operators symbolizes their dependence of the parameter. Differentiate Neumann series (2.22):

$$\frac{\partial}{\partial a}(\Psi_a \, B_a) = \frac{\partial}{\partial a}(\Psi_a \, \mathbf{q}_a) + \frac{\partial}{\partial a}(\Psi_a \, K_a \, \mathbf{q}_a) + \ldots + \frac{\partial}{\partial a}(\Psi_a \, K_a^n \, \mathbf{q}_a) + \ldots \quad (5.10)$$

Use for brevity of the derivative specification the following form:

$$(\Psi_a \, B_a)' \equiv \frac{\partial}{\partial a}(\Psi_a \, B_a) \tag{5.11}$$

and assuming the symbolic writing of integral operators (2.20) obtain the following for the right part of series (5.10)

$$\left(\int \Psi_a(u) B_a(u) du \right)' = \int \Psi_a(u) B_a(u) \left(\frac{\Psi_a'(u)}{\Psi_a(u)} + \frac{B_a'(u)}{B_a(u)} \right) du$$

$$= \int \Psi_a(u) B_a(u) W_a(u) du , \tag{5.12}$$

where

$$W_a(u) = \frac{\Psi_a'(u)}{\Psi_a(u)} + \frac{B_a'(u)}{B_a(u)} . \tag{5.13}$$

The formula of the derivative is specially converted to form (5.12), (5.13). Written in this way, it looks as integral (2.20) directly calculated with the Monte-Carlo method according to (2.21).

$$\int \Psi_a(u)B_a(u)W_a(u)du = M_\xi(\Psi_a(\xi)W_a(\xi)) \qquad (5.14)$$

That is to say, the calculation of the derivatives according to (5.14) is reduced to the multiplying of the value written to the counter by a certain "weight" function $W_a(\xi)$ (Marchuk et al. 1980).

To construct the concrete algorithm of calculating $W_a(\xi)$ the derivative explicit form of the right part of series (5.10) is obtained. For that we are using the known expression of the derivative of the product through the sum of logarithm derivatives $(xyz\ldots)' = (xyz\ldots)(x'/x + y'/y + z'/z + \ldots)$. The following is obtained:

$$(\Psi_a\, q_a)' = \int \Psi_a(u)q_a(u)\left(\frac{\Psi_a'(u)}{\Psi_a(u)} + \frac{q_a'(u)}{q_a(u)}\right),$$

$$(\Psi_a\, K_a^n\, q_a)' = \int \ldots \int du\,du_1 \ldots du_n \Psi_a(u)q_a(u_1)K_a(u_1, u_2)\ldots K_a(u_n, u)$$

$$\times \left(\frac{\Psi_a'(u)}{\Psi_a(u)} + \frac{q_a'(u)}{q_a(u)} + \frac{K_a'(u_1, u_2)}{K_a(u_1, u_2)} + \ldots + \frac{K_a'(u_n, u)}{K_a(u_n, u)}\right).$$

$$(5.15)$$

After writing (5.15) to form (5.14) as it is more convenient for the Monte-Carlo method, finally derive:

$$(\Psi_a\, K_a^n\, q_a)' = \int \ldots \int du\,du_1 \ldots du_n \Psi_a(u)q_a(u_1)K_a(u_1, u_2)$$

$$\ldots K_a(u_n, u)W_a(u, u_1, u_2, \ldots, u_n),\qquad (5.16)$$

$$W_a(u, u_1, u_2, \ldots, u_n) = \frac{\Psi_a'(u)}{\Psi_a(u)} + \frac{q_a'(u)}{q_a(u)} + \frac{K_a'(u_1, u_2)}{K_a(u_1, u_2)} + \ldots + \frac{K_a'(u_n, u)}{K_a(u_n, u)}.$$

As it follows from (5.16), in the Monte-Carlo method the derivatives could be calculated using the same algorithms as desired values with multiplying value $\Psi(\xi)$ by special weight $W_a(\xi)$ during each writing to the counter. In addition, if value $\Psi(\xi)$ depends on the current magnitude of random value ξ only, i. e. of the current coordinates of the photon, then $W_a(\xi)$ is the sum and depends on the whole history of its trajectory.

Thus, to compute the derivatives of the irradiances, it is enough to differentiate the explicit expressions of functions $\Psi_a(u)$, $q_a(u)$ and $K_a(u, u')$ with respect to the retrieved parameters. Then the following elementary changes are introduced to the algorithm of irradiance calculations described in Sect. 2.1: the counting of values W_a (for entire set of parameters) at every modeling of the element of the photon trajectory with the writing to the special counters of the derivatives simultaneously with writing to the counters of the irradiances. Although the irradiances are calculated as integrals with respect to wavelength (5.4), the wavelength remains the fixed one while modeling every single trajectory. Hence, it is enough to consider the monochromatic case only during the differentiation and the derivative of integral (5.4) will be obtained automatically. It should be emphasized also that the optical thickness itself is the function of differentiated parameters. Thus, the atmospheric pressure is to be used as a vertical coordinate, while computing the derivatives. Nothing changes in the real modeling but for the derivation of (2.8) the photon free path probability from altitude level P_1 (in the pressure scale) to level P is written as:

$$
1 - \exp\left(-\frac{1}{\mu}\int_{P_1}^{P}\alpha(P')dP'\right),
$$

where $\alpha(P)$ is the extinction coefficient, then probability density (2.8) transforms to the following:

$$
\varrho(P) = -\frac{\alpha(P)}{|\mu|}\exp\left(-\frac{1}{\mu}\int_{P_1}^{P}\alpha(P')dP'\right). \tag{5.17}
$$

It is just (5.17), which is to be used as a probability density of the photon free path, while differentiating.

Now apply the algorithm of the irradiance calculation, described in Sect. 2.1 to the algorithm for the calculating of derivatives while taking into account the explicit form of the functions in (5.16).

Counters W_a are introduced for the whole set of parameters. Starting every trajectory of the counter $W_a := 0$ is assumed. While modeling every photon free path, the following value is assigned to the counter while taking into account (5.17):

$$
W_a := W_a + \frac{1}{\alpha(P_2)}\frac{\partial}{\partial a}(\alpha(P_2)) - \frac{1}{|\mu'|}\frac{\partial}{\partial a}(\Delta\tau'(P_1,P_2)), \tag{5.18}
$$

where $\Delta\tau'(P_1,P_2)$ is the photon free path from level P_1 to level P_2 (2.7). If the photon reaches the surface, then the item with value $\alpha(P_2)$ will be absent. While modeling every act of the interaction between the photon and atmosphere, i.e. while multiplying the photon weight by $\omega_0(\tau')$, the following value is written to the counter:

$$
W_a := W_a + \frac{1}{\omega_0(P')}\frac{\partial}{\partial a}(\omega_0(P')), \tag{5.19}
$$

where P' is the current coordinate (in the atmospheric pressure scale) corresponding to optical thickness τ'. Analogously, the values for the interaction of the photon with the surface is written to the counter in accordance with (2.23):

$$
W_a := W_a + \frac{1}{A}\frac{\partial}{\partial a}(A). \tag{5.20}
$$

The value is written to the counter at every step of modeling the photon scattering in the atmosphere according to (2.9):

$$W_a := W_a + \frac{1}{x(P', \chi)} \frac{\partial}{\partial a}(x(P', \chi)) \,. \tag{5.21}$$

Finally, the following value is written to the counter of the derivatives simultaneously with writing weight ψ to the counter of irradiances as per to (2.18):

$$\psi \left[W_a - \frac{1}{|\mu'|} \frac{\partial}{\partial a}(\Delta\tau(P', P)) \right] \,,$$

where P is the coordinate of the counter.

The obtained algorithm is essentially simplified while taking into account that the following sum is calculated simultaneously at the point of the scattering modeling $P' = P_2$

$$\frac{1}{\alpha(P')} \frac{\partial}{\partial a}(\alpha(P')) + \frac{1}{\omega_0(P')} \frac{\partial}{\partial a}(\omega_0(P')) + \frac{1}{x(P', \chi)} \frac{\partial}{\partial a}(x(P', \chi)) \,. \tag{5.22}$$

After substituting the expressions of the optical parameters through aerosol and molecular components (1.24) and (1.25) with the elementary algebraic manipulations, this sum is reduced to the following form:

$$\frac{\frac{\partial}{\partial a}[\sigma_m(P')x_m(\chi) + \sigma_a(P')x_a(P', \chi)]}{\sigma_m(P')x_m(\chi) + \sigma_a(P')x_a(P', \chi)} \,, \tag{5.23}$$

where σ_m, x_m, σ_a, x_a are the volume coefficients and phase functions of the molecular and aerosol scattering. In addition, remember that the phase function of the molecular scattering determined by (1.25) does not depend on optical parameters. Finally, the only value is written to the counter in the algorithm of the photon free path modeling:

$$W_a := W_a - \frac{1}{|\mu'|} \frac{\partial}{\partial a}(\Delta\tau'(P_1, P_2)) \,, \tag{5.24}$$

and after modeling the scattering angle the only value is:

$$W_a := W_a + \frac{x_m(\chi)\frac{\partial}{\partial a}\sigma_m(P') + x_a(P', \chi)\frac{\partial}{\partial a}\sigma_a(P') + \sigma_a(P')\frac{\partial}{\partial a}x_a(P', \chi)}{\sigma_m(P')x_m(\chi) + \sigma_a(P')x_a(P', \chi)} \,. \tag{5.25}$$

The explicit expressions of the above-mentioned derivatives through the desired parameters of the inverse problem are presented further. The total set of the retrieved parameters has been defined in the previous section. There are the vertical profile of air temperature $T(P_i)$, profiles of contents of four gases absorbing radiation $Q_{H_2O}(P_i)$, $Q_{O_3}(P_i)$, $Q_{NO_2}(P_i)$, $Q_{NO_3}(P_i)$ (The O_2 content is constant), volume coefficients of the aerosol absorption and scattering

$\sigma_a(P_i, \lambda_j)$, $\kappa_a(P_i, \lambda_j)$, and surface albedo $A(\lambda_j)$. The concentrations of the atmospheric gases will be expressed through the volume-mixing ratio that gives the simple relation for their counting concentrations:

$$n(P_i) = \frac{P_i Q(P_i)}{kT(P_i)} .$$

(5.26)

Let the sets of altitude levels P_i and wavelengths λ_j to be specified in a general form for the present, their concrete magnitudes will be obtained on the basis of the derivative analysis (Sect. 4.4). Note that in practice to simplify the derivatives computing (and to prevent the errors while programming) the derivatives are to be written as a chain of the simplest formulas using the rule of the composite function differentiation. It is also useful even if the substituting of the derivatives to the general formulas causes the simplification of the expressions. The other approach effectively simplifying the calculations is application of the expression of the derivative of the product through the logarithmic derivatives.

As intermediate values in the grids P_i and λ_j are computed with the linear interpolation according to the following:

$$F(u) = F(u_i)\frac{u_{i+1} - u}{u_{i+1} - u_i} + F(u_{i+1})\frac{u - u_i}{u_{i+1} - u_i} ,$$

the derivative of function $\partial F(u)/\partial F(u_i)$ is obtained as the following process:

After determining number n from condition $u_n \leq u \leq u_{n+1}$ the following equalities are correct:

$$\frac{\partial F(u)}{\partial F(u_i)} = 0 \quad \text{for} \quad i < n \quad \text{or} \quad i > n+1 ;$$

$$\frac{\partial F(u)}{\partial F(u_i)} = \frac{u_{i+1} - u}{u_{i+1} - u_i} \quad \text{for} \quad i = n ,$$

$$\frac{\partial F(u)}{\partial F(u_i)} = \frac{u - u_i}{u_{i+1} - u_i} \quad \text{for} \quad i = n+1 .$$

As the derivative depends on the argument only, specify it as $\partial F(u)/\partial F(u_i) \equiv L_i(u)$. Then the derivative with respect to the surface albedo is written as:

$$\frac{\partial}{\partial A(\lambda_j)}(A) = L_j(\lambda) .$$

The photon free path $\Delta\tau'(P_1 P_2)$, as per (2.1)–(2.4), is the quadratic function of volume extinction coefficient $\alpha(P_i)$. Hence the following algorithm is elaborated for computing derivative $\partial/\partial\alpha(P_i)(\Delta\tau'(P_1, P_2))$, where inequity $P_2 < P_1$ is assumed for the definiteness:

1. Finding numbers n_1 and n_2 from conditions $P_{n_1} \geq P_1 \geq P_{n_1+1}$, $P_{n_2} \geq P_2 \geq P_{n_2+1}$

2. Then three cases are considered depending on the magnitude of difference $n_2 - n_1$: $n_2 > n_1 + 1$

$$\frac{\partial}{\partial\alpha(P_i)}(\Delta\tau'(P_1,P_2)) = 0 \quad \text{for} \quad i < n_1 \quad \text{or} \quad i > n_2 + 1 ;$$

$$\frac{\partial}{\partial\alpha(P_i)}(\Delta\tau'(P_1,P_2)) = \frac{1}{2}\frac{(P_1 - P_{i+1})^2}{(P_i - P_{i+1})} , \quad \text{for} \quad i = n_1 ;$$

$$\frac{\partial}{\partial\alpha(P_i)}(\Delta\tau'(P_1,P_2)) = P_1 - P_i - \frac{1}{2}\frac{(P_1 - P_i)^2}{P_{i-1} - P_i} + \frac{1}{2}(P_i - P_{i+1}) ,$$
$$\text{for} \quad i = n_2 + 1 ; \tag{5.27}$$

$$\frac{\partial}{\partial\alpha(P_i)}(\Delta\tau'(P_1,P_2)) = \frac{1}{2}(P_{i-1} - P_{i+1}) , \quad \text{for} \quad n_1 + 2 \leq i \leq n_2 - 1 ;$$

$$\frac{\partial}{\partial\alpha(P_i)}(\Delta\tau'(P_1,P_2)) = P_i - P_2 - \frac{1}{2}\frac{(P_i - P_2)^2}{P_i - P_{i+1}} + \frac{1}{2}(P_{i-1} - P_i) , \quad \text{for} \quad i = n_2 ;$$

$$\frac{\partial}{\partial\alpha(P_i)}(\Delta\tau'(P_1,P_2)) = \frac{1}{2}\frac{(P_{i-1} - P_2)^2}{P_{i-1} - P_i} , \quad \text{for} \quad i = n_2 + 1 .$$

$n_2 = n_1 + 1$. This case differs from the latter by the derivative being equal to:

$$\frac{\partial}{\partial\alpha(P_i)}(\Delta\tau'(P_1,P_2)) = P_1 - P_2 - \frac{1}{2}\frac{(P_1 - P_i)^2}{P_{i-1} - P_i} - \frac{1}{2}\frac{(P_i - P_2)^2}{P_i - P_{i+1}} \tag{5.28}$$
$$\text{for} \quad i = n_1 + 1 = n_2$$

$n_2 = n_1$:

$$\frac{\partial}{\partial\alpha(P_i)}(\Delta\tau'(P_1,P_2)) = 0 \quad \text{for } i < n_1 \quad \text{or} \quad i > n_1 + 1 ;$$

$$\frac{\partial}{\partial\alpha(P_i)}(\Delta\tau'(P_1,P_2)) = P_i - P_2 + \frac{1}{2}\frac{(P_1 - P_{i+1})^2 - (P_i - P_2)^2}{P_i - P_{i+1}} ,$$
$$\text{for} \quad i = n_1 = n_2 , \tag{5.29}$$

$$\frac{\partial}{\partial\alpha(P_i)}(\Delta\tau'(P_1,P_2)) = P_1 - P_i + \frac{1}{2}\frac{(P_{i-1} - P_2)^2 - (P_1 - P_i)^2}{P_{i-1} - P_i} ,$$
$$\text{for} \quad i = n_1 + 1 = n_2 + 1 .$$

Note that, the volume extinction coefficient in the described algorithm is applied after recalculating per the pressure unit $\alpha_P(P_i)$, while it has been calculated per the altitude unit initially. After the differentiation of (5.1) we obtained:

$$\frac{\partial \alpha_P(P_i)}{\partial \alpha_z(P_i)} = \frac{RT(P_i)}{g(P_i)\mu(P_i)P_i} . \tag{5.30}$$

Owing to summarizing rules (1.24), analogous relations are also derived for the volume coefficient of the aerosol scattering.

Now the final formulas are presented for the derivatives of the radiative characteristics with respect to the desired parameters. The specifications, used in Chapter 1 and in the previous section are kept.

Derivatives with respect to contents of the gases absorbing radiation (excluding the water vapor). Volume coefficient of the molecular absorption $\kappa_m(P_i)$ depends on these contents and the volume extinction coefficient in its turn depends on the volume coefficient of the molecular absorption as per (1.24). Then specify the concrete gas with subscript k and obtain:

$$\frac{\partial (\Delta \tau'(P_1, P_2))}{\partial Q_k(P_i)} = \frac{\partial (\Delta \tau'(P_1, P_2))}{\partial \alpha_P(P_i)} \frac{\partial \alpha_P(P_i)}{\partial \alpha_z(P_i)} \frac{\partial \alpha_z(P_i)}{\partial \kappa_{m,k}(P_i)} \frac{\partial \kappa_{m,k}(P_i)}{\partial n_k(P_i)} \frac{\partial n_k(P_i)}{\partial Q_k(P_i)} , \tag{5.31}$$

where:

$$\frac{\partial \alpha_z(P_i)}{\partial \kappa_{m,k}(P_i)} = 1$$

according to (1.23) and (1.24),

$$\frac{\partial \kappa_{m,k}(P_i)}{\partial n_k(P_i)} = C_{a,k}$$

according to (1.22) and

$$\frac{\partial n_k(P_i)}{\partial Q_k(P_i)} = \frac{P_i}{kT(P_i)}$$

according to (5.18).

The cross-sections of the molecular absorption depending on wavelength and (only for ozone) on temperature are computed by the linear interpolation with (1.28) and (5.7). Certainly, the derivatives with respect to gases content are not equal to zero within the spectral regions of these gases absorption only (Table 5.1).

Derivative with respect to water vapor content. In addition to the volume coefficient of the molecular absorption, the volume coefficient of the molecular scattering also depends on H_2O content as per (1.27). It yields the following expression for the derivative of the free path:

$$\frac{\partial (\Delta \tau'(P_1, P_2))}{\partial Q_{H_2O}(P_i)} \tag{5.32}$$

$$= \frac{\partial (\Delta \tau'(P_1, P_2))}{\partial \alpha_P(P_i)} \frac{\partial \alpha_P(P_i)}{\partial \alpha_z(P_i)} \left(\frac{\partial \alpha_z(P_i)}{\partial \kappa_{m,k}(P_i)} \frac{\partial \kappa_{m,k}(P_i)}{\partial n_k(P_i)} \frac{\partial n_k(P_i)}{\partial Q_k(P_i)} + \frac{\partial \sigma_{z,m}(P_i)}{\partial Q_{H_2O}(P_i)} \right)$$

and the following is obtained for the derivative with respect to the coefficient of the molecular scattering:

$$\frac{\partial \sigma_{P,m}(P')}{\partial Q_{H_2O}(P_i)} = L_i(P') \frac{\partial \sigma_{P,m}(P_i)}{\partial \sigma_{z,m}(P_i)} \frac{\partial \sigma_{z,m}(P_i)}{\partial Q_{H_2O}(P_i)} . \tag{5.33}$$

The derivatives depending on volume coefficient of the molecular scattering have been calculated above, and the absorption cross-section for H_2O is computed with (5.8).

The expression for the derivative of the molecular scattering volume coefficient is obtained as follows:

$$\frac{\partial \sigma_{z,m}(P_i)}{\partial Q_{H_2O}(P_i)} = \frac{\partial \sigma_{z,m}(P_i)}{\partial m} \frac{\partial m}{\partial P_w} \frac{\partial P_w}{\partial Q_{H_2O}(P_i)} , \tag{5.34}$$

where:

$$\frac{\partial \sigma_{z,m}(P_i)}{\partial m} = \sigma_{z,m}(P_i) \frac{4m}{m^2 - 1}$$

$$\frac{\partial m}{\partial P_w} = 10^{-6} \frac{0.0624 - 0.00068\lambda^{-2}}{1 + 0.003661\, T(P_i)}$$

$$\frac{\partial P_w}{\partial Q_{H_2O}(P_i)} = 0.7501\, P_i .$$

Derivative with respect to volume coefficient of the aerosol absorption. The volume extinction coefficient only depends on volume coefficient of the aerosol absorption that directly yields:

$$\frac{\partial(\Delta \tau'(P_1, P_2))}{\partial \kappa_{z,a}(P_i, \lambda_j)} = \frac{\partial(\Delta \tau'(P_1, P_2))}{\partial \alpha_P(P_i)} \frac{\partial \alpha_P(P_i)}{\partial \alpha_z(P_i)} \frac{\partial \alpha_z(P_i)}{\partial \kappa_{z,u}(P_i)} L_j(\lambda) , \tag{5.35}$$

where $\partial \alpha_z(P_i)/(\partial \kappa_{z,a}(P_i)) = 1$ with taking into account (1.23) and (1.24).

Derivative with respect to volume coefficient of the aerosol scattering. The volume coefficients of the absorption and scattering and the phase function of the aerosol scattering as per (5.9) depend on the volume coefficient of the aerosol scattering. Therefore, we obtain:

$$\frac{\partial(\Delta \tau'(P_1, P_2))}{\partial \sigma_{z,a}(P_i, \lambda_j)} = \frac{\partial(\Delta \tau'(P_1, P_2))}{\partial \alpha_P(P_i)} \frac{\partial \alpha_P(P_i)}{\partial \alpha_z(P_i)} \frac{\partial \alpha_z(P_i)}{\partial \sigma_{z,a}(P_i)} L_j(\lambda) , \tag{5.36}$$

where $\partial \alpha_z(P_i)/\partial \sigma_{z,a}(P_i) = 1$ with taking into account (1.23) and (1.24).

Then we can write:

$$\frac{\partial \sigma_{P,a}(P')}{\partial \sigma_{z,a}(P_i, \lambda_j)} = L_i(P') \frac{\partial \sigma_{P,a}(P_i)}{\partial \sigma_{z,a}(P_i, \lambda)} L_j(\lambda) . \tag{5.37}$$

At last for the derivative of the phase function the following relations are correct:

$$\frac{\partial x_a(P',\chi)}{\partial \sigma_{z,a}(P_i,\lambda_j)} = L_i(P')\frac{\partial x_a(P_i,\chi,\lambda)}{\partial \sigma_{z,a}(P_i,\lambda)}L_j(\lambda) \tag{5.38}$$

and with accounting for (5.9) after simple transformations we obtain:

$$\frac{\partial x_a(P_i,\chi,\lambda)}{\partial \sigma_{z,a}(P_i,\lambda)} = \frac{x_a(P_i,\chi,\lambda)}{\sigma_{z,a}(P_i,\lambda)}\left(D(P_i,\chi,\lambda) - \frac{1}{2}\int_{-1}^{1} D(P_i,\chi',\lambda)x_a(P_i,\chi',\lambda)d\chi'\right),$$
$$\tag{5.39}$$

where

$$D(P_i,\chi,\lambda) = b_i(\chi,\lambda) + 2c_i(\chi,\lambda)\ln(\sigma_{a,z}(P_i,\lambda)) .$$

The derivative with respect to air temperature. A big quantity of values depends on temperature. Begin from the photon free path and obtain the following for it:

$$\frac{\partial(\Delta\tau'(P_1,P_2))}{\partial T(P_i)} = \frac{\partial(\Delta\tau'(P_1,P_2))}{\partial \alpha_P(P_i)}\frac{\partial \alpha_P(P_i)}{\partial T(P_i)} \tag{5.40}$$

and for the volume coefficient of the molecular scattering:

$$\frac{\partial \sigma_{P,m}(P')}{\partial T(P_i)} = L_i(P')\frac{\partial \sigma_{P,m}(P_i)}{\partial T(P_i)} . \tag{5.41}$$

An important feature of calculating the derivatives with respect to temperature is the necessity of accounting for the temperature dependence in the formula of the recalculation of the volume extinction coefficients in terms of atmospheric pressure (5.1). It is obtained as follows:

$$\frac{\partial \alpha_P(P_i)}{\partial T(P_i)} = \alpha_P(P_i)\left(\frac{1}{\alpha_z(P_i)}\frac{\partial \alpha_z(P_i)}{\partial T(P_i)} + \frac{1}{T(P_i)}\right) . \tag{5.42}$$

The analogous relation is written for derivative $\partial \sigma_{P,m}(P_i)/\partial T(P_i)$, and for the aerosol scattering volume coefficient the following is obtained:

$$\frac{\partial \sigma_{P,a}(P_i)}{\partial T(P_i)} = \frac{\sigma_{P,a}(P_i)}{T(P_i)} .$$

Now the expression for the extinction coefficient is derived:

$$\frac{\partial \alpha_z(P_i)}{\partial T(P_i)} = \frac{\partial \sigma_{z,m}(P_i)}{\partial T(P_i)} + \frac{\partial \kappa_{z,m}(P_i)}{\partial T(P_i)} . \tag{5.43}$$

Finally, the problem is reduced to the differentiation of the volume coefficients of the molecular scattering and absorption. The first coefficient is equal to the sum of the coefficients of absorbing gases (all, including O_2) by (1.22). The corresponding sum is inferred for the derivatives too. Specifying the concrete gas with subscript k, with accounting for (5.18) we get:

$$\frac{\partial \kappa_{m,k}(P_i)}{\partial T(P_i)} = \kappa_{m,k}(P_i) \left(-\frac{1}{T(P_i)} + \frac{1}{C_{a,k}} \frac{\partial C_{a,k}}{\partial T(P_i)} \right) . \tag{5.44}$$

The absorption cross-sections of gases NO_2, NO_3, O_3 within the range 426–848 nm don't depend on temperature, hence, equality $\partial C_{a,k}/(\partial T(P_i)) = 0$ is correct. The following is obtained from (5.7) for O_3 within the range 330–356 nm:

$$\frac{\partial C_{a,k}(\lambda, T(P_i))}{\partial T(P_i)} = C_1(\lambda) + 2C_2(\lambda)T(P_i) . \tag{5.45}$$

Equation (5.8) yields the following expression with taking into account the linear interpolation of cross-sections over wavelength:

$$\begin{aligned}
\frac{\partial C_{a,k}(\lambda, P_i, T(P_i))}{\partial T(P_i)} &= -C_{a,k}(\lambda_j, P, T(P_i)) \frac{C_2(\lambda_j)}{T(P_i)} \frac{\lambda_{j+1} - \lambda}{\lambda_{j+1} - \lambda_j} \\
&\quad - C_{a,k}(\lambda_{j+1}, P, T(P_i)) \frac{C_2(\lambda_{j+1})}{T(P_i)} \frac{\lambda - \lambda_j}{\lambda_{j+1} - \lambda_j} .
\end{aligned} \tag{5.46}$$

The following is obtained for the derivative of the volume coefficient of the molecular scattering with (1.25) and (1.26):

$$\frac{\partial \sigma_{z,m}(P_i)}{\partial T(P_i)} = \sigma_{z,m}(P_i) \left(\frac{4m}{m^2 - 1} \frac{\partial m}{\partial T(P_i)} + \frac{1}{T(P_i)} \right) , \tag{5.47}$$

and expression (1.27) yields the following:

$$\begin{aligned}
\frac{\partial m}{\partial T(P_i)} &= \frac{10^{-6}}{1 + 0.003661 T(P_i)} \Bigg\{ b(\lambda) \Bigg[2.178 \times 10^{-11} P_i^2 \\
&\quad - 5.079 \times 10^{-6} \frac{P_i(1 + 10^{-6} P_i(1.049 - 0.0157 T(P_i)))}{1 + 0.003661 T(P_i)} \Bigg] \\
&\quad + 10^{-4} P_w \frac{2.284 - 0.0249 \lambda^{-2}}{1 + 0.003661 T(P_i)} \Bigg\}
\end{aligned} \tag{5.48}$$

After analyzing the obtained derivatives with the methods described in Sect. 4.4 the concrete sets of altitudes and wavelengths are selected for the retrieval of the atmospheric parameters, namely:

- The grid over wavelengths: from 325 to 685 nm with step 20 nm and from 725 to 985 nm with step 40 nm (28 points in a whole).

- The grid over altitude: from 1000 to 800 mbar with step 10 mbar, from 800 to 500 mbar with step 20 mbar, 500 to 110 mbar with step 30 mbar, 90 to 10 mbar with step 10 mbar and levels 5.2 and 0.5 mbar (61 points as a whole).

The selection of the detailed grid in the lower atmospheric layers is caused by the irradiance sounding levels and have been measured with a step of 100 mbar. Note that the top of the atmosphere corresponding to 0.5 mbar (about 55 km) is in a good agreement with the altitude of the standard top atmospheric level, usually used in calculations of the radiative transfer in the shortwave region (Rozanov et al. 1995; Kneizis et al. 1996).

Consider briefly the specific features of the calculated derivatives of the irradiances and their magnitudes. This analysis allows estimating the mechanisms of the parameter influences on the measured characteristics of solar radiation and concluding the possibility of the retrieval of certain atmospheric parameters.

Dependence of the upwelling irradiance upon the surface albedo is well studied (Kondratyev et al. 1971, 1977). The inhomogeneous linear function ($y = ax + b$) has been proposed for its description, where the multiplicative item is the part of irradiance directly reflected from the surface proportional to albedo, and the additive item is connected with diffused radiation in the atmosphere. Correspondingly, the greater albedo is the stronger is the upwelling irradiance dependent on it. The dependence of the surface albedo is also elucidated in the downwelling irradiance (Sect. 3.4). The corresponding derivative is greater, when the albedo is greater and the scattering in the atmosphere is stronger. It could reach decimals of percent of the irradiance variation to one percent of the albedo variation as it follows from the calculations with the bright surfaces like snow. Thus, the influence of surface albedo on the downwelling irradiance could exceed the uncertainty of the irradiance observation.

Out of O_2 and H_2O absorption bands, *the dependence of the irradiance upon temperature* is extremely weak: it conserves close to the value of the observational uncertainty even if the a priori variations of the temperature are maximal. The same is valid in the case of the ozone absorption bands. Thus, the temperature dependence of the irradiances could be ignored out of the absorption bands and the corresponding derivatives could be assumed equal to zero. At the same time, the temperature dependence is essential within the O_2 and H_2O absorption bands including the weak bands also. In addition, within some spectral regions, for example in wavelength 932 nm in the center of the H_2O band, it is strong and reaches the percent of the irradiance variation to one-degree variation of the whole temperature profile.

Derivative with respect to water vapor content are also essential only within its absorption bands, hence the relationship between the volume coefficient of the molecular scattering and H_2O content could be neglected. These derivatives are maximal within the absorption band 910–980 nm, where the irradiance variation reaches 40% to the a priori variations of the vertical profile of H_2O content as a whole.

The derivatives with respect to ozone content reach the maximum in the stratospheric ozone layer. Note that the selection of the upper boundary at level 0.5 mbar is determined with the influence of the stratospheric ozone on the solar irradiance value because the influence of all other components (including the aerosols) is negligibly weak at the high altitudes. The maximal irradiance variation at wavelength 330 nm is about 5% to the range of the a priori ozone variations.

The values of *the derivatives with respect to* N_2O *content* are very low, and, even with accounting for the possible wide interval of its a priori variations, the retrieval of N_2O content is impossible. This conclusion does not contradict the results obtained in the previous section as we have used the extremely high values of the absorbing gases content there and have calculated the derivatives for the averaged model (Rozanov et al. 1995; Kneizis et al. 1996). The analogous situation is arising for NO_3 gas, although the derivatives with respect to N_2O content within the absorption maximums (bands 524 and 662 nm) are essentially greater and allow principally obtaining certain information about NO_3 contents with its high concentration.

The derivatives with respect to volume coefficient of the aerosol scattering are specified with the complicated vertical dependence. The volume coefficient of the aerosol scattering influences to the solar irradiances owing to two contrary processes: the irradiances are decreasing with the aerosol optical thickness growth and are increasing with the aerosol scattering growth. Thus, the profiles of the derivatives in question are sign-invertible: they have a positive maximum around the observational point, which is decreasing with holding away from this point and then they transform to the negative ones. Evidently, this obstacle is connected with the local character of the scattering yield to the irradiances: it is maximal around the point of the measurement. The absolute value of the derivatives with respect to volume coefficient of the aerosol scattering is quite high: the variations of the coefficient even in separate layers could cause the irradiance variations up to 10% and higher. The spectral behavior of the derivatives in question is weakly expressed. There is an approach for retrieval of the altitudinal dependence of the aerosol parameters from the remote measurements within the 760 nm oxygen absorption band (Badaev and Malkevitch 1978; Timofeyev et al. 1995). Indeed, there is a certain difference between the vertical profiles shape of the derivatives within this band and out of it but it is rather weak that is also provided by the conclusion of the study (Timofeyev et al. 1995). However, the vertical profile of the retrieved parameters is directly obtained from the airborne observations at different levels of the atmosphere.

The derivatives with respect to volume coefficient of the aerosol absorption greatly depend on the type of the selected aerosol model. The values are greater if the aerosol absorption is stronger. It is the reason why the retrieval of the aerosol absorption volume coefficient from the data of the observations above Ladoga Lake has turned out a difficult problem and it has been much more possible for the observations above the desert. The same conclusion is followed from the analysis of the irradiances accomplished in Sect. 3.3.

5.3
Results of the Retrieval of Parameters of the Atmosphere and the Surface

The inverse problem of the retrieval of atmospheric and surface parameters was solved with the method described in Sect. 4.3, i.e. with the method of statistical regularization as per (4.53) (Vasilyev A and Ivlev 1999). Before discussing the retrieval results, we are pointing at the selection of the a priori and covariance matrices of the desired parameters necessary for the inverse problem solving.

The corresponding a priori models of temperature, water vapor, and ozone were taken from the book by Zuev and Komarov (1986). Two cases: "mid-latitudinal winter" for the observations above the ice and snow and "mid-latitudinal summer" for the observations above the water and sand surfaces. These models were completed with the data from the study by Anderson et al. (1996) to expand them to the top of the atmosphere (0.5 mbar). While completing, the traditional exponential approximation was used for the covariance matrices (Biryulina 1981).

$$\mathrm{corr}(X(z_i), X(z_j)) = \exp(-|z_i - z_j|/r) \qquad (5.49)$$

where X is the temperature or content of the atmospheric gas; z_i, z_j are the altitudes, where the correlation is calculated, r is the correlation radius and the only scalar parameter, which the standard altitude of 5 km was used for (Biryulina 1981).

The mean profiles of NO_2 and NO_3 were adopted from the text of GOME-TRAN computer code (Rozanov et al. 1995; Vasilyev A et al 1998). The covariance matrices were modeled according to (5.49), and the a priori SD was assumed equal to 100%.

The mean values and the covariance matrices of the albedo of sand, snow, and pure lake water were calculated directly from the observations of the spectral brightness coefficient, presented in Sect. 3.4. In the approximation of the orthotropic surface, the albedo is equal to the spectral brightness coefficient.

Construction of the a priori aerosol models is the most difficult problem, because there is no data about the variations and correlation links of the aerosol parameters in the cited literature in spite of the significant amount of optical aerosol models. In addition, the known models are not intended for applying to the inverse problems solving and consists of not detailed enough grids over altitudes and wavelengths. Thus, the special aerosol models for the regions and seasons of the observations should be elaborated while taking into account the features of the problem.

While elaborating such models, in addition to the cited literature data, the results of the direct airborne observations of the number concentration and chemical composition of the aerosol particles were used as well. These observations were accomplished by the team of the Laboratory for Aerosols Physics of the Atmospheric Physics Department of the Physical Institute of the Leningrad University above the Kara-Kum Desert and Ladoga Lake (Dmokhovsky et al. 1972; Kondratyev and Ter-Markaryants 1976). The following approach, traditional for the modern modeling of the optical properties of the aerosols was used there. The aerosols microphysical parameters were specified and the total set of the desired optical parameters (in our case the aerosol absorption and scattering volume coefficients and the phase function of the aerosol scattering at the fixed altitude and spectral grids) were calculated with them. As the problem of the modeling was obtaining the a priori statistical parameters of the aerosols, they were calculated by the variations of the microphysical parameters. This methodology of modeling and the aerosol model itself are presented in detail in the study by Vasilyev and Ivlev (2000).

The considered inverse problem of the retrieval of atmospheric optical parameters from the data of solar irradiance observations has no analogs in contemporary literature. Thus, we aim the study at the principal possibility of the retrieval of atmospheric parameters from the data of the irradiance measurements and also to the revealing of the methodological algorithm shortcomings. Therefore, we are presenting the analysis of *all* retrieved parameters of the atmosphere including even the parameters whose obtaining from the observational data is of no practical interest (the profiles of temperature and humidity). Moreover, we are presenting some erroneous results, which are of interest from the point of elucidating methodological shortcomings of the algorithms. The results of the retrieval of the aerosol parameters are certainly the most important ones, especially from the aspect of constructing and improving the aerosol models of the atmosphere. However, it should be emphasized that, if the number of the accomplished experiments with the processed results is less than ten for every type of surface, it would not be enough for statistical analysis of the results and for presenting them as models. Nevertheless, it is possible to limit our consideration with the most typical results because they are the robust (statistically stable) estimation of the mean values of the aerosol parameters for constructing the aerosol models. The obtained results are presented in Tables A.8–A.11 of Appendix A.

Figure 5.4 illustrates the examples of retrieving the temperature vertical profile. The specific features of the profiles, particularly, the strong maximum at the level 500 mbar, hardly correspond to the real altitudinal temperature behavior in the atmosphere, so they have been caused by the essential systematic uncertainty during the retrieval of the temperature profile. It is easy to explain with the significant temperature dependence of the irradiance within molecular absorption bands. In particular, it concerns oxygen narrow band 760 nm. However, as has been mentioned in Chap. 3, while describing the observations with the K-3 spectrometer the large systematic uncertainty could appear within the oxygen band connected with the shift of the wavelength scale owing to the mechanical scanning of the K-3 instrument. Besides, the instrumental function obtained from the measurements in the VD spectral region can (moreover, from the properties of spectral instruments, it has to) show the relationship between its halfwidth and spectral region and can be wider in the NIR region. Note that both specific features are clearly seen in comparison of the observations and calculations, illustrated by Fig. 5.3. As the oxygen content is fixed, while solving the inverse problem, the temperature profile is the only parameter, which links with the absorption band shape and, which could be varied in the algorithm. The systematic uncertainties of the

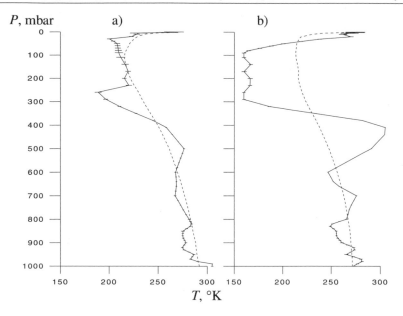

Fig. 5.4a,b. Results of the retrieval of the vertical temperature profile: **a** from the data of the airborne sounding 16th October 1983 above the Kara-Kum Desert, **b** from the data of the airborne sounding 29th April 1985 above Ladoga Lake. *Dotted line* indicates the a priori profile

temperature profile are inevitable because of the existence of the observational uncertainties within the oxygen band.

In this connection, the question of the possibility of using the radiosounding data for the irradiance data processing was discussed even in the 70th, while accomplishing the described experiments (e. g. Kondratyev et al. 1977). However, the geographical regions of the experiments differed with their microclimatic properties. The weather and atmospheric conditions above Ladoga Lake varied from those above the shore points, where the radiosounding was accomplished. While carrying out the observations above the Kara-Kum Desert, the nearest point of the radiosounding was the city Krasnovodsk at the Caspian Sea shore, where the weather and atmospheric conditions were essentially different than in the center of the desert (200 km from Krasnovodsk). Therefore, it was decided not to use the data of the direct measurements of the temperature and humidity profiles in the nearest points to the sites of the observations.

Figure 5.5 illustrates the examples of the retrieved water vapor vertical profiles. As follows from the previous analysis of the derivatives, H_2O absorption bands together with the oxygen absorption bands are the only spectral regions, where the essential temperature dependence of the irradiance exists. Thus, the significant uncertainties mentioned above could affect only H_2O content profile. However, as has been mentioned above, the retrieval of the temperature and humidity is not of practical interest, and the pointed systematic uncer-

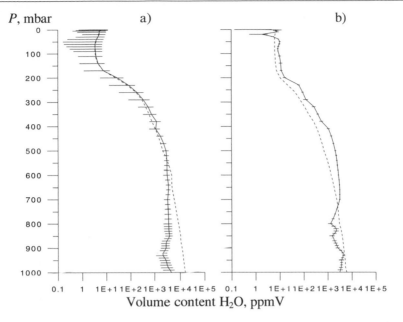

P, mbar

Fig. 5.5a,b. Results of the retrieval of the volume H_2O content vertical profile: **a** from the data of the airborne sounding 16th October 1983 above the Kara-Kum Desert; **b** from the data of the airborne sounding 29th April 1985 above Ladoga Lake. *Dotted line* indicates the a priori profile

tainties could be ignored. It should be emphasized that there is no significant contradiction in the results of H_2O content profile. In particular, H_2O content at the ground level retrieved from the observations above the desert is less than the a priori magnitude for mid-latitudes, as it should be in accordance with logic.

The results of the ozone content retrieval are presented in Fig. 5.6. It is seen that the retrieved profiles weakly differ from the a priori ones, though O_3 content above the desert is rather higher than the a priori content.

As for the results of N_2O and NO_3 contents, their uncertainties are close to the a priori ones, so it is better to discuss the correct accounting of the a priori indefinites of their content assignments but not the results of the vertical profiles of these gases.

Consider the most interesting components of the vector of the retrieved parameters, namely the optical parameters of the atmospheric aerosols. The examples of retrieving the vertical profiles of the aerosol scattering and absorption volume coefficients are presented in Figs. 5.7–5.9 and in Tables A.8–A.11 of Appendix A. Note that they are significantly lower than the a priori ones in the lower troposphere that points out the necessity of correcting the a priori models to decrease the aerosol particles content in the corresponding altitudinal zones. In this connection, the known effect of the strong dependence of the results upon zeroth approximation selection should be stressed (Zuev and

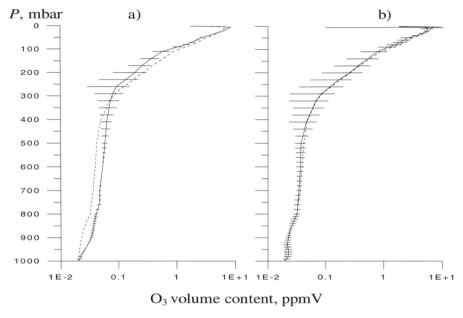

Fig. 5.6a,b. Results of the retrieval of the vertical volume ozone content profile: **a** from the data of the airborne sounding 16th October 1983 above the Kara-Kum Desert; **b** from the data of the airborne sounding 29th April 1985 above Ladoga Lake. *Dotted line* indicates the a priori profile

Naats 1990; Vasilyev O and Vasilyev A 1994). Thus, the retrieved results could be changed after correcting the aerosol model.

The systematic uncertainties of the instrument calibration strongly affect the results of the vertical profiles of the coefficients in question (Vasilyev A and Ivlev 1999). We illustrate this influence with the simplest example. Let the measured value of the downwelling irradiance at the level 500 mbar be systematically underestimated only for 1–2% (Sect. 3.3). The only way to adjust the direct problem solution to this observational data is introducing the extinction aerosol layer to the model at the altitude higher than 500 mbar. Taking into account small a priori aerosol content at these altitudes, the introduced aerosol layer must be sufficiently thick to cause the extinction of the downwelling irradiance to 1–2%. Thus, even with the low systematic uncertainty in the observed irradiances the algorithm of the inverse problem solving could cause the false conclusion about the existence of the aerosol layers in the upper troposphere and in the stratosphere. Hence, the results of the retrieval of the aerosol scattering and absorption volume coefficients obtained in altitudinal diapason of the airborne observations 500–950 mbar are much more reliable, because only the relative values of the solar irradiances are essential there. The corresponding profiles are presented in Fig. 5.8. The calibrating factor is likely to be introduced to the vector of the parameters for retrieval though it is make the retrieval accuracy worse.

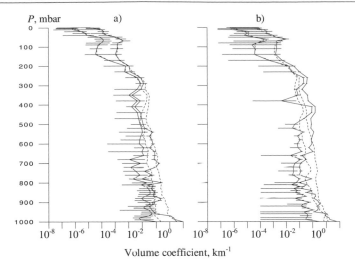

Fig. 5.7a,b. Results of the retrieval of the vertical profiles volume coefficients of the aerosol scattering (*right curves*) and absorption (*left curves*) at wavelength 545 nm: **a** from the data of the airborne sounding 16th October 1983 above the Kara-Kum Desert; **b** from the data of the airborne sounding 29th April 1985 above Ladoga Lake. *Dashed lines* indicate the relevant profiles of the a priori models

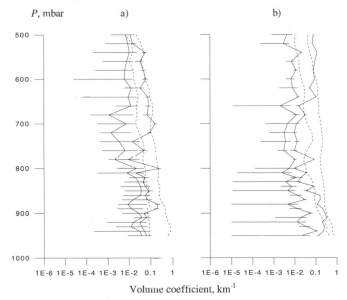

Fig. 5.8a,b. Results of the retrieval of the vertical profiles of the volume coefficient of the aerosol scattering (*right curves*) and absorption (*left curves*) at wavelength 545 nm and at the altitude levels corresponding to atmospheric pressure 500–950 mbar: **a** from the data of the airborne sounding 16th October 1983 above the Kara-Kum Desert; **b** from the data of the airborne sounding 29th April 1985 above Ladoga Lake. *Dashed lines* indicate the relevant profiles of the a priori models

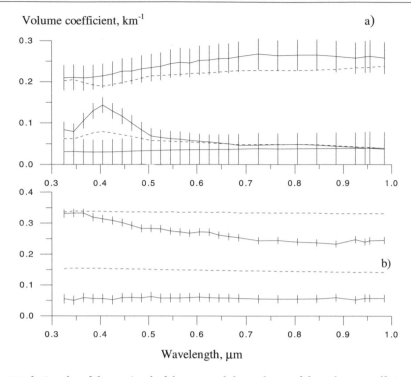

Fig. 5.9a,b. Results of the retrieval of the spectral dependence of the volume coefficients of the aerosol scattering (*upper curves*) and absorption (*lower curves*) at the altitude levels corresponding to atmospheric pressure 850 mbar: **a** from the data of the airborne sounding 16th October 1983 above the Kara-Kum Desert, **b** from the data of the airborne sounding 29th April 1985 above Ladoga Lake. *Dashed lines* indicate the relevant profiles of the a priori models. (*Middle curve* in Fig. 5.9a) – The aerosol absorption volume coefficient from the airborne sounding 12th October 1983 under sand storm conditions

The irregular, indent shape of the vertical profiles was obtained in the other studies (for example Krekov and Zvenigiriodsky 1990; Polyakov et al. 2001) from the remote sounding processing and it was also obtained from the airborne direct measurements of the aerosols particle concentrations even after the statistical smoothing over a big volume of data (Hudson and Yonghong 1999). Thus, the retrieved serrated profiles of the optical parameters of the atmospheric aerosols are not to be explained as an effect of only systematic errors of the calibration and altitudinal conjunction, and they are likely to reflect the real profile of the aerosol content in the atmosphere. The altitudes of the most probable formation of cloudiness correspond to the local maximums in curves of Fig. 5.8 (Hudson and Yonghong 1999). As is well known, the process of the cloudiness formation is connected with the presence of atmospheric aerosols as they serve the condensation nuclei.

In particular, the local maximum of the volume coefficients of the aerosol scattering and absorption at altitude 1900 m (corresponding to 800 mbar) is evident in Fig. 5.8 and could be explained with the atmospheric aerosols presence at this altitude. It corresponds to the most probable altitude of the cloud formation of the lower level.

The examples of the spectral dependence of the volume aerosol scattering and absorption coefficients are demonstrated in Fig. 5.9. We should mention the essentially different character of the spectral dependence of the scattering coefficient in question above the desert and above Ladoga Lake. In the first case, there is no spectral dependence of the scattering coefficient or there is a weak growth with wavelength. It might be explained by the rather high amount of large particles in the atmospheric aerosols above the desert.

Figure 5.9 illustrates the results of the volume coefficient of the aerosol absorption obtained from the sounding data above the desert under pure atmospheric conditions (the weak absorption) and after a sand storm (the strong absorption). The latter case demonstrates the evident absorption band of the hematite, which appears even in the spectra of the solar radiative flux divergence (Sect. 3.3). The second case illustrates the apparent decreasing of the aerosol scattering coefficient with wavelength.

The examples of retrieving the spectral values of the surface albedo are presented in Fig. 5.10. The deviation of the spectrum of the snow surface from the monotonic behavior (Fig. 5.10b) is likely caused by the surface inhomogeneity (the snow was melting on 29th April). The surface inhomogeneity has been smoothed during the second stage of the data processing, but the spectral distortions of the upwelling irradiances have remained and they cause the systematic uncertainty of the retrieved albedo, which does not exceed the interval of three SD and statistically can be assumed the insignificant one.

Note, that the spectral albedo is retrieved with the relative uncertainty about 1–3% that is much more accurate than in the case of direct dividing the upwelling irradiance by the downwelling (Sect. 3.4). In addition, the retrieved albedo is exactly correspondent to the notion of albedo used in the radiative transfer theory (Sect. 1.4) and it has no distortion connected with the gases absorption bands. Thus, the airborne experiments accomplished with the sounding scheme and the following inverse problem solving could be used for obtaining the surface albedo values with high accuracy.

We should mention that the uncertainties of the retrieved atmospheric parameters are greatly affected by the information content of the results of solar irradiance observations (Sect. 3.3) and by the spectral resolution within the absorption bands of gases, while retrieving their content. Thus, the uncertainties of the retrieval from different soundings data are essentially different. It is seen from the presented figures, where the posterior SDs of the retrieved parameters are shown. The uncertainties of the retrieval in the lower troposphere are 10–50% on the average for the volume aerosol scattering coefficient; are 50–100% for the volume aerosol absorption coefficient (that, however, is less than the a priori uncertainty); are 20–30% for ozone content and are 20–50% for H_2O content. We should point out that the higher the aerosol content in the atmosphere the higher the accuracy of the aerosol parameters is.

The discreteness of the registration during the measurements (Sect. 4.3) is not accounted for in the formal scheme of the inverse problem solving

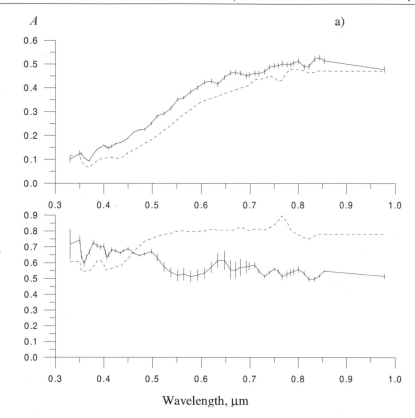

Fig. 5.10a,b. Results of the retrieval of the spectral dependence of the surface albedo: **a** from the data of the airborne sounding 16th October 1983 above the Kara-Kum Desert, sand surface; **b** from the data of the airborne sounding 29th April 1985 above Ladoga Lake, snow surface. *Dashed lines* indicate the relevant profiles of the a priori models

[consequence 4 from (4.38)]. However, the digitizing of the signal during the observations with the K-3 instrument has been accomplished with an accuracy to 10 binary, i.e. 3 decimal orders. This means that after averaging the results of about 100 measurements the accuracy of the mean value could exceed the accuracy of the instrument signal registration (Otnes and Enochson 1978). The ratio about 1/100 appears between the number of the independent retrieved parameters (during the transformation to the basis of the a priori covariance matrix) and the number of observations. Certainly not only the averaging but more complicated data processing is carried out during the inverse problem solving, but it does not matter and the obtained SD of the retrieved parameters could turn out lower than the real values. Especially it appears during the surface spectral albedo retrieval because all observational results are used in this case and the spectral albedo is described with only several independent parameters owing to its strong autocorrelation. Therefore, the formal accuracy of the albedo retrieval turns out fantastically high. However, in reality, the albedo could not be obtained with the accuracy exceeding the instrument accuracy. Indeed, this very accuracy would be obtained without atmosphere. Taking this fact into account the relevant correction has been introduced to the SD value finally attributed to the spectral albedo, for it is not lower than the random SD of K-3 instrument observations (Table 3.1).

It could be ascertained from the first experience of the inverse problem solving that the problem in question is quite solvable. There is sufficiently high susceptibility of the downwelling and upwelling irradiances to the variations of the gas and aerosol composition of the atmosphere and surface albedo. The strong relationship between the retrieval results and systematic uncertainties of calibration, graduation, and instrumental function is revealed. To diminish this relationship and to account for the calibration parameters correctly, they are to be included to the vector of the retrieved parameters, while the algorithm is improving. Thus, it is seen that the presented method allows the full and correct extraction of the information about the aerosol and gaseous composition of the atmosphere from the large arrays of the accumulated data of the field observations. It is doubtless that the elements of this method could be used in the processing algorithms of the contemporary satellite data of the scattered and reflected solar radiation in the shortwave spectral range (Vasilyev A et al. 1998).

References

Anderson GP, Clough SA, Kneizys FX et al. (1996) AFGL atmospheric constituent profiles (0–120 km). Environmental Research Paper 954, Air Force Geophysics Laboratory, Hanscom, Massachusetts

Badaev VV, Malkevitch MS (1978) The possibility of obtaining vertical profiles of aerosol extinction by satellite observations of reflected radiation within the band O_2 0.76 μ. Izv Acad Sc USSR, Atmosphere and Ocean Physics 14:1022–1030 (Bilingual)

Barteneva OD, Dovgyallo EN, Polyakova EA (1967) Experimental studies of optical properties of ground layer of the atmosphere. Main Geophysical Observatory Studies 220 (in Russian)

Barteneva OD, Laktionova AG, AdnashkinVN, Veselova LK (1978) Phase functions of light scattering in the ground layer above the Ocean. In: Atmospheric Physics Problems, Leningrad University Press, Leningrad, Vol. 15, pp 27–43 (in Russian)

Bass AM, Paur RJ (1984) The ultraviolet cross-section of ozone. In: Zerofs CS, Chazi AP (eds) The measurements of atmospheric ozone. Reidel, Dordrecht, pp 606–610

Biryulina MS (1981) Modeling of a priori ensemble of the inverse problem solution and stability of optimal planes of the ozone satellite experiment. Meteorology Hydrology 4:45–51 (in Russian)

Borovin GK, Komarov MM, Yaroshevsky VS (1987) Errors-traps by programming in Fortran. Nauka, Moscow (in Russian)

Dmokhovsky VI, Ivlev LS, Ivanov VN (1972) Airborne observations of the vertical structure of atmospheric aerosols according to the CAENEX program. In: Main geophysical observatory studies, 276, pp 37–42 (in Russian)

Gorchakov GI, Isakov AA (1974) Halo phase functions of the gaze. Izv. RAS, Atmosphere and Ocean Physics 10:504–511 (Bilingual)

Gorchakov GI, Isakov AA, Sviridenkov MA (1976) Statistical links between scattering co-efficient and coefficient of the directional light scattering in the angle ranges 0.5–165°. Izv. Acad. Sci USSR, Atmosphere and Ocean Physics 12:1261–1268 (Bilingual)

Hudson JG, Yonghong X (1999) Vertical distributions of cloud condensation nuclei spectra over the summertime northeast Pacific and Atlantic Oceans. J Geophysical Res 104(D23):30219–30229

Ivlev LS, Vasilyev AV (1998) Refined interpretation of the spectral behavior of optical thickness and residual atmospheric absorption in the short–wavelength region of spectrum. SPIE, Vol. 3583, pp 35–38

Kneizis FX, Robertson PC, Abreu LW et al. (1996) The Modtran 2/3. Report and Lowtran 7 model. Phillips Laboratory, Hanscom, Massachusetts

Kondratyev KYa, Timofeyev YuM (1970) Thermal sounding of the atmosphere from satellites. Gydrometeoizdat, Leningrad (in Russian)

Kondratyev KYa, Ter-Markaryants NE (eds) (1976) Complex radiation experiment. Gydrometeoizdat, Leningrad (in Russian)

Kondratyev KYa, Buznikov AA, Vasilyev OB et al. (1971) Certain results of combined complex under satellite geophysical experiment. Doklady Acad. Sci. USSR, Ser. Mathematics and Physics 196:1333–1336 (in Russian)

Kondratyev KYa, Buznikov AA, Vasilyev OB, Smoktiy OI (1977) Atmosphere influence on albedo by aero-cosmic survey of the Earth in visual spectral region. Izv. Acad. Sci USSR, Atmosphere and Ocean Physics 13:471–478 (Bilingual)

Krekov GM, Rakhimov RF (1986) Optical models of atmospheric aerosols. Tomsk Department of Siberian Branch Acad Sci USSR Press, Tomsk (in Russian)

Krekov GM, Zvenigorodsky SG (1990) Optical model of the middle atmosphere. Nauka, Novosibirsk (in Russian)

Lenoble J (ed) (1985) Radiative transfer in scattering and absorbing atmospheres: standard computational procedures. A. DEEPAK Publishing, Hampton, Virginia

Marchuk GI, Mikhailov GA, Nazaraliev NA et al. (1980) The Monte-Carlo method in the atmosphere optics. Springer-Verlag, New York

Minin IN (1988) The theory of radiation transfer in the planets atmospheres. Nauka, Moscow (in Russian)

Otnes RK, Enochson L (1978) Applied Time-Series Analysis. Toronto. Wiley, New York

Pokrovsky AG (1967) The methodology of calculation of spectral absorption of the infrared radiation in the atmosphere. In: Atmospheric Physics Problems, Iss.5. Leningrad, Leningrad University Press, pp 85–110 (in Russian)

Polyakov AV, Timofeyev YuM, Poberovsky AV, Vasilyev AV (2001) Retrieval of vertical profiles of coefficient of the aerosol extinction in the stratosphere by results of observations with instruments "Ozone-Mir" (DOS Mir). Izv. RAS, ser. Atmosphere and Ocean Physics 37:213–222 (Bilingual)

Rozanov VV, Timofeyev YuM, Barrows JP (1995) The information content of observations of outgoing UV, visual and near infrared solar radiation (instruments GOME). Earth Observations and Remote Sensing 6:29–39 (Bilingual)

Rudich Y, Talukder RK, Ravishankara AR (1998) Multiphase chemistry of NO_3 in the remote troposphere. J Geophys Res 103(D13):16133–16143

Timofeyev YuM, Vasilyev AV, Rozanov VV (1995) Information content of the spectral measurements of the 0.76 μm O_2 outgoing radiation with respect to the vertical aerosol properties. Advances of Space Research 16:91–94 (Bilingual)

Tvorogov SD (1994) Certain aspects of the problem of the function presentation by the exponent series. Atmosphere and Ocean Optics 7:793–798 (Bilingual)

Vasilyev AV (1996) "Vertical" is the collection of gas models of the Earth atmosphere. In: The Herald of the St. Petersburg University. Ser. 4, Physics, Chemistry 4:87–90 (in Russian)

Vasilyev AV, Ivlev LS (1995) Numerical modeling of optical characteristics of poly-disperse spherical particles. Atmosphere and Ocean Optics 8:921–928 (Bilingual)

Vasilyev AV, Ivlev LS (1996) Numerical modeling of spectral aerosol phase function of the light scattering. Atmosphere and Ocean Optics 9:129–133 (Bilingual)

Vasilyev AV, Ivlev LS (1999) Determination of parameters of gas and aerosol composition of the atmosphere by airborne observations of spectral irradiance. In: Ivlev LS (ed) Natural and anthropogenic aerosols, Collection of articles, St. Petersburg. Chemistry Institute, St. Petersburg University Press, pp 97–103 (in Russian)

Vasilyev AV, Ivlev LS (2000) Optical statistical model of the atmosphere for the region of Ladoga Lake. Atmosphere and Ocean Optics 13:198–203 (Bilingual)

Vasilyev AV, Rozanov VV, Timofeyev YuM (1998) The analysis of the informative content of observations of outgoing reflected and diffused solar radiation within the spectral region 240–700 nm. Earth Observations and Remote Sensing 2:51–58 (Bilingual)

Vasilyev OB, Vasilyev AV (1989) Information content of obtaining optical parameters of atmospheric layers by observations spectral irradiances at different levels in the atmosphere. I. Problem statement and results of calculations for the separate layer. Atmosphere and Ocean Optics 2:428–433 (Bilingual)

Vasilyev OB, Vasilyev AV (1989) Information content of obtaining optical parameters of atmospheric layers by observations spectral irradiances at different levels in the atmosphere. II. Estimation of the informatic content of observations in the multi-layer atmosphere. Atmosphere and Ocean Optics 2:433–437 (Bilingual)

Vasilyev OB, Vasilyev AV (1994a) Information content of obtaining optical parameters of atmospheric layers by observations spectral irradiances at different levels in the atmosphere. III. Obtaining optical parameters of layers in the inhomogeneous multi-layer atmosphere (numerical experiment). Atmosphere and Ocean Optics 7:625–632 (Bilingual)

Vasilyev OB, Vasilyev AV (1994b) Two-parametric model of the phase function. Atmosphere and Ocean Optics 7:76–89 (Bilingual)

Vasilyev OB, Contreras AL, Velazques AM et al. (1995) Spectral optical properties of the polluted atmosphere of Mexico City (spring–summer 1992). J Geophysical Res 100:D12, 26027–26044

Virolainen YaA, Polyakov AV (1999) The algorithm of the direct calculation of transmission functions in problems of ground remote sounding of the atmosphere. In: The Herald of the St. Petersburg University, ser. 4, Physics, Chem 1:25–31 (in Russian)

Weaver A, Solomon S, Sanders RW, Arpag K, Muller HL Jr (1996) Atmospheric NO_3 off-axis measurements at sunrise: estimates of tropospheric NO_3 at 40°N. J Geophys Res 101(D13):18605–18612

Zuev BE, Komarov VS (1986) Statistical models of temperature and gaseous components of the atmosphere. (Contemporary problems of the atmospheric optics, vol. 1). Gydrometeoizdat, Leningrad (in Russian)

Zuev VE, Naats IE (1990) Inverse problems of the atmospheric optics. (Contemporary problems of the atmospheric optics, vol. 7). Gydrometeoizdat, Leningrad (in Russian)

CHAPTER 6

Analytical Method of Inverse Problem Solution for Cloudy Atmospheres

6.1
Single Scattering Albedo and Optical Thickness Retrieval from Data of Radiative Observation

The approach of the numerical solving of the inverse problem of atmospheric optics has been presented in Chaps. 4 and 5. In addition, the direct problem solution compared with the values of the observed radiative characteristics has been obtained with the universal numerical Monte-Carlo method. In some cases we succeeded to find the solution of the direct problem in the analytical form (Sects. 2.2 and 2.3), then the procedure of computing the derivatives of the irradiances with respect to the atmospheric parameters becomes faster and simpler. Moreover, the possibility of the analytical expressions of the radiative characteristics suggest an idea to convert these expressions and to obtain the inverse formulas for the retrieval of the desired parameters after substituting the measured values of the radiative characteristics. The first studies in this field assumed either the infinitely thick or the conservative scattering atmosphere to exclude one of the unknown parameters. Thus, we are citing here only the studies, which have presented some analytical expression for finding the optical parameters but not the studies where the optical parameters have been obtained with a simple comparison of calculations and observational results.

The authors of the study by Rozenberg et al. (1974) took the first step by using the observation of the reflected solar radiation from satellites for obtaining the small parameter connected with the single scattering albedo ω_0 of the cloud while assuming its infinite optical thickness and the expansion analogous to (2.29). Only the first power of the expansion was taken into account, and the optical thickness of the cloud layer was not analyzed. In the study by Yanovitskij (1972), the expression for spherical albedo of the infinite atmosphere was inferred and applied to the clouds of Venus for defining the single scattering albedo with the same assumption about the optically infinite atmosphere (proven to be more correct than the assumption of the study by Rozenberg (1974) for terrestrial clouds). The spectral values of ω_0 for six wavelengths from the data of the astronomical observations of the atmosphere of Venus were evaluated there as well.

The expressions for the retrieval of optical thickness from the radiance observations above the cloud layer and within it were firstly proposed in several studies (King 1987, 1993; King et al. 1990). Regretfully the authors of the mentioned studies applied the formulas for the case of conservative scattering only to obtain the optical thickness in the VD spectral region. As the radiation absorption of the cloud layer was not accounted for, the significant errors (the unknown a priori ones) could occur if there was radiation absorption in the clouds. The problem of optical thickness retrieval from solar radiance measurements in several wavelength channels within the cloud layer was solved in the study by King et al. (1990) again with the assumption of the conservative scattering.

The important exact expression for scaled optical thickness $\tau' = 3(1 - g)\tau_0$ through the reflected radiance was derived in the study by King (1987) as a result of transforming the first of (2.24). Regretfully the author of the study (King 1987) continued the further application of the obtained formula assuming the conservative scattering of radiation only.

The approach based on using the ratio of the radiances or irradiances at different levels within the cloud layer was proposed in several studies (Duracz and McCormick 1986; McCormick and Leathers 1996) and the corresponding analytical formulas was derived for the realization of the approach. Unfortunately, we have found no results of its application to the analysis of the observational data.

The parameters of the optically thick atmosphere were determined on the basis of applying the irradiance gradients to the observations of automatic interplanetary station "Venera" in the atmosphere of Venus. The high accuracy of the measurements (1) and sufficiently high variations of the correspondent radiative characteristics with altitude (2) are readily needed to reach the acceptable accuracy of the retrieval of optical parameters. While the second condition is fulfilled in the atmosphere of Venus due to its large optical thickness, the high observational accuracy is easier to reach in Earth's atmosphere. Anyway, calculating the derivative of the radiative characteristics with respect to altitude with high accuracy is difficult. Various studies (Germogenova et al 1977; Ustinov 1977; Konovalov and Lukashevitch 1981; Konovalov 1982) have considered the approach of retrieving the optical parameters of the atmosphere of Venus from irradiance observations basing on the asymptotic formulas of the transfer theory.

In this connection the study by Zege and Kokhanovsky (1994) should be mentioned, where the relations for optical parameters of the cloudy atmosphere were deduced. The expansions over the parameter similar to the parameter used in the study by Rozenberg (1974), (with taking only the first power of the expansion), were convoluted together with the asymptotic formulas with respect to τ_0. This approach provided certain advantages but impeded the analysis of its applicability region to the single scattering albedo and to the optical thickness separately. The algorithm of the cloud retrieval for this method was presented in the study by Kohanovsky et al. (2003). In spite of the advantages of the method itself, the algorithm was elaborated with certain shortcoming assumptions: the conservative scattering in the VD spectral region, the invariant optical thickness with respect to wavelength, together with the usage of the insufficient number of the expansion terms (only the first power). Moreover, the authors have not presented yet any example of the application of this approach.

It should be emphasized that these assumptions are very often used because it is believed in the model of the pure water clouds that nothing is absorbing solar radiation out of the molecular bands. However, the existence of significant absorption follows even from the results of the study by Rozenberg et al. (1974). Note, that introducing the atmospheric aerosols to the cloud model provides certain radiation absorption in the calculations of the radiative characteristics (direct problem solving), but this is rather weak. Hence, many investigators have been still assuming the conservative scattering of solar radiation in clouds within the VD spectral region, though these assumptions strongly restrict the problem solution. Further, in Chap. 7 it will be shown that both assumptions (the conservative scattering in the VD spectral region, the invariant optical thickness with respect to wavelength) are false in most cases of extended clouds.

Let us return to Sects. 2.2–2.5, where the optical model of the extended cloudiness and radiative characteristics observed during the experiments have been considered. The radiation scattering in the layers above and below the cloud is neglected at the first stage. In this chapter, we will derive the formulas for the cloud optical parameters from the observed solar radiances or irradiances values. Expansions (2.29), (2.30), and (2.44) over small parameter $s = \sqrt{(1 - \omega_0)/(3 - 3g)}$ that is characterized the radiation absorption and phase function influence are substituted to (2.24), (2.26), (2.41) (2.43), and (2.45) describing the diffused radiation in the cloudy atmosphere. Then operations of multiplying and dividing are to be accomplished by exploiting the rules of algebra of series while taking into account the terms till the second power of the small parameter. Only the observed values of solar irradiances $F^{\uparrow}$ and $F^{\downarrow}$ or radiance I and values of functions $K_0(\mu_0)$, $K_2(\mu_0)$ and $a_2(\mu_0)$ for the fixed incident and viewing (for radiance) angles are included to the final equations.

The approach for obtaining the cloud optical parameters, presented here is useful for the interpretation of the airborne, ground and satellite observational data, but it needs careful analysis of the accuracy of the desired parameters retrieval, using the numerical models. This analysis has been accomplished for the set of single scattering albedo ω_0 and optical thickness τ_0 (Melnikova 1991, 1992) and will be briefly presented in Sect. 6.3.

Relation $\kappa = s^2\tau'/\Delta z$ expresses the absorption volume coefficient and no assumption concerning the phase function is needed. Obtaining scattering volume coefficient $\alpha = \tau'(3 - 3g)/\Delta z - \kappa$ demands the assignment of phase function parameter g or its determining using an independent methodology. Fortunately, the phase function parameter does not vary strongly with wavelength within the stratus clouds; hence, we assume the spectral values of mean cosine of the scattering angle g in accordance with the results of the study by Stephens (1979). If geometrical thickness Δz stays unknown then it is only possible to determine the optical thickness and single scattering albedo, but not the scattering and absorption volume coefficients.

6.1.1
Problem Solution in the Case of the Observations of the Characteristics of Solar Radiation at the Top and Bottom of the Cloud Optically Thick Layer

Now appeal to (2.25) describing the upwelling and downwelling irradiance scattered by the cloud layer at its top and bottom, which could be measured from an aircraft board. Let the cloud layer adjoin the surface with albedo $A = F^\uparrow(\tau_0)/F^\downarrow(\tau_0)$. Function $K(\mu)$ and constants m, l, $n$and k in (2.25) depend on true radiation absorption and phase function extension, which in their turn are defined by microphysical properties of cloud medium and described by expansions over parameter s (2.29). Substituting (2.29) to (2.25), after the algebraic transformations we obtain two equations with two unknowns – parameter s and scaled optical thickness τ' (Melnikova 1991, 1992; Melnikova and Mikhailov 1993, 1994). The corresponding manipulations are presented in Appendix 2. Thus, for determining values s and τ' using the data of the irradiances measurements, the following expressions are deduced:

$$s_2 = \frac{F(0)^2 - F(\tau_0)^2}{16[K_0(\mu_0)^2 - F^\uparrow(\tau_0)^2] - 2a_2(\mu_0)F(0) - 24q'F^\uparrow(\tau_0)F(\tau_0)},$$

$$\tau'_0 = 3\tau_0(1 - g) = \frac{1}{2s}\ln\left[\bar{l}\left(1 + \frac{mnK(\mu_0)}{a(\mu_0) - F_0^\uparrow}\right)\right].$$

(6.1)

In the case of using the observational data of the solar radiances, the expressions keep the similar structure, but they become more awkward:

$$s^2 = \frac{\bar{K}_0(\mu)^2(\varrho_0 - \varrho)^2 - K_0(\mu)^2\sigma^2}{16K_0(\mu)^2\left[\bar{K}_0(\mu)^2 K_0(\mu_0)^2 - (\frac{A\sigma}{1-A})^2\right] - J},$$

$$\tau'_0 = \frac{1}{2s}\ln\left[\bar{l}\left(1 + \frac{mK(\mu)K(\mu_0)}{(\varrho_0 - \varrho)}\right)\right].$$

(6.2)

where the following is specified for brevity:

$$J = \frac{24q'AK_0(\mu)}{1 - A}\left[\bar{K}_0(\mu)(\varrho_0 - \varrho)^2 - \frac{AK_0(\mu)\sigma^2}{1 - A}\right]$$
$$+ \frac{2A\bar{K}_0(\mu)}{1 - A}\left[a_2(\mu) + n_2 - \frac{K_2(\mu)}{K_0(\mu)}\right](\varrho_0 - \varrho)^2 + \frac{a_2(\mu)a_2(\mu_0)}{6q'}\bar{K}_0(\mu)(\varrho_0 - \varrho)$$

$F(0) = F^\downarrow(0) - F^\uparrow(0)$ and $F(\tau_0) = F^\downarrow(\tau_0) - F^\uparrow(\tau_0)$ are the net solar fluxes at the top ($\tau = 0$) and at the bottom ($\tau = \tau_0$) of the cloud layer; $a_2(\mu_0)$ is the second coefficient in the expansion for the plane albedo of the semi-infinite atmosphere, a^∞ is the spherical albedo of the infinite atmosphere; $\varrho = \varrho(0, \mu, \mu_0)$ and $\sigma = \sigma(\tau_0, \mu, \mu_0)$ are the radiances (in relative units of the incident solar flux πS) measured during the radiative experiments: $I(0, \mu, \mu_0) = S\mu_0\varrho(0, \mu, \mu_0)$ and $I(\tau_0, \mu, \mu_0) = S\mu_0\sigma(\tau_0, \mu, \mu_0)$. Functions $K_0(\mu_0)$, $K_2(\mu_0)$ and

Table 6.1. Deviation of the approximation of zeroth harmonic $\varrho^0(0.67, 0.67)$ from unity

g	0.3	0.5	0.75	0.8	0.85	0.9
$|1 - \varrho^0(0.67, 0.67)|$	0.0037	0.024	0.021	0.0059	0.013	0.0046

$a_2(\mu_0)$ are completely defined by the cosine of the solar zenith angle, i. e. by the time and place of the experiment, and functions $K_0(\mu)$, $K_2(\mu)$ and $a_2(\mu)$ are defined by the cosine of the viewing angle. Their values could be taken from tables (Dlugach and Yanovitsky 1974; Hulst 1980; Minin 1988) or calculated with (2.31), (2.34) and (2.35) presented in Chap. 2, in addition to the expressions for constants m, l, n_2 and k (2.29).

We need the value of the surface albedo to calculate function $\bar{K}_0(\mu) = K_0(\mu) + A/(1 - A)$. In the case of using the solar irradiance observations, value A could be inferred by dividing the upwelling irradiance by the downwelling one at the bottom of the cloud. In the case of the radiance, the situation is not so simple. Therefore, we propose a convenient approach. We should mention the value of the cosine of the zenith angle $\mu = 0.67$ (it corresponds to $48°$), for which the zeroth harmonic of reflection function is close to unity, especially in the case of the Henyey-Greenstein phase function, the other harmonics are close to zero and escape function $K_0(\mu)$ is also equal to unity. Thus, the reflected radiance, measured at the viewing angles close to $48°$ is equal to the reflected irradiance (similar is also true for transmitted radiation). Both radiance and irradiance, measured at solar angle $48°$, approximately coincide with the spherical albedo of the cloud layer. The value of $|1 - \varrho^0(\mu, \mu_0)|$ in the case of recent approximations ((2.37), (2.39) and Tables 2.3 and 2.4) are presented in Table 6.1 for phase function parameter $g = 0.3-0.9$ and $\mu = \mu_0 = 0.67$. The small deviation from unity shows rather low error of the approach proposed here. The analysis of zeroth harmonic of reflection function $\varrho^0(\mu, 0.67)$ for different phase function parameter g demonstrates that the deviation from unity is about 8% for $g \leq 0.5$ and 10% for $g = 0.85-0.9$.

The reflection function has been calculated for the Mie phase function that corresponds to the model of the fair weather cumulus (FWC) clouds (King 1983). These results indicate that the reflection function differs from unity by 2–5% at the zenith angles in the range 47–50°. Hence, it is possible to conclude that the reflection function is close to unity at these zenith angles even for the complicated phase function. It has been shown (Kokhanovsky et al. 1998) that the impact of the phase function on the radiative forcing is almost the same for different phase functions, if the solar incident angle is about 45–50° ($\mu_0 = 0.643-0.707$). The slightest influence of particle size distribution on cloud phase function at scattering angles about 90° has been also obtained (Kokhanovsky et al. 1998); that approximately corresponds to cosines of zenith angle $\mu_0 = \mu = 0.67$. It is possible to explain these facts with the following: the reflection function (and the escape function) more weakly depends on the phase function at this angular range than at other angles. Thus, it is more suitable to accomplish measurements of the reflected radiation either at the zenith solar or at the viewing angle in the range 45–50° for the retrieval of the optical thickness and single scattering albedo. Otherwise, as has been pointed out earlier (Boucher 1998; Kokhanovsky 1998) it is better to use other zenith angles for estimating the phase function parameter, at which the radiance is susceptible to the phase function.

The viewing direction during observations of the solar radiance reflected from the ground surface is not of great importance for the orthotropic surface, but the directions close to nadir are preferable for maximal excluding an effect of the radiation scattering within the atmospheric layer below the cloud.

6.1.2
Problem Solution in the Case of Solar Radiation Observation Within the Cloud Layer of Large Optical Thickness

In the general case, the stratus cloud is known to be vertically inhomogeneous. The determination of the vertical structure of the stratiform clouds is of great interest from the point of studying its physical characteristics and the processes of its forming. Here we are presenting the approach for the optical parameters retrieval from the data of the airborne radiative observations within the cloud, which could also serve to study the aerosol content and distribution.

While studying the scattered radiation field within clouds, it is necessary to distinguish two essentially different cases: (1) a single optically thick cloud layer with the optical properties, varying with vertical direction; and (2) a cloud system consisting of several cloud layers separated from each other with clear atmosphere.

In the case of vertically inhomogeneous clouds, the calculation uncertainties of applying the formulas for the homogeneous clouds to the inhomogeneous ones turns out to be sufficiently low because the information about the remote points is "forgotten" due to the multiple scattering within the cloud. Therefore, the photon, registered by the instrument, brings information only about the last collision. Thus, while measuring the diffused irradiance, the information collected by the instrument mainly concerns the points, remote from the instrument for the photon free path. In the case of stratus cloud this value is equal to ~ 20–50 m. Soundings within the stratus cloud is usually accomplished at every 100 m, so it is possible to apply the results obtained below for the interpretation of the airborne observations within the stratus cloud (Melnikova and Mikhailov 2001).

It is possible to divide the process of the radiative transfer through the optically thick cloud layer into three stages (Ivanov 1976), namely:

1. the transfer of radiation through the upper boundary sublayer adjusting the cloud top $\tau = 0$ (*pumping*);

2. the transfer through the inner layers (*diffusion*);

3. the transfer through the layer adjusting the cloud bottom $\tau = \tau_0$ (*escaping*).

Within the optically thick cloud, the processes of pumping, diffusion, and escaping could be assumed independently. Consider consequently these stages.

Suppose the optical thickness of the cloud is equal to $\tau_0 = \Sigma(\tau_i - \tau_{i-1})$, where τ_i is the optical deepness of the i-th observational level. The values of $\varepsilon_i = (\tau_i - \tau_{i-1})/(z_i - z_{i-1})$, $\varepsilon_i = \kappa_i + \alpha_i$ are the volume coefficients of extinction, absorption, and scattering, where z_i is the geometrical deepness of the i-th level. The value of $\omega_{0,i} = \alpha_i/\varepsilon_i$ is the single scattering albedo in the i-th sublayer (situated between $i - 1^{\text{st}}$ and i-th levels). In the short-wave spectral range one can assume $1 - \omega_{0,i} \ll 1$. The phase function is characterized with mean cosine of scattering angle γ: $g_i = \overline{\cos\gamma}$. Let the number of the analyzed sublayers be equal to N.

We should mention the main points of deriving the formulas defining parameters s^2 and τ' from *the irradiances observations* within the cloud. The cloud sublayer adjusting the cloud top is described by the first group of the formulas, the inner sublayers are described by the second group and adjusting the cloud bottom sublayer is described by the third group. Consider (2.26) together with (2.42) and expansions (2.29) and (2.30) to infer the corresponding formulas. The equations linear in value s^2 can be obtained after the algebraic manipulation while keeping the terms proportional to s^2. The results are considered in detail elsewhere (Melnikova 1998; Melnikova and Fedorova 1996; Melnikova and Mikhailov 2001). The formulas for the optical thickness of the sublayer $(\tau_i - \tau_{i-1})$ between measurement levels $i - 1$ and i are deduced from the combination of (2.26) and (2.42). The following expressions are obtained:

- for the cloud top sublayer (level numbers are 0, 1)

$$s_1^2 = \frac{F(0)^2 - F(\tau_1)^2}{16K_0(\mu_0)^2 - 4(F_1^{\downarrow} + F_1^{\uparrow})^2 + F(\tau_1)^2(2n_2 - 9q'^2) - 2F(0)a_2(\mu_0)} \tag{6.3}$$

$$3(1 - g)\tau_1 = \frac{1}{2s}\ln\left[\frac{(F(\tau_1) - 4F_1^{\downarrow}s(1 - 2s))(F(0) + s(4K_0(\mu_0) + a_2(\mu_0)s))}{(F(\tau_1) + 4F_1^{\uparrow}s(1 - 2s))(F(0) - s(4K_0(\mu_0) - a_2(\mu_0)s))}\right]$$

- for the internal sublayers (level numbers are $i - 1$, i)

$$s_i^2 = \frac{[F(\tau_{i-1})^2 - F(\tau_i)^2]F^2(\tau_{i-1})}{16[F(\tau_i)^2 F^{\downarrow}(\tau_{i-1})F^{\uparrow}(\tau_{i-1}) - F(\tau_{i-1})^2 F^{\downarrow}(\tau_i)F^{\uparrow}(\tau_i)]} \tag{6.4}$$

$$3(1 - g)(\tau_{i-1} - \tau_i) = \frac{1}{2s}\ln\left[\frac{(F(\tau_{i-1}) + 4F_{i-1}^{\uparrow}s(1 - 2s))(F(\tau_i) - 4F_i^{\downarrow}s(1 - 2s))}{(F(\tau_{i-1}) - 4F_{i-1}^{\downarrow}s(1 - 2s))(F(\tau_i) + 4F_i^{\uparrow}s(1 - 2s))}\right]$$

- for the cloud bottom sublayer (level numbers are $N - 1$, N)

$$s_N^2 = \frac{F(\tau_{N-1})^2 - F(\tau_N)^2}{4(F_{N-1}^{\downarrow} + F_{N-1}^{\uparrow})^2 - 16F_N^{\uparrow 2} - F(\tau_{N-1})^2(2n_2 - 9q'^2) - 24q'F_N^{\uparrow}F(\tau_N)}$$

$$\tau_N' - \tau_{N-1}' = 3(1 - g)(\tau_N - \tau_{N-1}) = \tag{6.5}$$

$$= \frac{1}{2s}\ln\left[1\frac{(F(\tau_{N-1}) + 4F_{N-1}^{\uparrow}s(1 - 2s))(F(\tau_0) - 4F_N^{\uparrow}s(1 + 3q's))}{(F(\tau_{N-1}) - 4F_{N-1}^{\downarrow}s(1 - 2s))(F(\tau_0) + 4F_N^{\downarrow}s(1 - 3q's))}\right]$$

The expressions in the numerators of the first relations in (6.3)–(6.5) are the differences of squares of the net fluxes, measured at corresponding altitude levels. The following nominations are used for brevity: $F_0^{\downarrow\uparrow} = F^{\downarrow\uparrow}(0, \mu_0)$, $F_i^{\downarrow\uparrow} = F^{\downarrow\uparrow}(\tau_i, \mu_0)$. It is important that the dependence upon ground albedo appears in the case of the bottom sublayer only (through value $F_N^{\uparrow}$) and the dependence upon incident solar angle (through functions $K_0(\mu_0)$ and $a_2(\mu_0)$) appears at the upper sublayer only. Meanwhile, the results for the internal sublayers are independent of these parameters.

Expressions (2.24) for the reflection and transmission functions together with (2.40) and expansion (2.41) serve the base for inferring the inverse formulas allowing the retrieval of values s^2 and τ' from radiance observations within clouds. The formulas are divided into three groups in the case of observations of the radiance analogously to the case of the irradiance dependence upon sublayer location.

The following modification of (2.40) is also used similar to (2.44):

$$J(\mu) = I(\mu) - I(-\mu) = \frac{1}{2}mK_0(\mu_0)\frac{\left[1 + \bar{l}e^{-2k(\tau_0-\tau)}\right]}{1 - \bar{l}\bar{l}e^{-2k\tau_0}},$$

$$\frac{I(-\mu)}{I(\mu)} = \frac{b^{\infty}(\mu) - \bar{l}e^{-2k(\tau_0-\tau)}}{1 - b^{\infty}(\mu)\bar{l}e^{-2k(\tau_0-\tau)}}.$$

(6.6)

where the nomination is specified analogously to value b^{∞}

– for the cloud top sublayer (level numbers are 0, 1)

$$s^2 = \frac{(\varrho_0 - \varrho)^2 - J(\tau_1)^2}{16K_0(\mu_0)^2K_0(\mu)^2 + J(\tau_1)^2[2Q_2w(\mu) - 9q'^2] - 9\mu^2[I(\mu) + I(-\mu)]^2 - \Xi}$$

where specified $\Xi = (\varrho_0 - \varrho)\dfrac{a_2(\mu)a_2(\mu_0)}{6q'}$,

(6.7)

$$\tau' = \frac{1}{2s}\ln\left[l\frac{[J(\tau_1) - 6\mu I(\mu)s(1 - 3\mu s)]}{[J(\tau_1) + 6\mu I(-\mu)s(1 - 3\mu s)]}\right]$$

$$\times\left[\frac{(\varrho^{\infty} - \varrho) + 4K_0(\mu_0)K_0(\mu)s + \frac{a_2(\mu_0)a_2(\mu)}{12q'}}{(\varrho^{\infty} - \varrho) - 4K_0(\mu_0)K_0(\mu)s + \frac{a_2(\mu_0)a_2(\mu)}{12q'}}\right].$$

– for the internal sublayers (level numbers are $i - 1, i$)

$$s^2 = \frac{[J(\tau_{i-1})^2 - J(\tau_i)^2]J(\tau_{i-1})^2}{36\mu^2[J(\tau_i)^2I_{i-1}(\mu)I_{i-1}(-\mu) - J(\tau_{i-1})^2I_i(\mu)I_i(-\mu)]},$$

(6.8)

$$\tau_i' - \tau_{i-1}' = \frac{1}{2s}\ln\left[\frac{[J(\tau_{i-1}) - 6I_{i-1}(\mu)\mu s(1 - 3\mu s)][J(\tau_i) + 6I_i(-\mu)\mu s(1 - 3\mu s)]}{[J(\tau_{i-1}) + 6I_{i-1}(-\mu)\mu s(1 - 3\mu s)][J(\tau_i) - 6I_i(\mu)\mu s(1 - 3\mu s)]}\right],$$

– for the cloud bottom sublayer (level numbers are $N - 1, N$)

$$s^2 = \frac{\bar{K}_0(\mu)^2 J(\tau_{N-1})^2 - \sigma^2}{\bar{K}_0(\mu)^2 \{9\mu^2 [I_{N-1}(\mu) + I_{N-1}(-\mu)]^2 - J^2(\tau_{N-1})(2n_2 - 9q'^2) - W\}} \; ,$$

where specified

$$W = 2\frac{K_0(\mu)}{\bar{K}_0(\mu)} J^2(\tau_{N-1}) \left(n_2(w(\mu) - 1) + \frac{3Ah(\mu)}{1 - A} \right) - 24q'^2 \frac{A(\bar{q}' + q')}{(1 - A)} \; ,$$

$$\tau'_N - \tau'_{Nn-1} = \frac{1}{2s} \ln \left[l \frac{J(\tau_{N-1}) + 6I_{N-1}(-\mu)\mu s(1 - 3\mu s)}{J(\tau_{N-1}) - 6I_{N-1}(\mu)\mu s(1 - 3\mu s)} \right]$$

$$\times \left[\frac{1 - A - 4As - 12Aq's^2}{1 - A + 4As - 12Aq's^2} \right] \; ,$$

(6.9)

function $w(\mu)$ has been specified in Sect. 2.2 with (2.34).

Following notation are introduced in (6.9) for brevity:

$$h(\mu) = \frac{3.8\mu - 2.7}{1 + g} \quad \text{and} \quad \bar{q}' = q' + \frac{4A}{3(1 - A)} \; ,$$

(6.10)

It should be noted that the formulas for value s^2 are characterized with the same properties, as the formulas for the irradiances are.

The expressions proposed here, constitute the set of formulas for solving the inverse problem in the case of radiation observations within the cloud layer. They allow more correct data interpretation than has been done in the study by King (1987), because these expressions are derived while taking into account the absorption of solar radiation and provide the data processing independently at every wavelength.

6.1.3
Problem Solution in the Case of Observations of Solar Radiation Reflected or Transmitted by the Cloud Layer

The possibilities of the interpretation of the radiation data from airborne observations have been analyzed hereinbefore. In addition to these data, there is a huge volume of data either from satellite observations or from the remote observations from the ground, which are conducted more regularly than the airborne ones. The satellite observations have been implemented continuously during the last decades while making use of various instruments, and the ground remote observations are cheaper and easier to carry out from the methodological and technical points. Besides, the remote observations provide the averaged characteristics necessary for using in the climatic models. However, not all volume of the data from satellite and remote observations are suitable for applying the approach developed here. The principal restrictions are put on space homogeneity and the temporal stability of cloud fields.

It should be pointed out that the interpretation of the radiation observations based on the monochromatic radiative transfer theory is available with the spectral measurements only. Applying the methodology to the observational data of total radiation needs the special analysis of uncertainties appearing, while integrating the formulas over wavelength. The values and functions in the asymptotic formulas of the radiative transfer theory depend on single scattering albedo and optical thickness, which in their turn are greatly varying with wavelength. Regretfully, this fact is neither mentioned nor analyzed in the many studies dealing with the observational data of total radiation.

The data of both the radiance and irradiance observations could be used for retrieval of the optical parameters. Interpretation of the irradiance data needs no high azimuthal harmonics of reflected radiances and the calculating errors of these harmonics neither included to the result.

The reflected and transmitted solar irradiance for the optically thick and weakly absorbing cloud layer are described by formulas (2.25). Consider these expressions for two values of cosine of the incident solar angle $\mu_{0,1}$, $\mu_{0,2}$ corresponding to the observations accomplished at two moments. The expressions for parameter s^2 and for scaled optical thickness $\tau' = 3\tau_0(1 - g)$ are easy to derive taking the ratios of the reflected (transmitted) irradiances for two different values of the cosine of the incident solar angle as has been shown in Melnikova and Domnin (1997) and Melnikova et al. (1998, 2000). Here they are:

- for the reflected irradiance

$$
s^2 = \frac{\left[\frac{(a(\mu_{0,1})-F_1^\uparrow)K_0(\mu_{0,2})}{(a(\mu_{0,2})-F_2^\uparrow)K_0(\mu_{0,1})} - 1 \right]}{n_2(w(\mu_{0,1}) - w(\mu_{0,2}))},
$$

$$
\tau' = \frac{1}{2s} \ln \left\{ \frac{mn\bar{l}K(\mu_{0,i})}{a(\mu_{0,i}) - F^\uparrow} + \bar{l}\bar{l} \right\},
$$

(6.11)

where function $w(\mu)$ is defined with (2.34) for function $K_2(\mu)$, and subscript i indicates that any of two values $\mu_{0,1}$, $\mu_{0,2}$ could be substituted to the second of (6.11). It is convenient to apply these expressions for the data processing of satellite observations of the reflected solar irradiance.

- and for transmitted irradiance:

$$
s^2 = \frac{\left[\frac{F_1^\downarrow K_0(\mu_{0,2})}{F_2^\downarrow K_0(\mu_{0,1})} - 1 \right]}{n_2(w(\mu_{0,1}) - w(\mu_{0,2}))},
$$

$$
\tau' = s^{-1} \ln \left[\frac{\sqrt{(4F^{\downarrow 2}\bar{l}\bar{l} + m^2\bar{n}^2K(\mu_{0,i})^2)} + m\bar{n}K(\mu_{0,i})}{2F^\downarrow \bar{l}\bar{l}} \right],
$$

(6.12)

where subscript i indicates that any of two values $\mu_{0,1}$, $\mu_{0,2}$ could be substituted to the second of (6.12). The positive value of the square root is chosen, owing to the demand of the logarithm argument positiveness.

Consider the observations of reflected radiance ϱ_1 and ϱ_2 at two viewing angles: $\arccos \mu_1$ and $\arccos \mu_2$. The first of (2.24) gives difference $[\varrho_\infty(\mu, \mu_0) - \varrho]$, where the arguments of measured value ϱ are omitted. The ratio of differences $[\varrho_\infty(\mu_1, \mu_0) - \varrho_1]/[\varrho_\infty(\mu_2, \mu_0) - \varrho_2]$ for different μ_1 and μ_2 provides the following expressions for values s and $\tau' = 3(1-g)\tau_0$ after the algebraic manipulations (Melnikova and Domnin 1997; Melnikova et al. 1998, 2000):

$$s^2 = \frac{[\varrho_0(\varphi, \mu_1\mu_0) - \varrho_1]K_0(\mu_2) - [\varrho_0(\varphi, \mu_2,\mu_0) - \varrho_2]K_0(\mu_1)}{[\varrho_0(\varphi, \mu_2,\mu_0) - \varrho_2]K_0(\mu_1)\left(\frac{K_2(\mu_1)}{K_0(\mu_1)} - \frac{K_2(\mu_2)}{K_0(\mu_2)}\right) - R},$$

where specified

(6.13)

$$R = \frac{0.955a_2(\mu_0)K_0(\mu_1)K_0(\mu_2)}{q'(1 + g)}[\mu_1 - \mu_2],$$

$$\tau' = (2s)^{-1}\ln\left\{\frac{m\bar{l}K(\mu_i)K(\mu_0)}{\varrho_\infty(\varphi, \mu_i, \mu_0) - \varrho_1} + \bar{l}\bar{l}\right\}$$

where φ is the viewing azimuth relative to the Sun's direction. It is possible to use these formulas for processing the multi-directional satellite observational data of the reflected solar radiance.

The couples of different pixels of the satellite image are characterized with different solar and viewing angles. Let the cosines of the zenith solar and viewing angles $\mu_{0,1}$, μ_1 relate to the first pixel and $\mu_{0,2}$, μ_2 relate to the second pixel. It is suitable to apply this approach for the one-directional satellite observations of the reflected solar radiance. Then the following expression of parameter s^2 is derived from the ratio of the radiances:

$$s^2 = \frac{\begin{array}{c}[\varrho_0(\varphi_1, \mu_1, \mu_{0,1}) - \varrho_1]K_0(\mu_2)K_0(\mu_{0,2}) \\ -[\varrho_0(\varphi_2, \mu_2, \mu_{0,2}) - \varrho_2]K_0(\mu_1)K_0(\mu_{0,1})\end{array}}{\begin{array}{c}K_0(\mu_1)K_0(\mu_{0,1}) \\ \times\left[[\varrho(\varphi_2, \mu_2, \mu_{0,2}) - \varrho_2]\left(\frac{K_2(\mu_1)}{K_0(\mu_1)} - \frac{K_2(\mu_2)}{K_0(\mu_{0,2})}\right) + \frac{a_2(\mu_2)a_2(\mu_{0,2})}{12q'}\right] - R_1\end{array}}$$

where specified

$$R_1 = K_0(\mu_2)K_0(\mu_{0,2})$$

(6.14)

$$\times\left[[\varrho(\varphi_1, \mu_1, \mu_{0,1}) - \varrho_1]\left(\frac{K_2(\mu_2)}{K_0(\mu_{0,2})} - \frac{K_2(\mu_1)}{K_0(\mu_1)}\right) + \frac{a_2(\mu_1)a_2(\mu_{0,1})}{12q'}\right]$$

With the very big magnitudes of optical thickness, the atmosphere is considered as a *semi-infinite* one. In this case, difference $[\varrho_\infty(\mu, \mu_0) - \varrho]$ tends to zero and reduce the numerator to zero. Thus, (6.11), (6.13) and (6.14) become inappropriate and another formulas are necessary to use. The closeness of the numerator to zero is defined by the expression

$$\frac{mn\bar{l}K(\mu_0)\exp(-2k\tau)}{1 - \bar{l}l\exp(-2k\tau)} \xrightarrow[\tau\to\infty]{} C\exp(-2k\tau)$$

that is about 0.02 for τ_0 equal to 100. The optical thickness is preliminarily estimated approximately while assuming the conservative scattering as has been proposed for example in the work by King (1987) and Kokhanovsky et al. (2003). Then, if $\tau_0 \geq 100$, the quadratic equations with respect to parameter s^2 are derived using the expression of $a(\mu_0)$ and $\varrho_\infty(\mu, \mu_0)$ (2.30) taken with the items proportional to s^2:

$$a_2(\mu_0)s^2 - 4K_0(\mu_0)s + 1 - F^\uparrow(\mu_0) = 0$$

$$\frac{a_2(\mu_0)a_2(\mu)}{12q'}s^2 - 4K_0(\mu_0)K_0(\mu)s + [\varrho_0(\mu, \mu_0, \varphi) - \varrho] = 0$$

Its solution is trivial:

$$s = \frac{2K_0(\mu_0) - \sqrt{4K_0(\mu_0)^2 - a_2(\mu_0)\left(1 - F^\uparrow(\mu_0)\right)}}{a_2(\mu_0)}. \tag{6.15}$$

And the similar expression for case of the reflected radiance:

$$s = \frac{2K_0(\mu)K_0(\mu_0) - \sqrt{4[K_0(\mu_0)K_0(\mu)]^2 - \frac{a_2(\mu_0)a_2(\mu)}{12q'}[\varrho_0(\mu, \mu_0, \varphi) - \varrho]}}{\frac{a_2(\mu_0)a_2(\mu)}{12q'}}. \tag{6.16}$$

Problem of choosing the sign before the radicals is the consequence of the ambiguity of the inverse problem solution, and it needs the special analysis of the concrete data. It is easy to demonstrate that just minus has to be chosen here. Indeed, in the case of the conservative scattering the equalities $\varrho = \varrho_0(\mu, \mu_0, \varphi)$ and $s^2 = 0$ are satisfied only with minus before the radical.

In the case of using the transmitted radiance, the corresponding equation for the values of parameter s^2 and scaled optical thickness τ' are similar to (6.12):

$$s^2 = \left[\frac{\sigma_1\bar{K}_0(\mu_2)}{\sigma_2\bar{K}_0(\mu_1)} - 1\right]\frac{1}{\frac{\bar{K}_2(\mu_1)}{\bar{K}_0(\mu_1)} - \frac{\bar{K}_2(\mu_2)}{\bar{K}_0(\mu_2)}}, \tag{6.17}$$

$$\tau' = s^{-1}\ln\left[\frac{\sqrt{4\sigma(\tau, \mu_{1,2}, \mu_0)^2\bar{l}l + m^2\bar{K}(\mu_{1,2})^2K(\mu_0)^2} + m\bar{K}(\mu_{1,2})K(\mu_0)}{2\sigma(\tau, \mu_{1,2}, \mu_0)\bar{l}l}\right],$$

where functions $\bar{K}_0(\mu)$ and $\bar{K}_2(\mu)$ are defined with formulas (2.35). The positive value of the square root is chosen, owing to the demand of the logarithm argument positiveness.

Any of the values of σ_1 or σ_2 (ϱ_1 or ϱ_2) corresponding to cosines of the viewing angles μ_1 or μ_2 could be substituted to the expressions of the scaled optical thickness. However, for better accuracy we recommend the use of the observations for all available viewing angles and then to average the retrieved values. We should mention that if the data of radiation measured in arbitrary units is enough for the parameter s^2 retrieval it will be necessary to use these data in relative units of the incident solar flux at the top of the atmosphere for the scaled optical thickness retrieval.

It is necessary to point out that the rigorous demand of the cloud field stability is suggested in the case of the approach applied to the transmitted irradiance observations because this approach needs carrying out the measurements at several time moments. Using different pixels of the satellite images [as per (6.14)] needs the horizontal homogeneity of the cloud field, which is checked out at the initial stage of the approximate retrieval of the optical thickness with assumption of the conservative scattering. The likewise demand is advanced, while using the transmitted radiance at different viewing angles, where the verification of the horizontal homogeneity is provided with the observations at several azimuth angles.

6.1.4
Inverse Problem Solution in the Case of the Cloud Layer of Arbitrary Optical Thickness

The case of the cloudiness with arbitrary optical thickness (not very thick clouds) is described by the formulas derived in the study by Dlugach and Yanovitskij (1974) and cited in Sect. 2 [(2.50)]. Applying the above-mentioned transformations to (2.50), we deduce the inverse formulas of the optical thickness and parameter s^2. The following is obtained for the nonreflecting surface:

$$s^2 = \frac{(1 - F^\uparrow)^2 - F^{\downarrow 2}}{16[u^2 - v^2]}, \tag{6.18}$$

$$3(1-g)\tau_0 = s^{-1} \ln \frac{tu + v \pm \sqrt{(u^2 - v^2)(t^2 - 1)}}{u + tv}, \quad \text{where} \quad t = \frac{1 - F^\uparrow}{F^\downarrow}.$$

The expression in the numerator of the first formula is the difference of squares of the net fluxes at the top and bottom of the cloud layer in units of the solar incident flux at the top, and value t is the ratio of the same net fluxes. The account of the surface reflection with albedo A transforms the functions and values in (6.18) as follows:

$$\bar{u} = u - A\bar{F}^\downarrow(p - 1), \quad \bar{v} = v + A\bar{F}^\downarrow p,$$

$F^\downarrow$ is changed to $(1 - A)\bar{F}^\downarrow$ and t is changed to $\bar{t} = \frac{1 - \bar{F}^\uparrow}{(1 - A)\bar{F}^\downarrow}.$ \qquad (6.19)

The obtained expressions would be suitable for the optical parameters retrieval but there is one obstacle complicating the solution. Namely, functions $u(\mu_0, \tau_0)$ and $v(\mu_0, \tau_0)$ depend not only on the cosine of the solar zenith angle μ_0 but also on optical thickness τ_0, therefore (6.18) is inconvenient in this case. We propose two ways for getting round this difficulty:

1. The problem is solved with successive approximation. To begin with, the optical thickness is estimated from other approaches (e.g. with the assumption of the conservative scattering) then the values of functions $u(\mu_0, \tau_0)$ and $v(\mu_0, \tau_0)$ are taken from the look-up tables. After that parameter s^2 is calculated and τ_0 is defined precisely using the observational data of semispherical irradiances $F^{\downarrow}, F^{\uparrow}$ at the cloud top and bottom. The process is repeated, and it is broken after the preliminary fixed difference between the values of the desired parameters obtained at the neighbor steps is reached.

2. Otherwise the analytical approximation of functions $u(\mu_0, \tau_0)$ and $v(\mu_0, \tau_0)$ together with the approximation of value p included in (6.22) should be derived. Thus, it is necessary to deduce the formulas similar to (6.18).

6.1.5
Inverse Problem Solution for the Case of Multilayer Cloudiness

The cloudy system consisting of the separate cloud layers has been discussed in Sect. 2.3, and the model of multilayer cloudiness together with the set of the formulas solving the direct problem (2.54), (2.57) for irradiances and (2.55) for radiances has also been presented there. The inversion of these formulas for the optical parameters retrieval is analogous to the above-described procedures. The expressions for the upper cloud layer ($i = 1$) is similar to those for the one-layer cloud with surface albedo $A = A_1$. In formulas for all below layers ($i > 1$), escape function $K_{0,i}(\mu_0)$ is substituted with $F^{\downarrow}(\tau_{i-1})$ and second coefficient of the plane albedo $a_2(\mu_0)$ is substituted with value $12q'$ (Melnikova and Zhanabaeva 1996a). The derivation of the expressions using the observational data of the irradiance has been presented in Melnikova and Fedorova (1996) and Melnikova and Zhanabaeva 1996a,b), which yields the following for parameter s^2:

$$s_1^2 = \frac{F(0)^2 - F(\tau_1)^2}{16[K_0(\mu_0)^2 - F^{\uparrow}(\tau_1)^2] - 2a_2(\mu_0)F(0) - 24q'F^{\uparrow}(\tau_1)F(\tau_1)}, \quad \text{for} \quad i = 1,$$

$$s_i^2 = \frac{F(\tau_{i-1})^2 - F(\tau_i)^2}{\begin{array}{c} 16[F^{\downarrow}(\tau_{i-1})^2 - F^{\uparrow}(\tau_i)^2] + 24q'[F^{\downarrow}(\tau_{i-1})^2 - F^{\downarrow}(\tau_i)^2] \\ \times [F^{\downarrow}(\tau_{i-1})F^{\uparrow}(\tau_{i-1}) - F^{\downarrow}(\tau_i)F^{\uparrow}(\tau_i)] \end{array}}, \quad \text{for} \quad i > 1,$$

$$(6.20)$$

where $F(0) = 1 - F^{\uparrow}(0)$ and $F(\tau_i) = F^{\downarrow}(\tau_{i-1}) - F^{\uparrow}(\tau_i)$ are the net fluxes at the top of the whole cloud system and at the layer boundaries correspondingly.

The expressions for $\tau_i' = 3\tau_i(1 - g_i)$ look like

$$\tau_1' = \frac{1}{2s_1} \ln \left[l_1^2 \left(1 + \frac{2K_0(\mu_0)s_1(4 - 9s_1^2)}{a(\mu_0) - F^\uparrow(0)} \right) \left(1 - \frac{8A_1s_1}{1 - A_1 a^\infty} \right) \right] , \quad i = 1 ,$$

$$\tau_i' = \frac{1}{2s_i} \ln \left[l_i^2 \left(1 + \frac{2s_i(4 - 9s_i^2)}{a_i^\infty - A_{i-1}} \right) \left(1 - \frac{8A_i s_i}{1 - A_i a_i^\infty} \right) \right] ,$$

$$(6.21)$$

where $a(\mu_0)$ and a^∞ are the plane and spherical albedo of the upper layer and a_i^∞ is the spherical albedo of the i-th layer.

For the data of the radiance observations the expressions for parameter s^2 are the following:

- for the upper layer ($i = 1$)

$$s^2 = \frac{\bar{K}_0(\mu)^2(\varrho_0 - \varrho_1)^2 - \bar{K}_0(\mu)^2 \sigma_1^2}{16K_0(\mu)^2 \left[\bar{K}_0(\mu)^2 K_0(\mu_0)^2 - \sigma_1^2 \left(\frac{A_1}{1-A_1} \right)^2 \right] - J} , \tag{6.22}$$

where J is specified as following

$$J = \frac{2A_1}{1 - A_1} [a_2(\mu) + n_2(1 - w(\mu))]\bar{K}_0(\mu)(\varrho_0 - \varrho_1)^2$$

$$+ \frac{a_2(\mu)a_2(\mu_0)\bar{K}_0(\mu)^2(\varrho_0 - \varrho_1)}{6q'}$$

$$- 24q' \frac{A_1}{1 - A_1} K_0(\mu) \left[\bar{K}_0(\mu)(\varrho_0 - \varrho_1)^2 - \frac{A_1}{1 - A_1} K_0(\mu)\sigma_1^2 \right]$$

- for the layer with number $i > 1$

$$s_i^2 = \frac{\bar{K}_0(\mu)^2(\sigma_{i-1} - \varrho_i)^2 - \bar{K}_0(\mu)^2 \sigma_i^2}{16K_0(\mu)^2 \left[\bar{K}_0(\mu)^2 \sigma_{i-1}^2 - \sigma_i^2 \left(\frac{A_i}{1-A_i} \right)^2 \right] - J} ,$$

$$J = \frac{2A_i}{1 - A_i} [a_i(\mu) + n_2(1 - w(\mu))]\bar{K}_0(\mu)(\sigma_{i-1} - \varrho_i)^2 \tag{6.23}$$

$$+ 2a_2(\mu)\bar{K}_0(\mu)^2(\sigma_i - \varrho_i)$$

$$- 24q' \frac{A_i}{1 - A_i} K_0(\mu) \left[\bar{K}_0(\mu)(\sigma_i - \varrho_i)^2 - \frac{A_i}{1 - A_i} K_0(\mu)\sigma_i^2 \right] .$$

Functions $a_2(\mu)$, $K_0(\mu)$ and $w(\mu)$ and value n_2 are calculated for phase function parameter g_i corresponding to the properties of the i-th layer. The subscripts are omitted in the formula for brevity.

Remember here the above conclusion concerning the definition of albedo A_i. The ratio of the radiances observed at viewing angles $\vartheta_{1,2} = \arccos(\pm 0.67)$ at the boundaries between layers $i - 1$ and i defines the albedo corresponding to the boundary of the i-th layer: $\varrho_i(- 0.67)/\sigma_{i-1}$ (0.67).

Scaled optical thickness of separate layers $\tau'_i = 3(1 - g_i)\tau_i$ is described with the following formulas:

– for the upper layer: $i = 1$

$$\tau'_1 = \frac{1}{2s_1} \ln \left\{ l_1^2 \left(1 + \frac{2K_0(\mu)K_0(\mu_0)s_1(4 - 9s_1^2)}{(\varrho_\infty - \varrho_1)} \right) \left(1 - \frac{8A_1 s_1}{1 - A_1 a_1^\infty} \right) \right\},$$

(6.24)

– for the layer with number $i > 1$

$$\tau'_i = \frac{1}{2s_i} \ln \left\{ l_i^2 \left(1 + \frac{2K_0(\mu)\sigma_{i-1}s_i(4 - 9s_i^2)}{(a_i(\mu)\sigma_{i-1} - \varrho_i)} \right) \left(1 - \frac{8A_i s_i}{1 - A_i a_i^\infty} \right) \right\}.$$

(6.25)

The obtained expressions could be applied for the retrieval of the optical parameters of the cloud layer from the observations of solar radiation at the layer boundaries of the multilayer cloud system.

If the layers are not optically thick, it is possible to use the corresponding formulas:

– for the upper layer: $i = 1$

$$s_1^2 = \frac{(1 - \bar{F}_1^\uparrow)^2 - (1 - A_1)^2 \bar{F}_1^{\downarrow 2}}{16[\bar{u}_1^2 - \bar{v}_1^2]},$$

(6.26)

$$3(1 - g_1)\tau_1 = s_1^{-1} \ln \frac{r_1 \bar{u}_1 + \bar{v}_1 + \sqrt{(\bar{u}_1^2 - \bar{v}_1^2)(\bar{r}_1^2 - 1)}}{\bar{u}_1 + \bar{r}_1 \bar{v}_1},$$

where

$$\bar{r}_1 = \frac{1 - \bar{F}_1^\uparrow}{(1 - A_1)\bar{F}_1^\downarrow}, \quad \bar{u}_1 = u_1 - A_1 \bar{F}_1^\downarrow(p_1 - 1) \quad \text{and} \quad \bar{v}_1 = v_1 + A_1 \bar{F}_1^\downarrow p_1.$$

– for the layer with number $i > 1$

$$s_i^2 = \frac{(1 - \bar{F}_i^\uparrow)^2 - (1 - A_i)^2 \bar{F}_i^{\downarrow 2}}{16\bar{F}_{i-1}^{\downarrow 2}[\bar{p}_i^2 - \bar{q}_i^2]},$$

(6.27)

$$3(1 - g_i)\tau_i = s_i^{-1} \ln \frac{\bar{r}_i \bar{p}_i + \bar{q}_i + \sqrt{(\bar{p}_i^2 - \bar{q}_i^2)(\bar{r}_i^2 - 1)}}{\bar{p}_i + \bar{r}_i \bar{q}_i},$$

where

$$\bar{r}_i = \frac{1 - \bar{F}_i^\uparrow}{(1 - A_i)\bar{F}_i^\downarrow}, \quad \bar{p}_i = p_i - A_i \bar{F}_i^\downarrow q_i \quad \text{and} \quad \bar{q}_i = q_i + A_i \bar{F}_i^\downarrow p_i.$$

The latter group of formulas presupposed the same difficulties as (6.18) does, because functions $u(\mu, \tau_i)$, $v(\mu, \tau_i)$, $p(\tau_i)$ and $q(\tau_i)$ depend on optical thickness τ_i.

6.2
Some Possibilities of Estimating of Cloud Parameters

6.2.1
The Case of Conservative Scattering

Sometimes there is no true absorption of solar radiation by clouds at separate wavelengths, so the case of conservative scattering occurs. The single scattering albedo is equal to unity: $\omega_0 = 1$. Equations (2.45)–(2.49) describing the radiative characteristics are rather simple. The expressions of scaled optical thickness $3(1 - g)\tau_0$ are readily derived using (2.45) for the radiance data:

$$3(1 - g)\tau_0 = \frac{4K_0(\mu_0)K_0(\mu)}{\varrho_0(\mu, \mu_0) - \varrho} - \left(6q' + \frac{4A}{1 - A}\right),$$

$$3(1 - g)\tau_0 = \frac{4K_0(\mu_0)\bar{K}_0(\mu)}{\sigma} - \left(6q' + \frac{4A}{1 - A}\right),$$

(6.28)

and for the irradiance data using (2.46):

$$3(1 - g)\tau_0 = \frac{4K_0(\mu_0)}{1 - F^\uparrow(\tau)} - \left(6q' + \frac{4A}{1 - A}\right),$$

$$3(1 - g)\tau_0 = \frac{4K_0(\mu_0)}{F^\downarrow(\tau)(1 - A)} - \frac{(6q' + 4A)}{1 - A}$$

(6.29)

and for net flux data using (2.47):

$$3(1 - g)\tau_0 = \frac{4K_0(\mu_0)}{F(\tau)} - \left(6q' + \frac{4A}{1 - A}\right).$$

(6.30)

Thus, it is possible to retrieve the optical thickness of the conservative homogeneous layer measuring the data of net flux $F(\tau) = F^\downarrow(\tau) - F^\uparrow(\tau)$ at any level – within the cloud or at its boundaries – as the net flux is constant over altitude. The observation at one viewing direction only is enough for the case of conservative scattering.

It should be noted that the expression for the optical thickness using airborne radiance observations has been derived and applied in two studies (King 1987; King et al. 1990).

Remember that conservative scattering is a priori assumed in many studies concerning the deriving of optical thickness from radiation data (King 1987, 1993; King et al. 1990; Zege and Kokhanovsky 1994; Kokhanovsky et al. 2003). We present the result of analyzing the possible uncertainties of this approximation. The accuracy verification of applying (6.28)–(6.30) shows that they

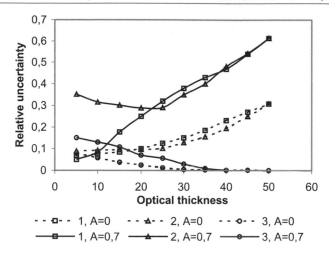

Fig. 6.1. Dependence of relative uncertainty $\Delta\tau_0/\tau_0$ upon optical thickness τ_0 with the value of $\omega_0 = 0.999$. *Solid lines* corresponds to $A = 0.7$, *dashed lines* corresponds to $A = 0.1$. 1 – for reflection irradiance; 2 – for transmitted irradiance; 3 – average values

are available even for $\tau_0 \geq 3$ and the relative error does not exceed 5% for $\omega_0 \geq 0.999$. The error of the retrieval of optical thickness strongly decreases with the increasing of radiation absorption. As is shown in Fig. 6.1 the error analysis using the numerical simulation indicates that the first formula from (6.29) provides the underestimation of value τ_0 for 20–50% while substituting the reflected irradiance at the cloud top, the second one overestimates value τ_0, while substituting the transmitted irradiance at the cloud bottom, and the average from these two values turns out to be rather close to real τ_0 (the relative error is about 10% for $\omega_0 \geq 0.990$).

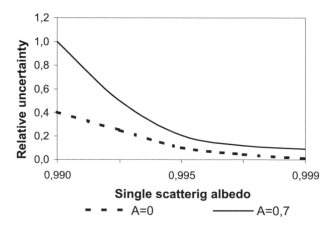

Fig. 6.2. Dependence of relative uncertainty $\Delta\tau_0/\tau_0$ upon ω_0 for mean value of τ_0, $(6 < \tau_0 < 25)$

The dependence of relative error $\Delta\tau_0/\tau_0$ of the average values of the optical thickness obtained from the reflected and transmitted irradiance assuming the conservative scattering versus to the single scattering albedo is demonstrated in Fig. 6.2. It is clear that the ground albedo strongly increases the uncertainty.

The interpretation of the irradiance observations within the conservative cloud layer is available using the formula readily derived from (2.46) and (2.49):

– the upper sublayer adjoins the cloud top

$$(1 - g)\tau_1 = \frac{4K_0(\mu_0) - 2(F_1^\downarrow + F_1^\uparrow)}{3F(\tau_1)} - q' , \tag{6.31}$$

– the sublayer within the cloud

$$(1 - g)(\tau_i - \tau_{i-1}) = \frac{4(F_{i-1}^\downarrow - F_i^\downarrow)}{3F(\tau_i)} , \tag{6.32}$$

– the sublayer adjoins the cloud bottom

$$(1 - g)(\tau_N - \tau_{N-1}) = \frac{2(F_{N-1}^\downarrow + F_{N-1}^\uparrow)}{3F(\tau_{N-1})} - \left(q' + \frac{4A}{3(1 - A)} \right) , \tag{6.33}$$

where N is the number of sublayers and $\tau_N = \tau_0$.

6.2.2
Estimation of Phase Function Parameter g

All the above-presented expressions retrieve the *scaled* optical thickness, so phase function parameter g is needed to obtain the optical thickness. The inferring of phase function parameter g (asymmetry factor) of ice clouds has been made in the 90th by measuring the radiative fluxes, calculating the radiative transfer models, and selecting parameter g for the best coincidence with the observations. However, the methodology of *selecting parameters* is ambiguous as has been shown in Chap. 4 and needs careful error analysis. Probably, it is the reason for inconsistent results. Besides, parameter g dramatically influences the calculation of reflection function $\varrho_\infty(\mu, \mu_0)$, thus it has to be obtained from measurements for the adequate interpretation of the satellite radiation observations.

The attempts to obtain parameter g from observations has been made in two studies (Gerber et al. 2000; Garrett et al. 2001) using the nephelometer measurements, and the values of parameter g is revealed to be equal to 0.85 for stratiform liquid clouds, to 0.81 for convective clouds, and to 0.73 for nonconvective ice clouds. It is seen that the variation of the asymmetry factor is significant and it is desirable to retrieve parameter g and the other optical parameters together during one experiment.

Here we propose a way of estimating phase function parameter g for the optically thick cloud from radiative observations as other optical parameters.

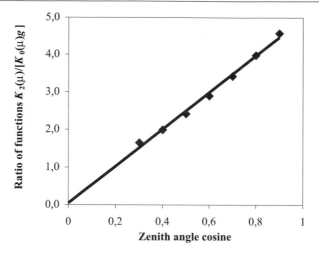

Fig. 6.3. Dependence of the ratio of $K_2(\mu_0)/[K_0(\mu_0)g]$ upon solar zenith angle μ_0; The *points* indicate the calculated values; the *solid line* is the linear approximation

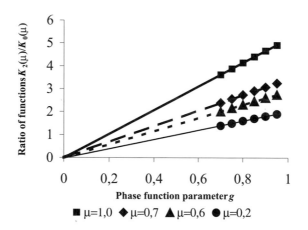

Fig. 6.4. Dependence of the ratio of $K_2(\mu_0)/K_0(\mu_0)$ upon the value of g for different μ_0

The analysis of the two-moment observation of the irradiances (two values of solar zenith angle) indicates that the dependence of difference $K_2(\mu_1)/K_0(\mu_1) - K_2(\mu_2)/K_0(\mu_2)$ upon parameter g is the linear one as is shown in Fig. 6.3 for different zenith angles (see also Fig. 6.4). Then parameter g may be empirically expressed as follows:

$$g = \frac{1}{5.57(\mu_1 - \mu_2)} \left[\frac{K_2(\mu_1)}{K_0(\mu_1)} - \frac{K_2(\mu_2)}{K_0(\mu_2)} \right]. \qquad (6.34)$$

However, in spite of the simplicity of (6.34), there is a problem in applying it. It is impossible to obtain parameter g from the reflected or transmitted radiance because the system of (6.34) with (6.11) or (6.12) for irradiance (6.13) or (6.20) for radiance turns out to be homogeneous. There is a way to obtain parameter s^2 with another approach for example from the airborne observations with (6.1) or (6.2). Then difference $K_2(\mu_1)/K_0(\mu_1) - K_2(\mu_2)/K_0(\mu_2)$ is expressed through parameter s^2 and through the observational data of the transmitted irradiance or radiance using (6.34). Finally, parameter g is estimated using one of the following expressions:

$$g = \frac{\left[\frac{(\varrho_0-\varrho_1)K_0(\mu_2)}{(\varrho_0-\varrho_2)K_0(\mu_1)} - 1\right]}{[5.57(\mu_1 - \mu_2)s^2]} \quad g = \frac{\left[\frac{\upsilon_1\bar{K}_0(\mu_2)}{\sigma_2\bar{K}_0(\mu_1)} - 1\right]}{[5.57(\mu_1 - \mu_2)s^2]} \tag{6.35}$$

Here the expressions are written for the case of the radiance observational data with demand of the horizontal homogeneity of the cloud field. The irradiance data need the temporal stability because of using the two-moment observations, and the formulas of the irradiances are almost likewise, excluding value $F^\downarrow$, which is substituted with value σ, and $(a(\mu_0) - F^\uparrow)$, which is substituted with $(\varrho_0 - \varrho)$. The evident advantages and disadvantages are seen, while using the reflected or transmitted radiance, or the irradiance observations. Thus, value $\varrho^\infty(\mu, \mu_0)$ strongly depends on phase function. The dependence of the plane albedo is weaker so using the reflected irradiance or transmitted radiance is more preferable than using the reflected radiance. Using the transmitted radiance is strongly influenced by the ground albedo, thus the transmitted irradiance provides the better accuracy for the cloud above the snow surface.

Now obtain the cloud optical parameters using the numerical model of the radiative characteristics, calculated with the doubling and adding method. Value s^2 and scaled optical thickness τ' are retrieved from $F^\downarrow$ and $F^\uparrow$ data. Then parameter g is obtained for the pair of radiances with (6.35), and single scattering albedo and optical thickness are calculated. Table 6.2 presents the obtained results.

Table 6.2. Retrieval of the optical parameters of the cloud layer from the model values of the radiative characteristics

Value	Model magnitudes		Retrieved magnitudes	Uncertainty (%)
$F^\downarrow$	0.3051			
$F^\uparrow$	0.6398			
τ_0	25		28.55	14
ω_0	0.99900		0.99919	0.2
g	0.850		0.872	2.5
s^2	0.002222		0.002227	0.2
μ	1.0	0.846		
$I^\downarrow(\mu)$	0.3866	0.3499		
$K_0(\mu)$	1.272	1.153		

Even the small uncertainty of value g causes a significant error of the optical thickness as per expression $\tau_0 = \tau'/[3(1 - g)]$ and is seen from Table 6.2. Model value $g = 0.85$ allows obtaining $\tau_0 = 24.36$ with the uncertainty equal to 2.6%, while retrieved value g leads to the uncertainty equal to 14%. Hence, the necessity of an accurate value of g is evident.

It is important to mention that a similar approach for the phase function parameter has been considered in the book by Yanovitskij (1997) for the case of conservative scattering on the basis of the rigorous theory. The approach for obtaining parameter g has also been proposed in the study by Konovalov (1997) with the approximation of the reflection function.

6.2.3
Parameterization of Cloud Horizontal Inhomogeneity

The simple approximate parameterization of the cloud top heterogeneity was proposed earlier in the study by Melnikova and Minin (1977). The rough cloud top causes an increase of the diffused radiation part in the incident flux. Therefore, this obstacle turns out to be an essential one for calculating the radiative characteristics depending on solar incident angle. Both the escape and reflection functions describe this dependence for the reflected radiance, and the escape function together with the plane albedo of semi-infinite atmosphere describe this dependence for the reflected irradiance. Thus, it was proposed (Melnikova and Minin 1977) to replace all functions depending on incident angle cosine μ_0 with their modifications according to expressions:

$$\varrho_0(\mu, \mu_0) = \varrho_0(\mu, \mu_0)(1 - r) + ra(\mu) \, ,$$

$$K(\mu_0) = K(\mu_0)(1 - r) + rn \, , \tag{6.36}$$

$$a(\mu_0) = a(\mu_0)(1 - r) + ra^\infty \, ,$$

where spherical albedo a^∞, plane albedo $a(\mu_0)$ and value of n are defined with (2.27).

$$a^\infty = 2 \int_0^1 a(\mu_0)\mu_0 d\mu_0 = 4 \int_0^1 \mu_0 d\mu_0 \int_0^1 \varrho^0(\mu, \mu_0)\mu d\mu$$

$$n = 2 \int_0^1 K(\mu_0)\mu_0 d\mu_0 \tag{6.37}$$

and parameter r describes the diffused part of light in the incident flux.

The influence of the overlying atmospheric layers (including high thin clouds), the difference between the reflection functions of the real cloud (described by the Mie phase function) and model cloud (described by the Henyey-Greenstein phase function), and other factors impacting the angular dependence of radiation, are also partly corrected by parameter r.

Let us consider the numerical and analytical results concerning the cloud heterogeneity. There have been many studies in this field lately (Tarabukhina 1987; Loeb and Davis 1997; Galinsky and Ramanathan 1998; Marshak et al. 1998). It was shown that the influence of geometrical variations of the cloud parameters is by an order of magnitude greater than the internal variations (Titov 1998). The analytical solutions (Tarabukhina 1987; Galinsky and Ramanathan 1998) emphasize that the cloud heterogeneity greatly impacts the radiance and irradiance, and this obstacle is actually described with modifying the escape function (or the analogous functions) as per the expression similar to (6.26).

There are different estimations of the role, which this impact plays, while simulating the radiative transfer within clouds. In our case it is expressed with the value of parameter r and the analysis of above-mentioned studies (Tarabukhina 1987; Galinsky and Ramanathan 1998) allows us to let $r \sim$ 0.01–0.1. Most results also show that the minimal disturbance in the radiation field caused by the cloud heterogeneity is at the solar angle equal to 48–49°. As has been mentioned above, all functions depending on incident angle are approximately equal to the integrals over this angle. That is why parameter r does not influence the result if the measurement is accomplished at this incident angle.

Parameter r can be estimated from radiance or irradiance measurements in the stable overcast conditions with the following approach. The ground-based and satellite observations indicate that the measured radiance or irradiance dependence upon solar incident angle is weaker than the dependences of the calculated radiance and irradiance upon viewing and incident angles (Loeb and Davis 1997), and it is called *the violation of the directional reciprocity* for the reflected radiation. Both the incident and viewing angle cosine dependences of the radiation escaped from the optically thick layer is described with the escape function $K(\mu_0)$. Thus, the data set measured during several hours could give us the solar incident angle dependence of the escape function. If it differs from the radiance dependence upon viewing angle, it is possible to obtain the value of r as follows:

$$r = \frac{I(\mu_1, \mu_2) - I(\mu_2, \mu_1)}{1 - I(\mu_1, 0.67)} \frac{K_0(\mu_1)}{K_0(\mu_1) - K_0(\mu_2)} . \tag{6.38}$$

In this expression $I(\mu_0, \mu)$ is the observed (reflected or transmitted) radiance. In addition, the assumption of $\varrho^0(\mu, 0.67) = K_0(0.67) = 1$ is used here. The radiation absorption influencing the escape function as per expression $(1-3q's)$ is divided out in the ratio. Certainly, this way needs high stability of clouds that is possible sometimes (but not often) especially in the North Regions. This method seems preferable for ground-based observations.

There is another method for parameter r estimation from the multi-directional radiance measurements (e.g. from the measurements by POLDER instrument). The approximate values of the optical thickness of the cloud layer are obtained for every available viewing direction and for every pixel assuming the conservative scattering at the first stage of data processing and (2.24). Then the average value of the optical thickness is calculated for every pixel. The relative deviations of the optical thickness obtained for every direction from the average one could be taken as a measure of the deviation of the cloud top from the plane. It is necessary to have in mind that parameter r also includes the influence of the radiation scattering by the above atmospheric layers and thin semitransparent above clouds. Then the following is proposed for the evaluation of parameter r:

$$r = \frac{1}{N\bar{\tau}} \sum_{i=1}^{N} |\bar{\tau} - \tau_i| \,, \tag{6.39}$$

where N is the number of viewing directions for every pixel and $\bar{\tau}$ is the average optical thickness over viewing directions. This methodology was applied to POLDER (Polarization and Directionality of the Earth's Reflectance) Level-2 data containing the reflected radiance at 14 directions (Melnikova and Nakajima 2000).

6.3
Analysis of Correctness and Stability of the Inverse Problem Solution

The above-proposed set of formulas is the solution of the inverse problem of atmospheric optics for the accepted cloud model. According to the book by Prasolov (1995) the range of the continuality of the obtained functions is to be analyzed for testing the solution correctness.

In the case of (6.1) the analysis of continuality and positiveness of function $s^2(F^\downarrow, F^\uparrow, \mu_0)$ taking into account evident condition $F(0) \geq F(\tau_0)$ yields the following inequalities:

– For cosine of solar incident angle $\mu_0 > 0.3$

$$1 > 2[F^{\uparrow 2}(0) + F^{\downarrow 2}(\tau_0) - F^{\downarrow}(\tau_0)F^{\uparrow}(\tau_0)] + F(0) \,,$$
$$s[8.0 + 0.2(1 - A)] > 0.54(1 - A)^2 + 0.3(1 - A) \,, \tag{6.40}$$

– For cosine of solar incident angle $\mu_0 > 0.9$

$$1 > 0.7[F^{\uparrow 2}(0) + F^{\downarrow 2}(\tau_0) - 0.7F^{\downarrow}(\tau_0)F^{\uparrow}(\tau_0)] + 1.1F(0) \,,$$
$$s[8.0 - 0.2(1 - A)] > 0.54(1 - A)^2 - 0.2(1 - A) \,. \tag{6.41}$$

The concrete numerical magnitudes of the parameters providing continuality and positiveness of function $s^2(F^\downarrow, F^\uparrow, \mu_0)$ are different for every observed pair of upwelling and downwelling irradiances at the single level and wavelength. Thus, the experimental data have to be tested for satisfying these inequalities before applying (6.1) to the observational results. Corresponding procedures are provided in the algorithms of the observational data processing. The analogous inequality could be easily derived for all cases considered hereinbefore and the corresponding analysis is included to the processing algorithms.

6.3.1
Uncertainties of Derived Formulas

There are four main sources of uncertainties, while using the proposed formulas for the retrieval of the cloud optical parameters:

1. observational uncertainties;

2. a priori specification of parameter g;

3. breakdown of the applicability region of the asymptotic formulas;

4. inhomogeneity of the cloud layer, while the derived expressions are assuming the cloud homogeneity (while consideration of the observations within the cloud layer).

It is easy to deduce the corresponding formulas for relative uncertainties $\Delta s/s$ and $\Delta \tau_0/\tau_0$ caused by observational uncertainty, as we have the analytical expressions for the calculation of the optical parameters using the approach described in Sect. 4.3, namely, if the vector of observations $y = f(x_1, x_2, \ldots, x_n)$, then:

$$\Delta y \leq \left| \frac{\partial f}{\partial x_1} \right| \Delta x_1 + \left| \frac{\partial f}{\partial x_2} \right| \Delta x_2 + \ldots + \left| \frac{\partial f}{\partial x_n} \right| \Delta x_n ,$$

where Δx_i is the mean square deviation caused by the observational uncertainty or interpolation of the functions over look-up tables.

In particular, if irradiances $F^\uparrow$ and $F^\downarrow$ have been measured with uncertainty ΔF and the optical parameters have been calculated with (6.1), the expression of the relative uncertainties are the following (Melnikova 1992; Melnikova and Mikhailov 1994):

$$\frac{\Delta s}{s} \leq \frac{\Delta F}{1 - F^\uparrow - F^\downarrow} + \frac{2\Delta F a_2(\mu_0) + 16 K_0(\mu_0)\Delta K_0 + F(0)\Delta a_2}{16 K_0(\mu_0) - 2F(0)a_2(\mu_0)} , \qquad (6.42)$$

and for relative uncertainty $\Delta \tau_0/\tau_0$:

$$\frac{\Delta \tau_0}{\tau_0} \leq \frac{1}{\tau_0} \left[30\Delta s + \frac{\Delta F}{F(0)^2} \right] + \frac{\Delta g}{1 - g} + \frac{\Delta s}{s} , \qquad (6.43)$$

where value $1 - F^\uparrow - F^\downarrow$ defines the radiative flux divergence in the cloud layer in relative units πS. In the short-wave range it is about 0.05–0.2. Then the first item provides the order of the magnitude of the uncertainty, namely $\Delta s/s \geq 4\%$ for $\Delta F \sim 1$–3 W/m^2.

The uncertainties of functions $\Delta K_0(\mu_0)$ and $\Delta a_2(\mu_0)$ are induced for two reasons: the inaccurate measuring of the incident angle and the income of partly scattered solar radiation to the cloud top. The first reason (measuring of solar incident angle $\arccos \mu_0$) could not give a significant error as the value of μ_0 is defined by the moment and geographical site of the observation and these parameters are known with sufficient accuracy. Concerning the second reason, we present the following consideration. According to the book by Minin (1988), the part of diffused radiation in the cloudless atmosphere depends on solar incident angle and wavelength, and this part is approximately equal to 0.3 of the total flux. Function $K_0(\mu_0)$ transforms to value n and function $a_2(\mu_0)$ transforms to value $12q' = 8.5$ for the fully diffused radiation, that yields $\Delta K_0 \sim 0.03$ and $\Delta a_2 \sim 0.25$, and these values are minimal for $\mu_0 = 0.6-0.7$. This condition should be provided during observations.

Relative uncertainty $\Delta \tau_0/\tau_0$ is defined mainly by uncertainties of the retrieval $\Delta s/s$ and $\Delta g/(1-g)$ as per (6.43), because the first item could be rather small in the case of large cloud optical thickness and could weakly influence the uncertainty. The value of $\Delta g/(1-g)$ is caused by the second uncertainty source and it depends on the consistency of the model value of parameter g to the real cloud property. In accordance with the results of the study by Stephens (1979) where the spectral values of g have been calculated with Mie theory for eight cloud models, assuming $g = 0.85$, it is possible to conclude that the variations of parameter g in the short wavelength region are not exceeding 2%.

Uncertainty $\Delta s/s$ provided by (6.3)–(6.5) yields $\Delta s/s \leq 0.05$ after calculating the corresponding derivatives and substituting $\Delta F \sim 1-3\,\mathrm{W/m^2}$. The relative uncertainty of single scattering albedo ω_0 is derived from expression $1 - \omega_0 = 3s^2(1-g)$:

$$\Delta(1 - \omega_0)/(1 - \omega_0) = 2\Delta s/s + \Delta g/(1 - g) . \tag{6.44}$$

Assuming value $s \leq 0.05$, we have: $\Delta(1 - \omega_0)/(1 - \omega_0) \leq 0.12$.

Relative uncertainty $\Delta \tau_0/\tau_0$ provided by (6.6)–(6.8) is estimated according to the following expression (Melnikova and Mikhailov 2001):

$$\Delta\tau_0/\tau_0 \sim 2\Delta F \bigg/ \left[\Delta s(F^\downarrow - F^\uparrow)(1 - g)\right] + \Delta s/s + \Delta g/(1 - g) . \tag{6.45}$$

The values of the two first items in the sum defined by the observational uncertainty and by the uncertainty of the retrieval of parameter s is about 15%, the third item adds 2%, thus $\Delta\tau_0/\tau_0 \sim 17\%$.

The error analysis in the case of using the reflected or transmitted irradiance with (6.11) and (6.12) shows that the temporal stability of the cloud layer during observations is necessary. As has been demonstrated in Sect. 1.5, the existence of the overcast cloudiness during one hour is rather probable (about 80%). Uncertainties $\Delta s/s$ and $\Delta\tau_0/\tau_0$ are calculated in the case of using the reflected irradiance by the following expressions:

$$
\begin{aligned}
\frac{\Delta s}{s} &= \frac{\Delta K_0(a - F^\uparrow) + K_0(\mu_2)(\Delta a + \Delta F)}{K_0(\mu_2)(a(\mu_1) - F_1^\uparrow) - K_0(\mu_1)(a(\mu_2) - F_2^\uparrow)} \\
&\quad + \frac{\Delta K_0(a - F^\uparrow) + K_0(\Delta a + \Delta F)}{2K_0(\mu_{11})(a(\mu_2) - F_2^\uparrow)} + \frac{2\Delta w(\mu_1) + \Delta n_2}{(w(\mu_1) - w(\mu_2))n_2}
\end{aligned}
\tag{6.46}
$$

$$\frac{\Delta\tau_0}{\tau_0} = \frac{\Delta s}{s} + \left[\frac{\Delta\bar{l}}{\bar{l}} + \frac{mnK(\mu)(\frac{\Delta m}{m} + \frac{\Delta n}{n} + \frac{\Delta K}{K}) + \Delta l(a(\mu) - F^\uparrow) + l(\Delta a - \Delta F^\uparrow)}{mnK(\mu) + l(a(\mu) - F^\uparrow)}\right.$$

$$+ \left.\frac{\Delta a - \Delta F^\uparrow}{a(\mu) - F^\uparrow}\right]\frac{1}{\tau_0}.$$

And in case of using the transmitted irradiance:

$$\frac{\Delta s}{s} = \frac{2\Delta w + \Delta n_2}{(w(\mu_1) - w(\mu_2))Q_2} + \frac{\Delta F K_0 + \Lambda K_0 F^\downarrow}{F_1^\downarrow K_0(\mu_2) - \Gamma_2^\downarrow K_0(\mu_1)} + \frac{\Delta F K_0(\mu_1) + \Delta K_0 F^\downarrow}{2F_2^\downarrow K_0(\mu_1)},$$

$$\tag{6.47}$$

$$\frac{\Delta\tau_0}{\tau_0} = \frac{\Delta s}{s} + \left[\frac{\Delta r}{r} + \frac{\frac{r^2}{\bar{l}\bar{l}}\left(\frac{\Delta l}{l} + \frac{\Delta\bar{l}}{\bar{l}} + \frac{2\Delta r}{r}\right)}{\left(\sqrt{1 + \frac{r^2}{\bar{l}\bar{l}}} + 1\right)2\sqrt{1 + \frac{r^2}{\bar{l}\bar{l}}}}\right]\frac{1}{\tau_0},$$

where

$$\frac{\Delta r}{r} = \frac{\Delta F}{F} + \frac{\Delta m}{m} + \frac{\Delta\bar{n}}{\bar{n}} + \frac{\Delta K}{K} + \frac{\Delta l}{l} + \frac{\Delta\bar{l}}{\bar{l}}.$$

The error analysis as per (6.46)–(6.47) gives $\Delta s/s \sim 8\%$ and $\Delta\tau_0/\tau_0 \sim 10\%$ for reflected irradiance and for transmitted irradiance – 6% and 10% correspondingly, if the observational uncertainty is about 2%. In general, the irradiances data allow obtaining the optical parameters within the cloud more accurately than the radiances do, according to the study by McCormick and Leathers (1996).

6.3.2
The Applicability Region

As has been mentioned in Sect. 2.4, the main lower bound connecting with the diffusion domain is set on the optical thickness. The restriction on the true absorption arises due to expansions over the small parameter for the asymptotic constants. The applicability region of the inverse expressions for values s and τ' have been studied in several studies (Melnikova 1992, 1998; Melnikova et al. 2000) for the wide set of parameters. Calculation of the direct problem has been accomplished with the doubling and adding method, and the obtained radiative characteristics have served as measured values (Demyanikov and Melnikova 1986). The retrieved parameters have been compared with the model parameters of the direct problem for estimating the relative error. About 50 numerical models have been analyzed in total. The values of the relative uncertainties of $1 - \omega_0$ and τ_0 with fixed phase function parameter g are presented in Figs. 6.5 and 6.6 versus the single scattering albedo and optical thickness correspondingly. We should point out that the only uncertainties caused by the break of the applicability region have been studied in the above-mentioned

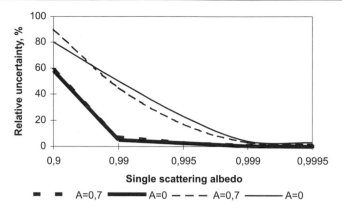

Fig. 6.5. Relative uncertainties $\Delta\tau_0/\tau_0$ (*solid line*) and $g\Delta(1-\omega_0)/(1-\omega_0)$ (*dashed line*) versus value ω_0 with fixed value $\tau_0 = 25$

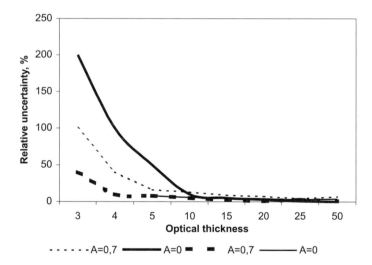

Fig. 6.6. Relative uncertainties $\Delta(1-\omega_0)/(1-\omega_0)$ (*solid line*) and $\Delta\tau_0/\tau_0$ (*dashed line*) versus value τ_0 with fixed value $\omega_0 = 0.999$

analysis, and the values of the radiative characteristics have been assumed as the exact ones.

The radiative characteristics of the inhomogeneous cloud layer have been calculated with the doubling and adding method for five sublayers with optical thickness $\tau_i = 5$ and single scattering albedo $\omega_{0,i}$ together with asymmetry factor g (two latter parameters varying for sublayers). The irradiances have been calculated at the boundaries of sublayers. Then the optical parameters have been retrieved with formulas (6.11) and (6.12). Table 6.3 demonstrates the obtained results and the uncertainties of these results. The results of the

Table 6.3. Influence of the vertical heterogeneity of the layer on the exactness of the optical parameters retrieval

i	G	ω_0	τ'_{model}	s^2_{model}	τ'	s^2	Δs^2 (%)	$\Delta \tau'$ (%)
Inhomogeneous layers								
1	0.85	0.999	2.25	0.00222	2.29	0.00235	4.0	4.8
2	0.85	0.999	2.25	0.02222	2.27	0.02287	2.7	3.6
3	0.85	0.970	2.25	0.06667	2.17	0.07042	5.8	5.2
4	0.85	0.950	2.25	0.10870	2.54	0.11620	7.3	9.4
5	0.85	0.930	2.25	0.15556	2.95	0.15732	8.8	15
Homogeneous layers								
1	0.85	0.999	4.59	0.00222	4.72	0.00228	3.0	3.1
3	0.85	0.970	4.59	0.06667	4.80	0.06349	5.4	5.0

analysis for two cases of the homogeneous layers with corresponding values of τ_0 and ω_0 are presented in the same table.

As is seen from the table, the uncertainty in the case of inhomogeneous layers is the same as in the case of the homogeneous layers and depends only on the applicability region of the used equations (magnitudes of the single scattering albedo and optical thickness). The high values of the uncertainties appear only for the high absorbing sublayers with values of the single scattering albedo $\omega_{0,i} = 0.95$ and 0.93, which provide significant errors both for the irradiance calculation (Fig. 2.2) and for the retrieved parameters (Fig. 6.5).

References

Boucher O (1998) On aerosol direct forcing and the Henyey-Greenstein phase function. J Atmos Sci 55:128–134

Demyanikov AI, Melnikova IN (1986) On the applicability region determination for asymptotic formulas of monochromatic radiation transfer theory. Izv Acad Sci USSR. Atmosphere and Ocean Physics 22:652–655 (Bilingual)

Dlugach JM, Yanovitskij EG (1974) The optical properties of Venus and Jovian planets. II. Methods and results of calculations of the intensity of radiation diffusely reflected from semi-infinite homogeneous atmospheres. Icarus 22:66–81

Duracz T, McCormick NJ (1986) Equation for Estimating the Similarity Parameter from Radiation Measurements within Weakly Absorbing Optically Thick Clouds. J Atmos Sci 43:486–492

Galinsky VL, Ramanathan V (1998) 3D Radiative transfer in weakly inhomogeneous medium. Part I: Diffusive approximation. J Atmos Sci 55:2946–2955

Garret T, Hobbs PV, Gerber H (2001) Shortwave, single scattering properties of arctic ice clouds. J Geoph Res 106(D14):15155–15172

Gerber H, Takano Y, Garret T, Hobbs PV (2000) Nephelometer measurements of the asymmetry parameter, volume extinction coefficient, and backscatter ratio in arctic clouds. J Atmos Sci 57:2320–2344

Germogenova TA, Konovalov NV, Lukashevitch NL, Feigelson EM (1977) Improval of the interpretation of optical observations from the board of automatic interspace station "Venera-8". Cosmic studies, XV: Iss 5. (in Russian)

Ivanov VV (1976) Radiative transfer in multi-layered optically thick atmosphere. Studies of the Astronomical Observatory (Mathematical Sci) Leningrad 32:3–23 (in Russian)

King MD (1983) Number of terms required in the Fourier expansion of the reflection function for optically thick atmospheres. J Quant Spectrosc Radiat Transfer 30:143–161

King MD (1987) Determination of the scaled optical thickness of cloud from reflected solar radiation measurements. J Atmos Sci 44:1734–1751

King MD (1993) Radiative properties of clouds. In: Hobbs V (ed) Aerosol-cloud-climate interactions. Academic Press, New York, pp 123–149

King MD, Radke L, Hobbs PV (1990) Determination of the spectral absorption of solar radiation by marine stratocumulus clouds from airborne measurements within clouds. J Atmos Sci 47:894–907

Kokhanovsky A (1998) Variability of the phase function of atmospheric aerosols at large scattering angles. J Atmos Sci 55:314–320

Kokhanovsky A, Nakajima T, Zege E (1998) Physically based parameterizations of the short-wave radiative characteristics of weakly absorbing optically media: application to liquid-water clouds. Appl Opt 37:335–342

Kokhanovsky AA, Rozanov VV, Zege EP, Bovensmann H, Burrows JP (2003) A semianalytical cloud retrieval algorithm using backscattered radiation in 0.4–2.4 μm spectral region. J Geophys Res 108(D1,4008,doi:10.1029/2001JD001543)

Konovalov NV (1982) Asymptotic properties of the transfer equation solution in plane-parallel layers. PhD Thesis, (KIAM) RAS, Moscow (in Russian)

Konovalov NV (1997) Certain properties of the reflection function of optically dense layers. Preprint of Keldysh Institute for Applied Mathematics (KIAM) RAS, Moscow (in Russian)

Konovalov NV, Lukashevitch NL (1981) The inverse problem of the interpretation of optical observations within the Venus atmosphere from the station "Venera-10". Preprint of Keldysh Institute for Applied Mathematics (KIAM) RAS, Iss 15 Moscow (in Russian)

Loeb NG, Davies R (1997) Angular dependence of observed reflectance: a comparison with plane parallel theory. J Geophys Res 102:6865–6881

Marshak A, Davis A, Wiscomb W, Cahalan R (1998) Radiative effects of sub-mean free path liquid water variability observed in stratiform clouds. J Geoph Res 103(D16):19557–19567

McCormick NJ, Leathers RA (1996) Radiative Transfer in the Near-Asymptotic Regime. In: IRS'96: Current Problems in Atmospheric Radiation:826–829

Melnikova IN (1991) Spectral coefficients of scattering and absorption in strati clouds. Atmospheric Optics 4:25–32 (Bilingual)

Melnikova IN (1992a) Analytical formulas for obtaining optical parameters of cloud layer from measured characteristics of solar radiation field. I. Theory. Atmosphere and Ocean Optics 5:169–177 (Bilingual)

Melnikova IN (1998) Vertical profile of spectral scattering and absorption coefficients of stratus clouds. I. Theory. Atmospheric and Ocean Optics 11:5–11 (Bilingual)

Melnikova IN, Minin IN (1977) To the transfer theory of monochromatic radiation in cloud layers. Izv Acad Sci USSR Atmosphere and Ocean Physics 13:254–263 (Bilingual)

Melnikova IN, Mikhailov VV (1993) Obtaining of optical characteristics of cloud layers. Doklady of Russian Academy of Sciences 328:319–321 (Bilingual)

Melnikova IN, Mikhailov VV (1994) Spectral scattering and absorption coefficients in strati derived from aircraft measurements. J Atmos Sci 51:925–931

Melnikova IN, Fedorova EYu (1996) Vertical profile of optical parameters inside cloud layer. In: Problems of Atmospheric Physics, St.Petersburg State University Press, St.Petersburg 6:261–272 (in Russian)

Melnikova IN, Zshanabaeva SS (1996) Evaluation of uncertainty of approximate methodology of accounting the vertical stratus structure in direct and inverse problems of atmospheric optics. International Aerosol Conference, December, Moscow (in Russian)

Melnikova IN, Zshanabaeva SS (1996) Exactness of method for calculation of solar irradiances in vertical inhomogeneous scattering layers. Intern Symp "Geokosmos", June, St.Petersburg (in Russian)

Melnikova IN, Domnin PI (1997) Determination of optical parameters of homogeneous optically thick cloud layer. Atmosphere and Ocean Optics 10:734–740 (Bilingual)

Melnikova IN, Nakajima T (2000) Single scattering albedo and optical thickness of stratus clouds obtained from "POLDER" measurements of reflected radiation. Earth Observations and Remote Sensing 3:1–16 (Bilingual)

Melnikova IN, Solovjeva SV (2000) Solution of direct and inverse problem in case of cloud layers of arbitrary optical thickness and quasi-conservative scattering. In: Ivlev LS (ed) Natural and anthropogenic aerosols, St. Petersburg, pp 86–90 (in Russian)

Melnikova IN, Mikhailov VV (2001) Vertical profiles of stratus clouds spectral optical parameters derived from airborne radiation measurements. J Geophys Res 106(D21):27465–27471

Melnikova IN, Domnin PI, Radionov VF (1998) Retrieval of optical thickness and single scattering albedo from measurements of reflected or transmitted solar radiation. Izv RAS Atmosphere and Ocean Physics 34:669–676 (Bilingual)

Melnikova IN, Dlugach ZhM, Nakajima T, Kawamoto K (2000) On reflected function calculation simplification in case of cloud layers. Appl Optics 39:541–551

Melnikova IN, Domnin PI, Radionov VF, Mikhailov VV (2000) Optical characteristics of clouds derived from measurements of reflected or transmitted solar radiation. J Atmos Sci 57:623–630

Minin IN (1988) The theory of the radiation transfer in the planets atmospheres. Nauka, Moscow (in Russian)

Minin IN, Tarabukhina IM (1990) To studying of optical properties of the Venus atmosphere (Bilingual). Izv Acad Sci USSR Atmosphere and Ocean Physics 26:837–840

Prasolov AV (1995) Analytical and numerical methods of dynamic processes studying. St. Petersburg State University Press, St. Petersburg (in Russian)

Rozenberg GV, Malkevitch MS, Malkova VS, Syachinov VI (1974) Determination of the optical characteristics of clouds from measurements of reflected solar radiation on the Kosmos-320 satellite. Izv. Acad. Sci. USSR. Atmosphere and Ocean Physics 10:14–24 (in Russian)

Stephens GL (1979) Optical properties of eight water cloud types. Technical Paper of CSIRO, Atmos Phys Division, Aspendale, Australia, No. 36:1–35

Tarabukhina IM (1987) On the reflection and transmission of light by a horizontally inhomogeneous optically thick layer. Izv RAS, Atmosphere and Ocean Physics 23:148–155 (Bilingual)

Titov GA (1998) Radiative horizontal transport and absorption in stratocumulus clouds. J Atmos Sci 55:549–2560

Ustinov EA (1977) The inverse problem of the multiple scattering theory and interpretation of the diffused radiation observations within the Venus atmosphere. Space Studies 15:768–775 (in Russian)

Van de Hulst HC (1980) Multiple Light Scattering. Tables, Formulas and Applications, Vol. 1 and 2. Academic Press, New York

Yanovitskij EG (1972) Spherical albedo of planet atmosphere. Astronomical J 49:844–849 (in Russian)

Yanovitskij EG (1997) Light scattering in inhomogeneous atmospheres. Springer, Berlin Heidelberg New York

Zege EP, Kokhanovsky AA (1994) Analytical solution for optical transfer function of a light scattering medium with large particles. Applied Optics 33:6547–6554

CHAPTER 7

Analysis of Radiative Observations in Cloudy Atmosphere

7.1
Optical Parameters of Stratus Cloudiness Retrieved from Airborne Radiative Experiments

The data of airborne experiments accomplished in 1970–1980-th within the range of research programs CAENEX, GATE, GARP have been presented in Sect. 3.3 with the results of the experiments under the overcast conditions being listed in Table 3.2. These results are used here for inferring spectral dependence of the optical parameters of cloud layers (optical thickness τ_0 and single scattering albedo ω_0), applying the approach described in Chap. 6 (Melnikova 1989, 1992; Melnikova and Mikhailov 1993,1992). The spectral values of phase function parameter g, needed for obtaining optical thickness τ_0, single scattering albedo ω_0, and the volume scattering coefficient are taken from the study by Stephens (1979). The procedure of retrieval is presented in detail elsewhere (Melnikova 1992, 1997; Melnikova and Mikhailov 1994).

7.1.1
Analysis of the Results of Radiation Observations in the Tropics

The observations were carried out as a part of the GATE experiment above the Atlantic Ocean close to the west coast of Africa (experiment No. 1: 12th July 1974, the latitude was 16°N, experiment No. 2: 4th August 1974, the latitude was 17°N). The cloud bottom and top were at altitudes 0.3–3.3 and 0.5–5.0 km for experiments 1 and 2 correspondingly. The uncertainties of the observations were about 5–7% depending on wavelength. The retrieval of the optical parameters was implemented for every wavelength independently using (6.1). The spectral values of optical thickness τ_0 and single scattering co-albedo $(1 - \omega_0)$ are shown in Figs. 7.1a and 7.2a correspondingly and the volume absorption and scattering coefficients are shown in Table A.12 of Appendix A. The oscillations in the curves presenting the optical thickness in Fig. 7.1a are explained with the high observational uncertainties; the smoothed curves are figured there as well. It should be mentioned that the high values of single scattering co-albedo $(1 - \omega_0)$ are explained with the strong flue sand escaping from the Sahara Desert to the observational site.

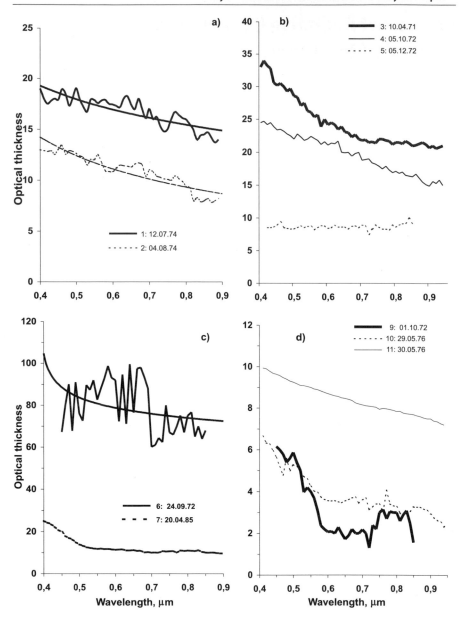

Fig. 7.1a–d. Spectral dependence of optical thickness τ_0 retrieved from the data of airborne radiative observations for different latitudinal zones: **a** 17°N; **b** 45°N; **c** 60°N; **d** 75°N

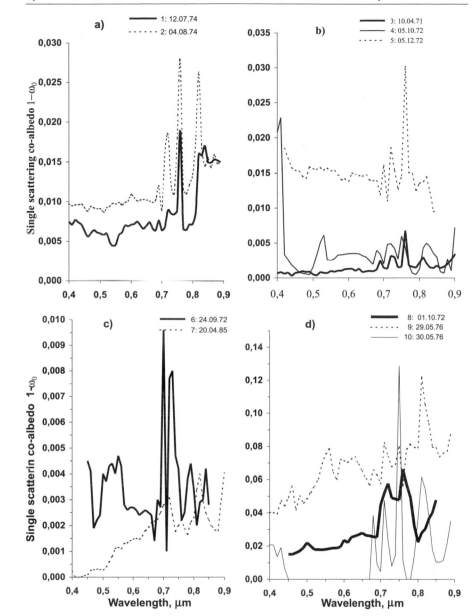

Fig. 7.2a–d. Spectral dependence of single scattering co-albedo $1 - \omega_0$ retrieved from the data of airborne radiative observations for different latitudinal zones: **a** 17°N; **b** 45°N; **c** 60°N; **d** 75°N

7.1.2
Analysis of the Results of Observations in the Middle Latitudes

The observations in the overcast sky were accomplished above the Black Sea (experiment No. 3: 10th April 1971, the latitude was 44°N) and the Azov Sea (experiment No. 4: 5th October 1971, the latitude was 47°N) within the range of the CAENEX program. The altitudes of the cloud bottom and top were 0.4–0.85 km in experiment 3 and 0.3–0.85 km in experiment 4. The retrieved results are presented in Fig. 7.1b and 7.2b and in Table A.12 of Appendix A. The observations above the Azov Sea were conducted under the conditions of strong industrial pollution of the atmosphere that is confirmed with the values of the volume absorption coefficient ($\kappa = 0.12\,\mathrm{km}^{-1}$) that are higher than the values of the observations above the Black Sea ($\kappa = 0.05\,\mathrm{km}^{-1}$).

Besides, the radiative experiment was accomplished above the ground surface in the suburb of Rustavi city, Georgia (experiment No.5: 5th December 1972, latitude 42°N). The cloud bottom and top were at altitudes 3.3–7.2 km correspondingly. The spectral optical parameters are also presented in Figs. 7.1b and 7.2b and in Table A.12 of Appendix A. The studied region had a heavy steel industry with significant air pollution. The cloud was generated just at the observational site and contained a lot of absorbing black carbon particles. The cloud volume scattering coefficient in the observations above the land turned out to be lower than in the case of the observations above the sea.

7.1.3
Analysis of the Results of Observations Above Ladoga Lake

The observations were accomplished at latitude 60°N under overcast conditions (experiment No. 6: 24th September 1972 and experiment No. 7: 20th April 1985). The two-layer cloudiness was observed in experiment 6 with the bottoms and tops at altitudes 0.3–2.5 km and 2.0–3.9 km correspondingly. These results have been obtained by considering the cloud system as a whole. The cloud bottom and top altitudes in experiment 7 were 0.9–1.4 km. The spectral dependence of optical thickness τ_0 and single scattering co-albedo $(1 - \omega_0)$ are presented in Figs. 7.1c and 7.2c. Value $(1 - \omega_0)$ obtained from the data of experiment 7 shows the conservative scattering at wavelength 0.42 μ. The oscillations in the curve illustrating the optical thickness in Fig. 7.1c are explained with the high observational uncertainties, and the smoothed curve is shown in the same figure. The observational accuracy of experiment 7 is better than the accuracy of the other cloud experiments. It appeared just during the data processing: every retrieved point was situated close to the neighbor points in the spectral curve (Fig. 7.1c), and the uncertainty of the processing were increasing (the oscillations of spectral curve of τ_0) only in regions of the molecular absorption bands due to the break of the applicability region. The values of volume scattering and absorption coefficients are presented in Table A.12 of Appendix A.

7.1.4
Analysis of the Results of Observations in the High Latitudes

The observations were accomplished above the Kara Sea at latitude 75°N (experiment No. 8: 1st October 1972, experiment No. 9: 29th May 1976 and experiment No. 10: 30th May 1976). The altitudes of the cloud bottom and top were 0.6–1.1 km in experiment 8, 5.0–8.0 km in experiment 9, 5.0–7.5 km and 8.0–9.0 km for two-layer cloudiness in experiment 10, considered as a whole layer. The spectral optical parameters are shown in Figs. 7.1d and 7.2d. The value of single scattering co-albedo $(1 - \omega_0)$ retrieved from the data of experiment 10 demonstrates the conservative scattering within wavelength interval 0.45–0.67 μm. The high absorption obtained from experiment 8 $(1-\omega_0 \sim 0.06)$ is caused by the source of the high pollution that has been situated at the lee side of the observational site.

7.2
Vertical Profile of Spectral Optical Parameters of Stratus Clouds

The retrieval of the vertical structure of stratus clouds is of interest for investigation of either the physical properties of the cloud or the study of distribution of pollution within clouds (Feigelson 1981; Marchuk et al. 1986; Kondratyev 1991). Hereinbefore the spectral dependence of volume scattering and absorption coefficients has been obtained for the cloud layer, considered as a whole. If the data of the airborne radiative measurements are available at several levels within the cloud, it is possible to derive the vertical structure of the cloud optical parameters. As has been discussed in Chap. 6, while studying the scattered radiation field inside the cloud, it is necessary to distinguish two essentially different cases: (1) a single optically thick cloud layer with the optical properties, which vary in vertical direction; and (2) cloud system consisting of several cloud layers separated by clear atmosphere.

The airborne radiative observations suitable for these two cases have been described in Sect. 3.4 (experiments 6 and 7). During these observations the spectral irradiance was measured at several levels within the cloud and at the cloud boundaries. The observational results provide the input data of the optical parameters retrieval.

The conditions of the experiments were the following:

Experiment 6. The measurements were carried out at five levels of the double-layer cloud: above the upper cloud layer at altitude $z = 4.1$ km, within the upper cloud at altitude $z = 3$ km, between the cloud layers at altitude $z = 1.6$ km, within the lower cloud at altitude $z = 0.6$ km and under the lower cloud at altitude $z = 0.05$ km;

Experiment 7. The measurements were carried out at six levels of the single-layer cloud: above the cloud top, within the cloud and under the cloud bottom at altitudes: 1.4, 1.3, 1.2, 1.1, 0.95 and 0.8 km (Table A.3 of Appendix A and Fig. 3.13).

The spectral range of experiment 6 is 0.45–0.85 μm, the data set contains only 41 wavelength points processed by hand; the spectral range in experiment 7 is 0.35–0.95 μm and the data set contains 180 wavelength points. The ground albedo were obtained from the measurements as $F^\uparrow(\tau_0)/F^\downarrow(\tau_0)$: experiment 6 – $A \sim 0.2$ (water surface) and experiment 7 – $A \sim 0.65$ (snow surface).

Equations (6.20) and (6.21) for the retrieval of values s^2 and τ' in the system of cloud layers have been applied to the data of experiment 6, and (6.3)–(6.5) for the retrieval of values s^2 and τ' in one inhomogeneous layer have been applied to the data of experiment 7. The results (single scattering albedo and optical thickness) are presented in Tables A.13 and A.14 of Appendix A for both experiments. There has been no problem with the interpretation of the data of experiment 6. The magnitudes of the optical thickness of four cloud sublayers have turned out to be high enough and the processing errors are less than the observational uncertainty. As a result, the errors of the retrieved optical parameter in experiment 6 are $\Delta s/s \sim 6\%$ and $\Delta\tau/\tau \sim 10\%$. It is necessary to point out that experiment 6 has been carried out with the earlier version of the spectrometer and experimental uncertainty ($\Delta F^{\uparrow\downarrow}/F^{\uparrow\downarrow} \sim 5$–7%) is higher than the uncertainty of experiment 7 ($\Delta F^{\uparrow\downarrow}/F^{\uparrow\downarrow} \sim 1$–2%), which has been carried out with the improved version of the spectrometer. This difference is expressed with the stronger oscillations on the curves of experiment 6 than they are on the curves of experiment 7. In experiment 7 for the bottom sublayer the negative values of s^2 have been obtained because the absorption was strong and the optical thickness was too small. Thus, the asymptotic formulas are off the applicability range in this case. It corresponds to the solution expressed through a discontinuous function that yields the absence of solution as per to Sect. 6.3.

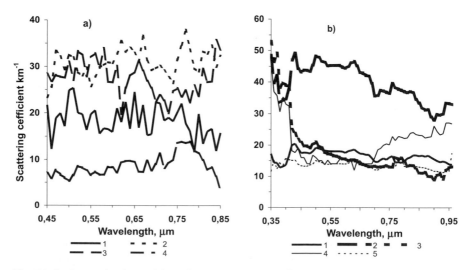

Fig. 7.3a,b. Spectral values of the volume scattering coefficient of the cloud layers between the measurement levels retrieved from experiment 6 –(**a**) and 7 – (**b**); digits in figures indicate layer's number: **a** 1 – layer 4.1–3 km; 2 – 3.0–1.6 km; 3 – 1.6–0.6 km; 4 – 0.6–0.05 km; **b** 1 – layer 1.4–1.3 km, 2 – 1.3–1.2 km, 3 – 1.2–1.1 km, 4 – 1.1–0.9 km, 5 – 0.9–0.8 km

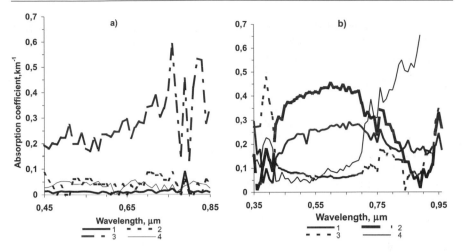

Fig. 7.4a,b. Spectral values of the volume absorption coefficient of the cloud layers between measurement levels retrieved from experiment 6 – (**a**) and 7 – (**b**); Digits at curves indicate the layer's number: **a** 1 – layer 4.1–3.0 km; 2 – 3.0–1.6 km; 3 – 1.6–0.6 km; 4 – 0.6–0.05 km; **b** 1 – layer 1.4–1.3 km, 2 – 1.3–1.2 km, 3 – 1.2–1.1 km, 4 – 1.1–0.9 km, 5 – 0.9–0.8 km

Comparison of the optical thickness, summarized over all layers, with the optical thickness retrieved from the measurements at the top and bottom of the same cloud layer (Kondratyev et al. 1998; Melnikova and Mikhailov 2001) as per Sect. 7.1, indicates that the difference does not exceed the errors due to the processing technique.

The spectral values of the volume scattering and absorption coefficients between the measurement levels have been obtained from the values of s^2 and τ', while taking into account the distance between the levels and the spectral dependence of g parameter according to (Stephens 1979). Figures 7.3a,b and 7.4a,b illustrate the spectral dependence of the retrieved optical parameters and Figs. 7.5a,b and 7.6a,b show their vertical dependence.

7.3
Optical Parameters of Stratus Cloudiness from Data of Ground and Satellite Observations

Here the analytical method of processing the reflected radiance measurements, elaborated in Sect. 6.1, is applied to the multi-angle ground observations of the transmitted radiance and to the satellite observations of the reflected radiance to obtain the cloud optical thickness and single scattering albedo. The observations have been described in Sect. 3.4 in Tables 3.3 and 3.4. The reflected and transmitted radiance in units of the solar incident flux (reflection and transmission functions) was measured in several viewing angles.

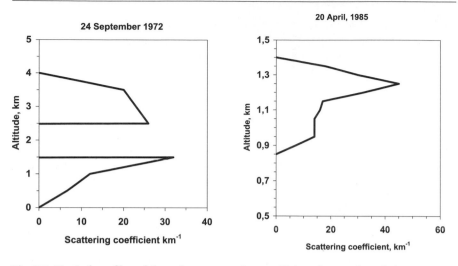

Fig. 7.5. Vertical profiles of the volume scattering coefficient for wavelength $\lambda = 0.55\,\mu m$ retrieved from airborne radiative experiments 6 and 7

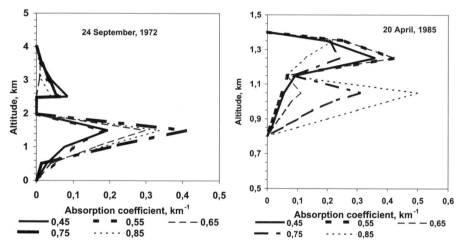

Fig. 7.6. Vertical profiles of the volume absorption coefficient retrieved from airborne radiative experiments 6 and 7 for wavelengths μm indicated in figure

7.3.1
Data Processing of Ground Observations

Two experiments of the ground observations have been processed with (6.17). The first experiment has been performed using spectrometer K-2 (see Sect. 3.2) under overcast conditions at drifting Arctic station SP-22 on the 13th August and 8th October 1979 with the error of the transmitted radiance measurements within 3% (Radionov et al. 1981). There were extended, horizontally homogeneous thick cloudiness during these observations, and the obtained data seemed suitable for applying the analytical method.

The second experiment was accomplished using the spectral instrument, which had spectral resolution 0.002 μm and spectral range 0.35–0.75 μm, under the overcast condition in St. Petersburg's suburb on 12th April 1996 (Melnikova et al. 1997).

In all these cases, the data were obtained for five viewing angles and for five azimuth angles. One set of the measurements took about 10 minutes. The measurements were conducted at noontime, when the solar zenith angle hardly varied during 10 minutes. The transmitted radiance at different azimuth angles with the same viewing angle varied in the range of the measurement error and it was averaged during the data processing.

During the Arctic experiment, the observations of the downwelling and up-welling irradiance were also accomplished and ground albedo A was obtained (Radianov et al. 1981). Different types of snow cover were studied (fresh snow, wet snow and so on), and in all cases, a slight spectral dependence of ground albedo A was observed. On 13th August 1979 the ground surface was covered with wet snow and ground albedo A was about 0.6. On 8th October 1979 there was fresh snow and ground albedo A was about 0.9. Thus, the data of 13th August 1979 are more favorable for the present interpretation because the lower value of ground albedo A decreases the errors of the optical cloud parameters retrieval. During the observation of 12th April 1996 there was heavy snowing, so the ground albedo was assumed equal to 0.9 (for fresh snow) as per (Radionov et al. 1981).

Besides, the observation of direct solar radiation was carried out in clear sky during the Arctic experiment of 1979. It gave the opportunity of calibrating the instrument in units of solar incident flux πS at the top of the atmosphere that is necessary for the retrieval of optical thickness τ_0 according to (6.17). The experiment of 12th April 1996 was accomplished likewise excluding the measurement of direct solar radiation in the clear sky, hence the instrument was not calibrated and optical thickness τ could not been obtained.

Each pair of the transmitted radiance is processed according to (6.20), and then values s^2 and τ' are averaged. The final spectral values of single scattering co-albedo $(1 - \omega_0)$ and optical thickness τ_0 are calculated accounting the spectral dependence of parameter g according to the book by Stephens (1979). The results are presented in Fig. 7.7 and in Table A.15 of Appendix A [the Arctic cases (a) and St. Petersburg's suburb case (b); Melnikova and Domnin 1997; Melnikova et al. 1998, 2000]. Out of the molecular absorption bands spectral values $(1 - \omega_0)$ are about 0.002–0.004, that corresponds to the values obtained earlier from the airborne experiments (Sect. 7.1). The optical thickness presented in Fig. 7.8 is typical for stratus clouds and demonstrates the apparent spectral dependence similar to the one, presented in Sect. 7.1.

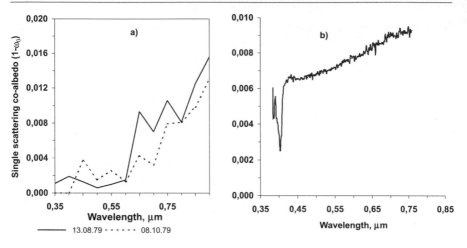

Fig. 7.7a,b. Spectral dependence of single scattering co-albedo $1 - \omega_0$ retrieved from the ground observation data: **a** in Arctic, 1979 and **b** in St. Petersburg suburb (city Petrodvorets), 1996

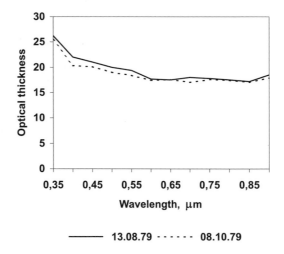

Fig. 7.8. Spectral dependence of optical thickness τ_0 retrieved from the data of the ground observations in Arctic: experiment 11 – 13 August 1979 and experiment 12 – 08 October 1979

7.3.2
Data Processing of Satellite Observations

Optical thickness τ_0 and single scattering co-albedo $1 - \omega_0$ for extended clouds were obtained with inverse asymptotic formulas [(6.13), (6.28)]. The approximate accounting of the horizontal inhomogeneity including the scattering of radiation by the upper atmospheric layers was accomplished with (6.36) and (6.39). Multidirectional reflected radiance measurements with the POLDER instrument were processed for the retrieval of cloud optical parameters. The pixels with the cloud amount exceeding 0.5 were only considered.

The following sequence of the procedures for every pixel is proposed for processing POLDER data:

1. At the first step the angular dependent functions are calculated.

2. The next step includes the calculation of the approximate optical thickness for every viewing direction with the simple formula, assuming the conservative scattering. The obtained values show the degree of the shadowing influence (or the influence of the cloud top deviation from the plane) and give the possibility to evaluate parameter r with (6.39). Besides, they allow choosing the pairs of viewing directions where the optical thickness is approximately equal.

3. The third stage consists of the parameter s^2 retrieval from the radiances at each pair of viewing directions with the equal optical thickness [(6.13)]. If the optical thickness defined at the previous stage without accounting of the absorption is more than 100, parameter s^2 is obtained according to (6.16). Then the averaging over all pairs of the viewing directions is accomplished, and the relative mean square deviation is estimated.

4. At the fourth stage optical thickness τ_0 is calculated for every viewing direction, assuming the true absorption, and the results are averaged.

5. Then, the similar procedure is repeated for every available wavelength.

6. At the sixth stage the results are prepared for mapping (inserting the missed pixels; inserting the values averaged over the neighbor pixels to the missed pixels or to the pixels with only one viewing direction; rejecting the edge pixels). The uncertainties are calculated for every pixel using the formulas similar to (6.46).

7. Finally, the images of the single scattering co-albedo and optical thickness are figured with the GRADS editor. The space distribution of single scattering co-albedo $(1 - \omega_0)$ is shown in Fig. 7.9, optical thickness τ_0 is shown in Fig. 7.10 (Melnikova and Nakajima 2000a,b). The values of $(1 - \omega_0)$ are in the range 0.001–0.010; the optical thickness is about 15–25 and can reach 100 in the Tropics. Black gaps in the images correspond to the pixels with the cloud amount less than 0.5. Four images are presented in Figs. 7.9 and 7.10, the upper picture join three images registered during the successive satellite pass with time interval about one hour (i.e. these images are presenting one cloud field). Figure 7.11 demonstrates the values of (a) – single scattering co-albedo $(1 - \omega_0)$, and (b) – optical thickness τ_0 and shadow parameter r multiplied by 10^2 in three spectral channels versus pixel numbers. The latter turns not to depend on wavelength, and in contrast the spectral dependence of the optical thickness decreases with wavelength for all (!) processed pixels. Please remember that the processing has been accomplished for every wavelength independently. The size of every pixel is about 60 km.

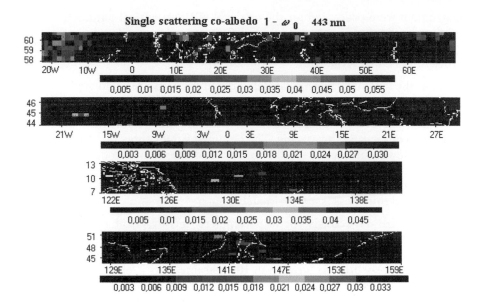

Fig. 7.9. Images of single scattering co-albedo $(1 - \omega_0)$ of the cloud pixels, retrieved from POLDER data

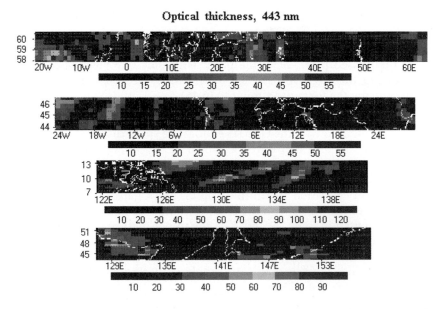

Fig. 7.10. Images of optical thickness τ_0 of the cloud pixels, retrieved from POLDER data

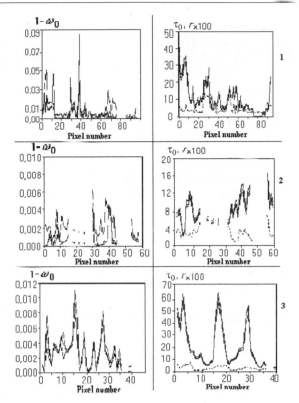

Fig. 7.11a,b. Cloud optical parameters versus pixel numbers: **a** – single scattering co-albedo $(1 - \omega_0)$, and **b** – optical thickness τ_0 (*solid line*) and shadow parameter $r \times 10^2$ (*dashed line*) for three wavelength channels 443 nm – black line; 670 nm – red line; 865 nm – blue line; 1 – latitude 58.75°N and longitude 23°W–75°; 2 – latitude 44.75°N and longitude 24°W–30°E; 3 – latitude 8.75°N and longitude 120°E–140°E

7.4
General Analysis of Retrieved Parameters of Stratus Cloudiness

7.4.1
Single Scattering Albedo and Volume Absorption Coefficient

Molecular absorption bands are apparent in the figures illustrating the spectral dependence of single scattering co-albedo $(1 - \omega_0)$ but they are expressed differently in different cloud layers. The molecular band at wavelength 0.42 μ appears in experiments 1, 2 and 4. It can be identified as an absorption by hematite (see Sect. 3.3, Fig. 3.14 and studies by Ivlev and Andreev 1986 and Sokolic and Toon 1999) contained in flue sand escapes from the Kara-Kum and Sahara deserts. One can see the weak bands of the aerosol absorption at wavelengths around 0.5 and 0.8 μm in the curves obtained from the data of experiments 3 and 4, accomplished above the sea surface. It could be attributed to sea salt (namely to NaCl) content in the atmospheric aerosols according to the study by Ivlev and Andreev (1986).

The atmosphere in the Arctic regions is purer – the conservative scattering becomes apparent within a large range of wavelength (Fig. 7.2d, experiment 11). Spectral values $(1-\omega_0)$ retrieved from airborne experiments 3 and 7 (Fig. 7.2b,c) and from the satellite experiments (certain parts of the curves in Fig. 7.11, 3) demonstrate a monotonic increase with wavelength that can be attributed to organic fuel combustion (Sokolic 1988). The values of single scattering co-albedo $(1-\omega_0)$ obtained from airborne experiments 1, 2 and 5 and most pixels of the satellite images show no spectral dependence, which is typical for the black carbon and dust aerosols.

Consideration of volume absorption coefficient κ of the separate cloud sublayers (Fig. 7.4) indicates strong vertical inhomogeneity. The upper curves in Fig. 7.4b demonstrate significant absorption by two upper cloud sublayers corresponding either to the oxygen and water vapor absorption bands (0.68, 0.72, 0.76 µm) or to the ozone Chappuis molecular absorption band (0.65 µm). Two lower sublayers show the opposite spectral dependence. It could be explained with the higher content of ozone in the upper tropospheric layers compared with the lower ones. The results of experiment 7 show the monotonic increase of the absorption coefficient with wavelength in the bottom layer (1.0–1.1 km). A similar result has been mentioned above for the cloud, considered as a whole layer.

In spite of significant uncertainties of the retrieval of values $(1 - \omega_{0,i})$ and especially τ_i the obtained result demonstrates the rather real magnitudes and spectral dependence coinciding with the results of considering the cloud layer as a whole. Using the spectral dependence of the irradiances promotes diminishing the uncertainties of the retrieval because the results obtained for the neighbor wavelengths do not distinguish strongly from each other. Smoothing over spectral values out of the absorption bands could be rather effective for obtaining the real values of the optical parameters.

Several pixels of the satellite images (in Fig. 7.11, 1) are characterized with magnitude 0.05 for value $(1 - \omega_0)$. It could be concluded that the observational errors increases at the edges of the image, especially for the single pixels with the strong absorption. However, the other parts consist of several pixels with the higher absorption and could correspond to the industrial regions with the increasing content of the soot aerosols. Only some rare pixels above the ocean are characterized with the conservative scattering of radiation.

7.4.2
Optical Thickness τ_0 and Volume Scattering Coefficient α

The values of volume scattering coefficient α vary strongly in different experiments. Spectral dependence $\alpha(\lambda)$ demonstrates the strong vertical inhomogeneity of the cloud, and both the magnitudes and the spectral dependence are different in different cloud sublayers. It reflects the inhomogeneity of the microphysical cloud structure. The volume scattering coefficient obtained for the cloud as a whole coincides with the averaged values obtained for the separate sublayers within the uncertainty range. The scattering coefficient is maximal for the inner sublayers close to the cloud top. The obtained vertical profile of the volume scattering coefficient is similar to the airborne results accomplished in stratus-cumulus cloudiness in the Southern hemisphere (Boers et al. 1996) and to the results of the FIRE experiment in the Arctic (Curry et al. 2000). The same values are cited in the book by Mazin and Khrgian (1989) for stratus clouds. Thus, our results could be assumed to be the quite real ones.

Figure 7.10 illustrates that most pixels are characterized with optical thickness τ_0 about 10–25, while in some regions consisting of several pixels the optical thickness reaches 70–80 and even 100 (in the Tropical latitudes). Space variations of the optical thickness seem rather monotonic in images obtained from the satellite data, and this obstacle points to the low enough uncertainty of either observations or data processing.

The presented results of the retrieval of optical thickness τ_0 and single scattering albedo ω_0 from the airborne, ground, and satellite radiative observations demonstrate the similar values and spectral features in spite of using different observational methods and different formulas. It shows the inverse asymptotic formulas to be quite suitable for obtaining the cloud optical parameters. The elaborated method has more advantages comparing with the other methods (Rosenberg et al. 1974; Asano 1994; Nakajima TY and Nakajima T 1995; Rublev et al. 1997) because it provides obtaining two parameters for every wavelength in the shortwave spectral range and for every pixel of the satellite images independently and with no additional restricting assumptions.

The approximate account of the cloud top inhomogeneity turns out to be rather effective either for inverse or for direct problems. The introduced shadow parameter turns out to take into account the upper atmospheric layers influence together with the uncertainty of the phase function approximation with the Henyey-Greenstein function. It will be promising to analyze the results of similar data processing in the global scale.

It should be mentioned that the more accurate presentation of the phase function would change the numerical magnitudes of the results because it has to retrieve the phase function parameter for substituting its real value instead of the model one to the formulas.

7.5
Influence of Multiple Light Scattering in Clouds on Radiation Absorption

7.5.1
Empirical Formulas for the Estimation of the Volume Scattering and Absorption

The results discussed in the previous section have common features, namely:

1. magnitudes of the single scattering albedo are lower than the values calculated with Mie theory,

2. and the existence of the spectral dependence of the optical thickness contradicted Mie theory results.

The interpretation of the UV radiation observations in the cloudy sky by (Mayer et al. 1998) also demonstrates the strong extinction: the cloud optical thickness in the UV region has been retrieved to be equal to several hundreds.

Mie theory calculations yield volume scattering coefficient α (and optical thickness τ_0) for ensemble of the particles with size $> 5\,\mu m$ independent of wavelength in the shortwave region, and the magnitude of the volume absorption coefficient in the cloud has to be in range 10^{-5}–10^{-8} (single scattering albedo ω_0 is about 0.99999–1.0).

Here we propose a possible explanation of this contradiction. It links with the multiple scattering within clouds. Qualitatively the similar assumption has been proposed in the book by Kondratyev and Binenko (1984), while considering the airborne observational data.

The cloud layer is considered to consist of droplets, sometimes with addition of aerosols within the droplet. The molecular scattering is accounted for with summarizing the scattering coefficients and as the molecular scattering coefficient is much lower (by a factor of 10^3) than the cloud scattering coefficient, its yield turns out to be negligible. It's known that the mean number of the scattering events in the cloud with optical thickness τ_0 is proportional to τ_0^2 owing to the multiple scattering (Minin 1981,1988; Yanovitskij 1997); for reflecting photons it is proportional to τ_0. Thus, the photon path within the optically thick cloud significantly increases compared to the photon path within the clear sky, and the number of collisions with air molecules (more rigorous with fluctuations of the molecular density) increases as well. The radiation absorption removes the part of photons and weakens the increasing effect of the molecular scattering. Since it is necessary to take into account that the cloud layer does not simply superpose to the molecular atmosphere, but it increases the molecular scattering. We should mention that the increasing of the molecular absorption within oxygen absorption band $\lambda = 0.76\,\mu m$ due to the increasing of the photon path within the cloud has been considered in various studies (Dianov-Klokov et al. 1973; Marshak et al. 1995; Kurosu et al. 1997; Pfeilsticker et al. 1997; Wagner et al. 1998; Pfeilsticker 1999). The same reasons are also valid for radiation scattering and absorption by the aerosol particles between droplets.

It is clear that the multiple scattering theory and the radiative transfer equation takes into account all processes of scattering and absorption, but it is right only, if they are accurately put in the model of scattering and absorbing medium. Usually the averaging values of scattering and absorption coefficients over the elementary volume are substituted to the transfer equation and then the solving is accomplished with one of the radiative transfer methods. However, from the physical point it is incorrect to average the initial parameters over the elementary volume before solving. The incorrectness is intensified with the essentially different scales of the elementary volumes for different particles (molecules, aerosols and droplets), whose sizes distinguish by an order of magnitude and much more (look Sect. 1.2) and the transfer equation is derived in a phenomenological way for this incorrect elementary volume. Strictly speaking, the equation of the radiative transfer for the complex multi-component medium is to be inferred from Maxwell equations accounting all its components. However, we don't aim here to consider the mathematical aspect of the problem, thus we propose the empirical approach, presented in several studies (Melnikova 1989, 1997; Kondratyev et al. 1997; Melnikova and Mikhailov 2000).

Usually the scattering or absorption coefficients of the whole medium are presented as a sum of the corresponding coefficients of separate components. Specify the optical parameters relating to the molecular component with M, relating to the aerosol component with A, and relating to the droplets with D. Then the usual notation looks like:

$$\alpha = \alpha_M + \alpha_A + \alpha_D ,$$
$$\kappa = \kappa_M + \kappa_A .$$
(7.1)

Accounting for the mutual influence of the scattering and absorption by different components, we propose the empirical relations:

$$\alpha = (\alpha_M + \alpha_A)C\tau_D^p\omega_0^q + \alpha_D ,$$
$$\kappa = (\kappa_M' + \kappa_A')C\tau_D^p\omega_0^q ,$$
(7.2)

where ω_0 is the single scattering albedo, C is the factor of proportionality, τ_D and α_D are the optical thickness and the volume scattering coefficient caused only by scattering by droplets (value of τ_0 in Fig. 7.1 and value of α in Fig. 7.12a for $\lambda > 0.8$ μm), α_M', α_A', κ_M', κ_A' are the values of scattering and absorption coefficients of molecules and aerosol particles in the clear sky (α_M' is a coefficient of Raleigh scattering) at corresponding wavelength and altitude of the atmosphere; p and q are the empiric coefficients, estimated in several studies (Melnikova 1989, 1992, 1997; Kondratyev et al. 1997; Melnikova and Mikhailov 2000). The coefficient of scattering by droplets α_D has no factor because the

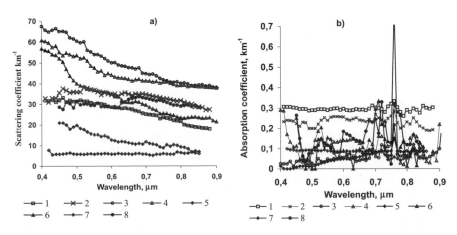

Fig. 7.12a,b. Spectral dependence of the volume coefficients (**a** – scattering and **b** – absorption) of the stratus cloud, retrieved from the data of the experiments, numbered as per Table 3.2

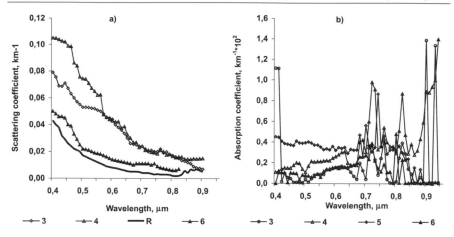

Fig. 7.13a,b. Volume coefficients of **a** – scattering and **b** – absorption, transformed using (7.2). The curve numbering corresponds to the experiments, listed in Table 3.2. The curve marked with letter R characterizes the molecular scattering at altitude 1 km

equation of radiative transfer and corresponding asymptotic formulas solving it are written for one component – droplet (in some cases for the droplet with the absorbing particle within it). Item $\kappa'_M \tau_D^p \omega_0^q$ in the second of (7.2) differs from zero only within the molecular absorption bands. Remember that the problem is considered only for $\tau_0 \gg 1$.

Factor C turns out to be equal to unity. Powers p and q are equal to: $p = 2$ and $q = \tau_0^2$, as per the estimations in several studies (Melnikova 1989, 1992, 1997; Kondratyev et al. 1997; Melnikova and Mikhailov 2000). These magnitudes correspond to the above-mentioned fact that the mean number of scattering events in the cloud of optical thickness τ_0 is proportional to τ_0^2 (Minin 1981; Yanovitskij 1997). We should point out that powers p and q were obtained from the analysis of the magnitudes of volume scattering and absorption coefficients for the data of two experiments at two wavelengths.

Transform values $[\alpha(\lambda) - \alpha(0.8)]$ and $\kappa(\lambda)$ (Tables A.8, Appendix A) using (7.2) leads to the values obtained with Mie theory and usually attributed to the cloud elementary volume (Grassl 1975; Nakajima et al. 1991). The spectral dependence of the transformed values of both difference $[\alpha(\lambda) - \alpha(0.8)]$ and the volume absorption coefficient is presented in Fig. 7.13a,b. It is seen that the magnitudes of the volume absorption coefficient demonstrated in Fig. 7.13b practically coincide with the ones usually calculated with Mie theory for cloud droplets (Grassl 1975). The molecular absorption bands become sharper. The values of the single scattering albedo corresponding to the absorption coefficients presented in Fig. 7.13b are about 0.99998 that is close to the standard magnitudes for the cloud layer. Difference $[\alpha(\lambda) - \alpha(0.8)]$ converted with (7.2) does not distinguish much from Raleigh scattering coefficient for the clear sky.

The presented consideration concerns the *external mixture*, i.e. the case, when aerosol particles are situated between the cloud droplets. When aerosol particles are situated within the droplets (the *internal mixture*) the aerosol absorption is correctly accounted for in calculation with the formulas for one-component medium. Basing on the obtained results one could conclude that the anomalous absorption by clouds points to the external mixture of the atmospheric aerosols and cloud droplets because in the opposite case the radiation absorption by clouds coincides with the theoretical values.

7.5.2
Multiple Scattering of Radiation as a Reason for Anomalous Absorption of Radiation by Clouds in the Shortwave Spectral Region

The aerosols consisting of hydrophobic particles such as sand, soot etc. could exist within the cloud between droplets with higher probability than the hydrophilic ones (salt, sulfates); hence, they increase the shortwave absorption of radiation by the cloud. Hydrophilic particles, being the nuclei of condensation increase the droplet number. This obstacle in turn increases the cloud optical thickness and causes the cloud cooling. The aerosol absorption by the cloud increasing up to 15% has been approximately estimated basing on the proposed mechanism with the mean values of the aerosol volume absorption coefficient equal to $0.08 \, \text{km}^{-1}$ and of the volume scattering coefficient equal to $30 \, \text{km}^{-1}$ with geometrical thickness $\Delta z = 1 \, \text{km}$ and within spectral range 0.4–1.0 µm. The molecule absorption within the ozone Chappuis band increases up to 6–10% and the molecule absorption within oxygen band 0.76 µm increases up to 10% that coincides with the results of the study by Dianov-Klokov et al. (1973). This effect turns out stronger for the thicker clouds, and it quantitatively explains the anomalous absorption by clouds.

Experimental studies (Boers et al. 1996; Bott et al. 1996) actually indicate the higher content of the carbonaceous and mineral compound in the atmospheric aerosols than has been assumed before together with their significant yield to forming the radiative regime of the atmosphere. The hydrophobic particles could be injected into the atmosphere as the result of industrial escapes, sand storms, volcanic eruptions, and fires. These sources do not seem enough to account for the cloud anomalous absorption displayed on a global scale, however the aerosols flue escapes extend up to 3000 km keeping their radiation activity in the optical range (Mazin and Khrgian 1989).

In the remainder of this chapter, we would like to point out that careful accounting of the optical properties of all atmospheric components is necessary for the construction of optical models (Vasilyev and Ivlev 2002).

References

Asano S (1994) Cloud and Radiation Studies in Japan. Cloud Radiation Interactions and Their Parameterization in Climate Models. In: WCRP-86 (WMO/TD No. 648), WMO, Geneva, pp 72–73

Binenko VI, Kondratyev KYa (1975) Vertical profiles of typical cloud forms. In: Main Geophysical Observatory Studies 331, pp 3–16 (in Russian)

Boers R, Jensen JB, Krummel PB, Gerber H (1996) Microphysical and short-wave radiative structure of wintertime stratocumulus clouds over the Southern Ocean. Q J R Meteorol Soc 122:1307–1339

Bott A, Trautmann T, Zdunkowski W (1996) A numerical model of the cloud-topped planetary boundary layer: Radiation, turbulence and spectral microphysics in marine stratus. Q J R Meteorol Soc 122:635–667

Curry JA, Hobbs PV, King MD, Randall DA, Minnis P, Isaac GA, Pinto JO, Uttal T, Bucholtz A, Cripe DG, Gerber H, Fairall CW, Garrett TJ, Hudson J, Intrieri JM, Jakob C, Jensen T, Lawson P, Marcotte D, Nguyen L, Pilewskie P, Rangno A, Rogers DC, Strawbridge KB, Valero FPJ, Willams AG, Wylie D (2000). FIRE Arctic Clouds Experiment. Bulletin of the American Meteorological Society 81:5–29

Dianov-Klokov BG, Grechko EI, Malkov GP (1973) Airborne measurements of the effective photon free pass from radiation reflected and transmitted by clouds in the oxygen band 0.76 μm. Izv Acad Sci USSR, Atmosphere and Ocean Physics 9:524–537 (in Russian)

Feigelson EM (ed) (1981) Radiation in the cloudy atmosphere. Gidrometeoizdat, Leningrad (in Russian)

Grassl H (1975) Albedo reduction and radiative heating of clouds by absorption aerosol particles. Beitr Phys Atmos 48:199–209

Ivlev LS, Andreev SD (1986) Optical properties of atmospheric aerosols. St. Petersburg University Press, St. Petersburg (in Russian)

Kondratyev KYa (ed) (1969) Radiative characteristics of the atmosphere and the surface. Gidrometeoizdat, Leningrad (in Russian)

Kondratyev KYa (ed) (1991) Aerosol and Climate. Gidrometeoizdat, Leningrad (in Russian)

Kondratyev KYa, Binenko VI (1984) Impact of Clouds on Radiation and Climate. Gidrometeoizdat, Leningrad (in Russian)

Kondratyev KYa, Binenko VI, Melnikova IN (1997) Absorption of solar radiation by clouds and aerosols in the visible wavelength region. Meteorology and Atmospheric Physics 0/319:1–10

Kondratyev KYa, Binenko VI, Melnikova IN (1998) Vertical profile of spectral scattering and absorption coefficients of stratus clouds. II. Application to data of airborne radiative observation. Atmosphere and Ocean Optics 11:381–387

Kurosu T, Rozanov VV, Burrows JP (1997) Parameterization schemes for terrestrial water clouds in the radiative transfer model GOMETRAN. J Geoph Res 102(D18):21809–21823

Marchuk GI, Kondratyev KYa, Kozoderov VV, Khvorostyanov VI (1986) Clouds and Climate. Gidrometeoizdat, Leningrad (in Russian)

Marshak A, Davis A, Wiscombe W, Cahalan R (1995) Radiative smoothing in fractal clouds. J Geoph Res 100(D18):26247–26261

Mayer B, Kylling A, Madronich S, Seckmeyer G (1998) Enhanced absorption of UV radiation due to multiple scattering in clouds: Experimental evidence and theoretical explanation. J Geophys Res 103(D23):31241–31254

Mazin IP, Khrgian AKh (eds) (1989) Clouds and cloudy atmosphere. Gidrometeoizdat, Leningrad (in Russian)

Melnikova IN (1989) Light absorption in cloud layers. In: Atmospheric Physics Problems. St.Petersburg State University Press, St.Petersburg, 20, pp. 18–25

Melnikova IN (1992) Spectral optical parameters of cloud layers. Theory, Part I. Atmosphere Optics 5:178–185 (Bilingual)

Melnikova IN (1992) Spectral optical parameters of cloud layers. Application to experimental data. Part II. Atmosphere Optics 5:178–185 (Bilingual)

Melnikova IN (1997) Determination of strati clouds optical parameters from measurement of reflected or transmitted solar radiation. In: IRS'96 Current problems in Atmospheric Radiation. Proceedings of the International Radiation Symposium, August 1996, Fairbanks, Alaska. A. Deepak Publishing, pp 210–213

Melnikova IN (1997) Studies of influence of the multiple scattering on the true light absorption in clouds with using calculations of the Monte-Carlo method. I International Conference Natural and anthropogenic aerosols, 29th September–4th October, St. Petersburg (in Russian)

Melnikova IN (1998) Vertical profile of spectral scattering and absorption coefficients of stratus clouds. I. Theory. Atmosphere and Ocean Optics 11:5–11 (Bilingual)

Melnikova IN, Mikhailov VV (1992) The optical parameters in strati on basis of aircraft spectral measurements. Thesis of International Radiation Symposium, August, Tallinn

Melnikova IN, Mikhailov VV (1993) Determination of optical characteristics of cloud layers. Dokl. RAS 328:319–321 (in Russian)

Melnikova IN, Mikhailov VV (1994) Spectral scattering and absorption coefficients in strati derived from aircraft measurements. J Atmos Sci 51:925–931

Melnikova IN, Fedorova EYu (1996) Vertical profile of optical parameters inside cloud layer. In: Problems of Atmospheric Physics. State University Press, St.Petersburg, 20, pp 261–272

Melnikova IN, Domnin PI (1997) Determination of optical parameters of homogeneous optically thick cloud layer. Atmosphere and Ocean Optics 10:734–740 (Bilingual)

Melnikova IN, Mikhailov VV (2000) Influence of multiple scattering of radiation on aerosols and molecular absorption and scattering into clouds. In: IRS'2000, Current problems in Atmospheric Radiation, Proceedings of the International Radiation Symposium, July, St. Petersburg, Russia, pp 326–328

Melnikova IN, Nakajima T (2000) Single scattering albedo and optical thickness of stratus clouds obtained from "POLDER" measurements of reflected radiation. Earth Observations and Remote Sensing 3:1–16 (Bilingual)

Melnikova IN, Nakajima T (2000) Space distribution of cloud optical parameters obtained from reflected radiation observations with the POLDER instrument on the board of the ADEOS satellite. In: Ivlev LS (ed) Natural and anthropogenic aerosols, St. Petersburg, pp 78–85 (in Russian)

Melnikova IN, Mikhailov VV (2001) Vertical profile of spectral optical parameters of stratus clouds from airborne radiative measurements. J Geophys Res 106(D21):27465–27471

Melnikova IN, Domnin PI, Varotsos C, Pivovarov SS (1997) Retrieval of optical properties of cloud layers from transmitted solar radiance data. In: Proceedings of SPIE, Vol. 3237, 23rd European Meeting on Atmospheric Studies by Optical Methods, September 1996, Kiev, Ukraine, pp 77–80

Melnikova IN, Domnin PI, Radionov VF (1998) Retrieval of optical thickness and single scattering albedo from measurements of reflected or transmitted solar radiation. Izv. RAS, Atmosphere and Ocean Physics 34:669–676 (Bilingual)

Melnikova IN, Mikhailov VV, Domnin PI, Radionov VF (2000) Optical characteristics of clouds derived from measurements of reflected or transmitted solar radiation. J Atmos Sci 57:2135–2143

Minin IN (1981) Leningrad School of the radiative transfer theory. Astrophysics. Academy of Sciences of Armenian Republic 17:585–618 (in Russian)

Nakajima T, King MD, Spinhirne JD, Radke LF (1991) Determination of the optical thickness and effective particle radius of clouds from reflected solar radiation measurements. II. Marine stratocumulus observations. J Atmos Sci 48:728–750

Nakajima TY, Nakajima T (1995) Wide-area determination of cloud microphysical properties from NOAA AVHRR Measurements for FIRE and ASTEX regions. J Atmos Sci 52:4043–4059

Pfeilsticker K (1999) First geometrical path length probability density function derivation of the skylight from high-resolution oxygen A-band spectroscopy. 2. Derivation of the Levy index for the skylight transmitted by midlatitude clouds. J Geoph Res 104(D43):4101–4116

Pfeilsticker K, Erle F, Platt U (1997) Absorption of solar radiation by atmospheric O_4. J Atmos Sci 54:934–939

Pfeilsticker K, Erle F, Funk O, Marquard L, Wagner T, Platt U (1998). Optical path modification due to tropospheric clouds: Implications for zenith sky measurements of stratospheric gases. J Geoph Resv 103(D19):35323–25335

Radionov VF, Sakunov GG, Grishechkin VS (1981) Spectral albedo of snow surface from measurements at drifting station SP-22. In: First global experiment FGGE, Vol. 2: Polar aerosols, extended cloudiness and radiation. Gidrometeoizdat, Leningrad, pp 89–91 (in Russian)

Rozenberg GV, Malkevitch MS, Malkova VS, Syachinov VI (1974) Determination of the optical characteristics of clouds from measurements of reflected solar radiation on the Kosmos-320 satellite. Izv. Acad. Sci. USSR, Atmosphere and Ocean Physics 10:14–24 (in Russian)

Rublev AN, Trotsenko AN, Romanov PYu (1997) Using data of the satellite radiometer AVHRR for cloud optical thickness determining. Izv. RAS, Atmosphere and Ocean Physics 33:670–675 (Bilingual)

Sokolic I, Toon OB (1999) Incorporation of mineralogical composition into models of the radiative properties of mineral aerosol from UV to IR wavelengths. J Geoph Res 104(D8):9423–9444

Sokolic IN (1988) Interpretation of measurements of optical characteristics of smoke aerosols. Izv Acad Sci USSR, Atmosphere and Ocean Physics 34:345–357 (Bilingual)

Stephens GL (1979) Optical properties of eight water cloud types. Technical Paper of CSIRO, Atmosph Phys Division, Aspendale, Australia, No. 36, pp 1–35

Vasilyev AV, Ivlev LS (2002) On optical properties of polluted clouds. Atmosphere and Ocean Optics 15:157–159 (Bilingual)

Vasilyev AV, Melnikova IN, Mikhailov VV (1994) Vertical profile of spectral fluxes of scattered solar radiation within stratus clouds from airborne measurements. Izv RAS, Atmosphere and Ocean Physics 30:630–635 (Bilingual)

Wagner T, Erle F, Marquard L, Otten C, Pfeilsticker K, Senne T, Stutz J, Platt U (1998) Cloudy sky optical paths as derived from differential optical absorption spectroscopy observations. J Geoph Res 103(D19):25307–25321

Yanovitskij EG (1997) Light scattering in inhomogeneous atmospheres. Springer, Berlin Heidelberg New York

CHAPTER 8

Conclusion

The authors have considered two effective methods for calculation of the solar radiance and irradiance under clear and cloudy conditions (the direct problem solving): the numerical one – the Monte-Carlo method for clear sky, and the analytical one – the method of the asymptotic formulas for overcast sky. The advantages of the methods during the calculation of the radiative characteristics have been shown. The methods have been presented in detail (including the algorithms) so that interested colleagues could directly use them. The uncertainties of these methods have been analyzed. In the beginning of the book (Chaps. 1 and 2) the physical characteristics and conceptions have been defined and the main physical principles of light propagation in the atmosphere have been explained.

While describing the experiments, the main emphasis has been put to the methodological details of observations for improving the exactness of measurements. Instruments are improved constantly, but the considered details of the accomplishment of radiation observations, as we hope, could be useful for specialists. The sources of observational and processing errors have been analyzed, and the possibilities for their minimization have been proposed. The elaborated algorithms of the experimental data processing are based on the methods of mathematical statistics and even if they could not be directly applied to the data of other experiments they would be useful to study because the common principles of processing a large volume of data are the fundamental ones.

The presented examples of the vertical profiles and spectral dependence of solar semispherical upward and downward fluxes are shown in figures and tables for using these data in radiative models under different atmospheric conditions or as the initial data of inverse problems. Here we have presented the examples of observational data for different atmospheric and meteorological conditions. For our colleagues who are interested in these data we would like to remind them that the database is extended enough.

The developed classification of different types of surfaces could be also mentioned. The obtained results allow effectively identifying the type of surface on the one hand and adequately taking into account the reflection of solar radiation from the surface in atmospheric optics on the other hand.

The numerical and analytical methods of the retrieval of the atmospheric parameters from the data of solar radiation measurements under clear and overcast sky conditions (the inverse problem solving), elaborated by the authors are described in detail. Significant attention is paid to the correctness of the inverse problem. Careful error analysis and study of the applicability range in every considered case is in fact the investigation of stability of the inverse problem solution. The detailed algorithms of the inverse problem solving and its analysis could be applied to other similar data.

The application of the elaborated methods to the interpretation of the experimental data allows the retrieval of new information: the spectral and vertical dependence of the optical parameters of the clear and cloudy atmosphere. The obtained examples of the vertical profiles and spectral dependence of the optical parameters of the atmosphere and surface are presented in figures and tables. There is a rich database of results similar to the examples presented here, which could be used as an optical model for different atmospheric conditions.

On the basis of cloud optical parameters obtained from observations, the mechanism of influence of the multiple scattering of radiation by cloud droplets on the increase of true absorption by atmospheric aerosols and on the molecular scattering and absorption by the cloudy atmosphere is proposed. The empirical formulas for taking into account this mechanism are inferred. They allow correcting numerical optical models. Numerically estimating validation of the obtained cloud optical parameters is accomplished.

This mechanism is applied to the multi-component medium (droplets, molecules, aerosols) and used for the explanation of the anomalous short-wave radiation absorption by clouds. Until now this effect has not had an adequate interpretation.

Appendix A: Tables of Radiative Characteristics and Optical Parameters of the Atmosphere

Table A.1. Semispherical solar irradiance ($mW\,cm^{-1}\,\mu m^{-1}$) reduced to solar incident angle $51°$ and to the levels of the atmospheric pressure from the results of processing the airborne sounding data 16 Oct. 1983 in the clear sky. Ground surface is the sand (*continued on next page*)

λ (nm)	Downwelling irradiance $mW\,cm^{-2}\,\mu m^{-1}$						Upwelling irradiance $mW\,cm^{-2}\,\mu m^{-1}$					
P (mbar)	1000	900	800	700	600	500	1000	900	800	700	600	500
350	21.6	24.0	26.7	29.8	33.3	37.0	1.48	3.41	5.31	7.20	9.06	10.9
360	25.9	28.1	35.0	33.3	36.3	39.5	2.09	4.10	6.10	8.08	10.0	12.0
370	42.6	45.7	47.8	51.6	54.6	57.5	3.96	7.09	10.1	13.0	16.0	18.9
380	43.4	46.1	48.9	51.7	54.5	57.3	4.87	7.64	10.4	13.2	16.0	18.8
390	46.0	48.6	51.2	53.9	56.6	59.2	5.70	8.31	10.9	13.6	16.3	19.0
400	64.6	67.1	69.9	73.1	76.5	80.2	7.86	10.4	12.9	15.5	18.1	20.7
410	68.8	71.4	74.3	77.5	81.0	84.8	8.30	10.9	13.5	16.2	18.8	21.5
420	76.5	79.3	82.2	85.3	88.6	92.2	9.33	12.2	15.0	17.9	20.8	23.8
430	55.7	58.3	61.2	64.4	67.9	71.7	7.24	9.47	11.7	14.0	16.3	18.7
440	69.2	72.2	75.6	79.3	83.3	87.6	9.97	12.5	15.1	17.8	20.5	23.2
450	86.3	89.5	93.0	96.8	101.0	106.0	13.7	16.6	19.5	22.5	25.5	28.7
460	88.1	91.0	94.3	97.8	102.0	106.0	15.0	17.6	20.3	23.1	26.0	28.9
470	87.5	90.2	93.2	96.5	100.0	104.0	15.5	17.9	20.5	23.1	25.8	28.5
480	92.9	95.6	98.5	102.0	105.0	109.0	17.4	19.8	22.2	24.7	27.3	30.0
490	86.8	89.2	92.0	95.0	98.3	102.0	16.9	19.1	21.3	23.6	26.0	28.4
500	85.8	88.1	90.7	93.5	96.7	100.0	17.7	19.7	21.8	23.9	26.2	28.4
510	90.0	92.2	94.7	97.4	100.0	104.0	19.8	21.8	23.8	25.9	28.1	30.3
520	83.3	85.4	87.7	90.3	93.1	96.2	19.4	21.1	22.9	24.8	26.8	28.8
530	89.4	91.5	93.8	96.4	99.2	102.0	22.1	23.8	25.6	27.5	29.4	31.4
540	86.6	88.4	90.5	92.9	95.5	98.3	22.7	24.2	25.8	27.5	29.3	31.1
550	89.2	90.9	92.9	95.2	97.7	101.0	24.6	26.1	27.7	29.4	31.1	32.9
560	87.1	88.8	90.8	93.0	95.5	98.2	25.4	26.8	28.2	29.7	31.3	33.0
570	87.7	89.5	91.4	93.6	96.1	98.7	26.6	27.9	29.2	30.7	32.2	33.7
580	88.1	89.8	91.8	94.1	96.5	99.2	27.8	29.0	30.2	31.6	33.0	34.5
590	83.5	85.2	87.1	89.2	91.6	94.2	27.2	28.3	29.4	30.6	31.9	33.3
600	86.2	87.7	89.3	91.2	93.4	95.7	28.6	29.7	30.8	32.0	33.3	34.6
610	86.1	87.4	88.9	90.7	92.7	94.9	29.3	30.3	31.4	32.6	33.8	35.2

Table A.1. (*continued*)

λ (nm)	Downwelling irradiance mW cm^{-2} µm^{-1}						Upwelling irradiance mW cm^{-2} µm^{-1}					
P (mbar)	1000	900	800	700	600	500	1000	900	800	700	600	500
620	84.4	85.8	87.4	89.2	91.3	93.5	29.3	30.3	31.4	32.5	33.7	35.0
630	80.7	82.2	83.9	85.9	88.0	90.4	28.6	29.4	30.3	31.2	32.3	33.3
640	80.4	81.7	83.2	85.0	86.9	89.1	28.7	29.6	30.5	31.5	32.5	33.6
650	78.3	79.5	80.9	82.6	84.4	86.5	28.3	29.1	29.9	30.9	31.8	32.9
660	76.4	77.4	78.7	80.1	81.8	83.6	27.7	28.5	29.4	30.3	31.3	32.4
670	77.7	78.7	79.9	81.3	83.0	84.8	28.6	29.4	30.3	31.2	32.2	33.3
680	75.4	76.4	77.6	79.0	80.7	82.6	28.1	28.9	29.7	30.6	31.6	32.6
690	66.7	68.1	69.8	71.7	73.8	76.2	25.5	25.9	26.4	27.0	27.6	28.2
700	67.5	68.8	70.3	72.0	73.9	76.1	25.3	25.8	26.5	27.1	27.8	28.6
710	65.3	66.5	67.9	69.6	71.5	73.6	25.1	25.6	26.2	26.8	27.5	28.2
720	59.7	61.3	63.1	65.2	67.4	69.8	22.5	22.9	23.2	23.7	24.1	24.6
730	59.9	61.3	62.8	64.6	66.6	68.8	23.1	23.5	23.9	24.4	24.9	25.4
740	61.2	62.3	63.6	65.1	66.8	68.7	24.8	25.2	25.6	26.1	26.7	27.3
750	59.6	60.7	62.0	63.5	65.2	67.1	24.6	25.0	25.5	26.0	26.5	27.1
760	42.2	45.0	48.0	51.1	54.5	58.0	18.5	18.5	18.6	18.6	18.7	18.8
770	53.7	55.0	56.5	58.3	60.2	62.4	22.8	23.1	23.4	23.7	24.1	24.6
780	56.2	57.2	58.4	59.7	61.3	63.1	23.9	24.3	24.7	25.1	25.6	26.2
790	54.4	55.4	56.6	58.0	59.5	61.3	23.3	23.6	24.0	24.4	24.8	25.3
800	53.0	54.0	55.2	56.5	58.0	59.8	22.8	23.1	23.5	23.9	24.3	24.8
810	49.3	50.5	51.8	53.4	55.1	57.0	20.9	21.1	21.4	21.7	22.0	22.3
820	46.1	47.4	49.0	50.7	52.5	54.6	19.3	19.5	19.7	19.9	20.2	20.5
830	45.5	46.7	48.0	49.5	51.3	53.2	19.5	19.7	19.9	20.2	20.5	20.8
840	46.8	47.7	48.8	50.0	51.4	53.0	20.6	20.9	21.1	21.4	21.8	22.2
850	45.0	45.9	46.9	48.1	49.5	51.0	20.3	20.5	20.8	21.1	21.4	21.8
860	43.7	44.5	45.5	46.7	48.0	49.6	19.8	20.1	20.3	20.6	20.9	21.3
870	45.5	46.2	47.1	48.1	49.3	50.6	20.4	20.7	21.0	21.3	21.6	22.0
880	44.3	45.1	46.1	47.2	48.5	49.9	19.9	20.1	20.4	20.8	21.1	21.5
890	41.8	42.7	43.7	45.0	46.4	48.0	18.4	18.6	18.9	19.2	19.4	19.8
900	38.3	39.4	40.8	42.3	44.0	45.8	16.3	16.5	16.6	16.8	17.0	17.2
910	35.1	36.5	38.1	39.8	41.7	43.7	14.8	14.9	14.9	15.0	15.0	15.1
920	36.6	37.7	39.1	40.5	42.2	43.9	15.6	15.7	15.9	16.0	16.1	16.3
930	21.3	23.7	26.3	29.1	32.1	35.3	8.35	8.34	8.31	8.28	8.23	8.18
940	21.1	23.5	26.1	28.8	31.7	34.8	8.19	8.12	8.05	7.97	7.89	7.80
950	21.5	23.7	26.1	28.7	31.4	34.3	8.52	8.44	8.36	8.27	8.19	8.10
960	5.71	6.56	8.22	10.6	13.7	17.5	3.73	3.40	3.09	2.81	2.55	2.32
970	0.16	1.11	1.11	1.11	9.00	9.43	0.104	1.05	1.05	1.05	0.80	1.23

Table A.2. Semispherical solar irradiance ($mW\,cm^{-1}\,\mu m^{-1}$) reduced to solar incident angle 48° and to the levels of the atmospheric pressure from the results of processing the airborne sounding 29 Apr. 1985 in the clear sky. Ground surface is the snow on ice (*continued on next page*)

λ (nm)	Downwelling irradiance $mW\,cm^{-2}\,\mu m^{-1}$						Upwelling irradiance $mW\,cm^{-2}\,\mu m^{-1}$					
P (mbar)	1000	900	800	700	600	500	1000	900	800	700	600	500
330	19.9	20.9	22.3	24.4	27.2	30.5	9.72	10.7	11.8	13.0	14.1	15.3
340	30.5	31.6	33.1	35.1	37.5	40.5	15.3	16.3	17.3	18.5	19.7	20.9
350	35.7	37.2	38.9	41.0	43.4	46.1	18.2	19.5	20.9	22.3	23.8	25.3
360	38.3	39.7	41.2	43.0	45.1	47.4	19.8	21.1	22.6	24.1	25.7	27.3
370	54.2	55.8	57.6	59.4	61.5	63.9	28.8	30.5	32.2	34.0	35.9	37.8
380	56.9	58.3	59.8	61.4	63.1	64.8	31.2	32.6	34.1	35.7	37.4	39.1
390	57.2	58.4	59.7	61.1	62.7	64.7	30.9	32.1	33.4	34.8	36.4	37.9
400	82.1	83.5	85.1	86.8	88.6	90.6	43.7	45.1	46.6	48.3	50.2	52.2
410	83.8	85.0	86.3	87.9	90.1	92.8	44.4	45.6	46.9	48.5	50.1	51.8
420	87.2	88.6	90.1	91.9	93.9	96.9	46.2	47.6	49.1	50.9	52.9	54.8
430	75.8	76.8	78.1	79.5	81.1	82.9	40.6	41.6	42.8	44.2	45.8	47.7
440	91.1	92.2	93.6	95.1	97.0	99.0	48.3	49.4	50.8	52.3	54.2	56.2
450	103.0	104.0	106.0	108.0	110.0	113.0	54.6	55.7	57.1	58.8	60.9	62.8
460	107.0	108.0	109.0	111.0	113.0	115.0	56.3	57.4	58.7	60.3	62.1	64.2
470	106.0	107.0	108.0	110.0	112.0	113.0	55.6	56.6	57.8	59.2	60.9	62.9
480	111.0	112.0	113.0	115.0	116.0	118.0	57.1	58.0	59.1	60.5	62.2	64.1
490	105.0	106.0	107.0	108.0	110.0	111.0	53.7	54.5	55.6	56.9	58.4	60.2
500	103.0	104.0	105.0	106.0	107.0	109.0	52.2	52.9	53.8	55.0	56.4	58.0
510	107.0	107.0	108.0	109.0	111.0	112.0	53.6	54.2	55.1	56.2	57.5	59.1
520	99.4	100.0	101.0	102.0	103.0	104.0	49.8	50.4	51.2	52.2	53.5	54.9
530	105.0	106.0	106.0	107.0	109.0	110.0	52.1	52.7	53.4	54.4	55.6	57.1
540	101.0	101.0	102.0	103.0	104.0	106.0	49.6	50.2	50.9	51.8	53.0	54.3
550	103.0	103.0	104.0	105.0	106.0	107.0	50.5	50.9	51.6	52.5	53.6	55.0
560	101.0	101.0	102.0	103.0	104.0	105.0	49.3	49.7	50.4	51.2	52.2	53.5
570	102.0	102.0	103.0	103.0	104.0	105.0	49.2	49.6	50.2	50.9	51.9	53.1
580	101.0	101.0	102.0	102.0	104.0	105.0	48.1	48.4	48.9	49.6	50.3	51.2
590	97.9	98.2	98.6	99.3	100.0	101.0	46.6	46.8	47.3	47.9	48.7	49.7
600	98.3	98.6	99.0	99.7	100.0	101.0	46.1	46.4	46.8	47.4	48.3	49.2
610	97.3	97.6	98.0	98.6	99.4	100.0	45.9	46.2	46.6	47.2	48.0	49.0
620	96.7	96.9	97.3	97.9	98.7	99.6	45.4	45.7	46.1	46.7	47.4	48.4
630	93.3	93.5	93.9	94.4	95.1	96.6	43.2	43.3	43.7	44.2	44.9	45.6
640	92.6	92.7	93.1	93.6	94.2	95.0	42.4	42.6	42.9	43.4	44.0	44.9
650	89.8	90.0	90.3	90.8	91.5	92.3	40.9	41.1	41.4	41.9	42.6	43.4
660	86.5	86.7	87.0	87.5	88.2	89.0	39.1	39.3	39.7	40.1	40.8	41.6
670	88.0	88.2	88.5	89.0	89.6	90.4	40.0	40.2	40.5	41.0	41.7	42.5
680	85.7	85.8	86.1	86.6	87.2	88.0	38.8	39.0	39.3	39.7	40.4	41.1
690	78.0	78.3	78.9	79.9	81.1	82.7	34.2	34.3	34.5	34.8	35.2	35.7
700	79.1	79.3	79.6	80.0	81.0	82.3	34.3	34.5	34.8	35.2	35.7	36.2
710	70.4	71.0	71.8	73.0	74.6	76.4	30.6	31.2	31.8	32.5	33.3	34.1
720	65.8	66.5	67.5	68.9	70.7	72.7	28.0	28.4	29.0	29.5	30.2	30.9

Table A.2. (*continued*)

λ (nm)	Downwelling irradiance mW cm^{-2} μm^{-1}						Upwelling irradiance mW cm^{-2} μm^{-1}					
P (mbar)	1000	900	800	700	600	500	1000	900	800	700	600	500
730	62.9	63.9	65.1	66.7	68.6	70.8	25.6	25.9	26.4	26.8	27.3	27.8
740	64.9	65.4	66.3	67.4	68.9	70.7	27.3	27.8	28.3	28.9	29.5	30.3
750	64.3	64.8	65.4	66.5	67.8	69.5	27.6	28.1	28.7	29.3	30.0	30.8
760	39.1	40.0	41.5	43.7	46.6	50.2	17.6	17.8	18.0	18.3	18.6	18.9
770	48.5	50.1	52.0	54.3	56.8	59.7	20.2	20.4	20.6	20.9	21.3	21.7
780	59.6	60.1	60.9	62.0	63.4	65.1	25.0	25.4	26.0	26.5	27.1	27.8
790	58.6	59.1	59.7	60.7	62.0	63.5	24.2	24.6	25.2	25.8	26.4	27.1
800	56.8	57.3	57.9	58.9	60.1	61.7	23.2	23.6	24.2	24.7	25.3	26.0
810	55.5	55.9	56.5	57.4	58.6	60.1	22.5	23.0	23.4	24.0	24.5	25.2
820	49.9	50.7	51.8	53.1	54.7	56.6	19.1	19.4	19.7	20.0	20.4	20.8
830	49.1	49.8	50.8	52.1	53.6	55.4	18.9	19.2	19.5	19.9	20.3	20.7
840	49.9	50.4	51.1	52.1	53.3	54.8	19.9	20.2	20.6	21.0	21.5	22.0
850	49.3	49.7	50.2	51.1	52.2	53.5	20.0	20.4	20.9	21.3	21.8	22.4
860	48.0	48.4	48.9	49.7	50.7	52.0	19.4	19.8	20.2	20.7	21.2	21.7
870	47.7	48.0	48.5	49.3	50.4	51.7	19.1	19.4	19.9	20.4	20.9	21.4
880	47.6	48.0	48.5	49.3	50.3	51.6	18.8	19.1	19.6	20.0	20.5	21.1
890	46.8	47.2	47.7	48.5	49.6	50.9	18.0	18.4	18.8	19.2	19.7	20.3
900	40.8	41.5	42.5	43.7	45.1	46.9	14.6	14.8	15.1	15.4	15.8	16.1
910	35.1	36.5	38.1	39.8	41.7	43.7	14.8	14.9	14.9	15.0	15.0	15.1
920	36.6	37.7	39.1	40.5	42.2	43.9	15.6	15.7	15.9	16.0	16.1	16.3
930	21.3	23.7	26.3	29.1	32.1	35.3	8.35	8.34	8.31	8.28	8.23	8.18
940	21.1	23.5	26.1	28.8	31.7	34.8	8.19	8.12	8.05	7.97	7.89	7.80
950	21.5	23.7	26.1	28.7	31.4	34.3	8.52	8.44	8.36	8.27	8.19	8.10
960	5.71	6.56	8.22	10.6	13.7	17.5	3.73	3.40	3.09	2.81	2.55	2.32
970	0.16	1.11	1.11	1.11	9.00	9.43	0.104	1.05	1.05	1.05	0.80	1.23

Table A.3. Semispherical solar irradiance ($mW\,cm^{-1}\,\mu m^{-1}$) reduced to solar incident angle 48° from the results of processing the airborne sounding 20 Apr. 1985 in the overcast sky. Ground surface is the snow on ice (*continued on next page*)

λ (nm)	Downwelling irradiance $mW\,cm^{-2}\,\mu m^{-1}$						Upwelling irradiance $mW\,cm^{-2}\,\mu m^{-1}$					
z (km)	1.4	1.3	1.2	1.1	0.9	0.8	1.4	1.3	1.2	1.1	0.9	0.8
350	62.01	57.70	47.16	37.50	31.49	29.52	39.26	37.62	30.34	25.44	21.09	20.38
360	68.89	64.81	55.08	44.37	37.39	35.43	44.17	42.26	33.76	28.05	23.92	22.30
370	85.71	81.02	68.52	54.42	47.03	44.75	57.19	54.86	43.40	36.24	30.92	29.03
380	76.88	72.38	60.97	49.09	42.24	40.56	47.65	43.54	34.68	30.78	27.60	27.71
390	79.70	74.86	62.74	50.75	43.97	42.00	47.00	43.40	36.02	33.63	30.97	27.62
400	111.72	104.84	87.95	71.48	61.52	58.99	67.47	61.7	49.86	43.96	39.19	39.95
410	113.57	106.09	89.42	73.66	63.86	61.34	71.80	65.24	51.99	45.81	40.88	41.26
420	109.40	99.78	78.82	70.72	65.59	62.34	65.91	59.76	49.13	44.1	39.98	40.56
430	110.17	100.11	78.01	70.32	65.19	62.05	66.05	59.87	49.00	44.22	40.35	40.66
440	115.66	105.34	82.80	74.70	68.04	65.84	68.98	62.47	51.36	46.15	41.88	43.06
450	130.51	118.52	91.65	82.68	76.18	73.28	76.02	69.04	57.18	51.28	46.47	47.78
460	139.61	126.44	96.97	87.83	81.15	77.80	80.75	73.32	60.26	54.13	49.18	49.79
470	138.56	125.49	96.20	87.15	80.93	77.41	80.04	72.79	59.96	53.74	48.74	49.59
480	139.61	126.31	96.51	87.67	81.09	77.86	79.33	71.96	58.89	52.76	47.88	49.09
490	133.62	120.97	92.39	83.72	77.02	73.94	75.12	68.30	56.17	50.26	45.54	46.48
500	132.58	119.99	91.43	82.91	76.94	73.61	74.09	67.24	55.40	49.47	44.71	45.77
510	129.45	117.02	88.77	80.72	74.77	71.64	72.02	65.24	53.08	47.58	43.20	44.17
520	126.84	114.41	86.46	78.82	72.77	69.82	70.27	63.70	52.11	46.71	42.39	43.36
530	125.80	113.35	85.19	77.86	72.53	69.429	69.32	62.66	50.78	45.47	41.22	42.36
540	129.19	116.44	87.83	80.50	74.43	71.28	70.91	64.08	52.19	46.72	42.32	43.46
550	130.23	117.31	88.20	80.74	75.00	71.44	71.23	64.25	51.99	46.69	42.42	43.46
560	129.44	116.67	87.89	80.70	74.66	71.52	70.78	63.94	51.78	46.53	42.29	43.36
570	125.59	113.06	84.77	77.80	71.81	68.57	67.91	61.03	49.29	44.29	40.25	41.25
580	123.97	111.50	83.30	76.52	70.33	67.36	66.51	59.62	47.88	43.02	39.12	40.05
590	123.97	111.46	83.34	76.68	70.14	67.13	66.23	59.32	47.57	42.78	38.93	39.75
600	122.14	109.80	81.47	74.67	68.48	65.39	64.36	57.51	45.81	41.10	37.32	38.14
610	119.8	107.64	80.42	74.24	68.43	65.22	63.51	56.86	45.40	40.93	37.32	38.14
620	116.68	104.79	78.12	72.16	66.39	63.29	61.61	55.21	44.03	39.56	35.95	36.54
630	112.77	101.39	75.77	69.91	64.14	61.10	59.48	53.17	42.12	37.84	34.43	34.93
640	111.73	100.41	74.86	69.08	63.59	60.39	58.36	52.07	41.25	37.16	33.86	34.23
650	109.90	98.89	74.03	68.31	62.42	59.41	57.38	51.06	40.15	36.04	32.77	33.43
660	107.30	96.32	71.73	66.32	60.93	57.72	55.51	49.48	38.85	34.97	31.89	32.42
670	106.78	96.10	72.11	66.60	60.84	57.77	55.58	49.59	38.85	35.07	32.07	32.32
680	105.48	94.79	71.10	65.91	60.32	57.01	55.23	49.24	38.32	34.51	31.48	32.02
690	95.50	85.71	63.65	58.42	53.04	49.64	48.87	43.26	33.16	29.59	26.80	27.11
700	96.44	86.96	65.86	60.104	52.40	48.70	49.80	44.09	32.33	28.73	25.98	26.70
710	96.56	87.12	65.95	59.88	52.25	48.42	49.57	43.97	32.44	28.71	25.83	26.55
720	94.23	84.81	63.69	58.02	50.11	46.72	48.56	42.81	31.08	27.55	24.87	25.50
730	87.89	78.82	59.04	53.55	45.25	41.87	44.76	38.84	26.93	23.70	21.32	21.88
740	89.65	80.79	61.24	55.46	47.57	43.83	45.52	40.07	28.93	25.3	22.59	23.39
750	88.95	80.24	61.85	56.55	48.30	44.82	46.37	40.91	29.55	26.09	23.48	24.39

Table A.3. (*continued*)

λ (nm)	Downwelling irradiance mW cm^{-2} μm^{-1}						Upwelling irradiance mW cm^{-2} μm^{-1}					
z (km)	1.4	1.3	1.2	1.1	0.9	0.8	1.4	1.3	1.2	1.1	0.9	0.8
760	83.27	75.06	57.38	52.71	44.81	42.15	44.58	39.14	27.75	24.40	21.85	22.98
770	61.52	55.77	41.63	36.80	30.57	28.23	29.55	25.74	17.81	15.30	13.52	14.45
780	84.39	76.70	58.12	52.02	43.64	40.48	42.39	37.28	26.23	22.90	20.42	21.08
790	85.67	77.98	59.26	52.80	44.54	40.97	42.79	37.53	26.09	22.75	20.30	21.08
800	81.98	74.42	56.44	50.39	42.18	38.93	41.35	36.06	24.64	21.39	19.03	19.48
810	80.27	72.56	55.58	50.12	41.58	38.46	40.73	35.50	24.04	20.95	18.71	19.27
822	72.60	65.31	49.62	44.66	36.33	33.10	36.54	31.35	19.98	17.20	15.27	15.96
830	71.06	63.83	48.59	43.65	35.70	32.42	35.67	30.64	19.90	16.98	14.91	15.86
840	71.72	64.44	50.54	45.91	37.62	34.16	37.37	32.45	21.21	18.24	16.16	17.16
850	71.07	63.87	50.67	46.42	38.37	34.70	37.56	32.86	21.99	18.96	16.79	17.56
860	67.77	60.88	48.59	44.53	36.74	33.26	35.78	31.09	20.44	17.62	15.61	16.96
870	67.95	61.02	48.98	44.96	36.66	33.34	36.18	31.46	20.62	17.70	15.62	16.46
880	67.37	60.34	48.16	44.28	35.91	32.63	35.49	30.68	19.58	16.88	15.01	15.66
890	66.18	59.10	47.06	43.31	35.00	31.69	35.05	30.18	18.94	16.15	14.23	14.56
900	59.91	53.33	41.36	37.24	28.65	25.84	30.00	25.15	14.28	12.12	10.75	11.04
910	58.60	52.16	40.23	36.14	27.43	24.47	29.12	24.30	13.36	11.32	10.06	10.44
920	57.80	51.44	39.21	34.84	26.50	23.40	28.09	23.36	12.77	10.52	9.09	8.98
930	57.78	51.43	39.79	36.03	27.76	24.80	29.08	24.29	13.53	11.26	9.83	10.74
940	41.78	36.74	25.88	22.41	15.81	13.52	18.78	14.77	6.33	5.27	4.75	4.80
950	42.19	37.32	26.40	22.60	15.38	13.78	18.09	14.38	6.70	5.38	4.61	4.38
960	44.77	39.64	27.87	23.83	16.79	14.83	19.23	15.33	7.16	5.91	5.16	4.89
970	49.87	44.31	31.63	27.24	20.09	17.23	21.84	17.31	7.54	6.48	5.93	6.32

Table A.4. Description of classes of the spectral brightness coefficients (SBC) of the water surface* (*continued on next page*)

Notation	N – number of spectra in the class (mean and root-mean-square values of SBC are calculated over them).
	H – altitude of the flight in meters, three values: minimum, arithmetic mean and maximum over all spectra of the class.
	Z – solar zenith angle in degrees, three values: minimum, arithmetic mean and maximum over all spectra of the class.
	C(cl) – total chlorophyll contents, attributed to the class (µg/l) for the water surface from the accompanied contact measurements
	C(ms) – mineral matter contents, attributed to the class (µg/l) for the water surface, from the accompanied contact measurements

Class 1.0	Pure lake water: central parts of the Ladoga and Onega Lakes, far from the coast and river mouths. C(cl) = 0.5 µg/l. C(ms) = 0.5 µg/l. N = 930, H = (200/292/300), Z = (37/44/64). Observation to nadir. Variation of weather conditions: clear sky, transparent cloudiness, overcast sky.
Class 2.0	East part of the Ladoga Lake, central part of the Rybinsk reservoir during the period before "water blossom". C(cl) = 1.5 µg/l, C(ms) = 1.5 µg/l. N = 55, H = (300/300/300), Z = (35/39/51). Observation to nadir. Weather conditions: clear sky.
Class 3.0	The Ladoga and Onega Lakes at the distance 10–15 km from the coast, central part of the Rybinsk reservoir. C(cl) = 2.5 µg/l, C(ms) = 1.0 µg/l, N = 226, H = (300/300/300), Z = (35/43/64). Observation to nadir. Variation of weather conditions: clear sky, overcast sky.
Class 4.0	The Ladoga Lake: areas of the Volkhov and Svir rivers mouths and Petrokrepost bay, the Rybinsk and Tsimlyansky reservoirs C(cl) = 2.5, C(ms) = 3.0, N = 182, H = (200/299/300), Z = (35/40/63). Observation to nadir. Variation of weather conditions: clear sky, overcast sky.
Class 5.0	The Ladoga Lake: areas of the Volkhov and Svir rivers mouths near the coast, the Rybinsk reservoir. C(cl) = 4.0 µg/l, C(ms) = 1.0 µ g/l, N = 165, H = (300/300/300), Z = (35/40/63). Observation to nadir. Variation of weather conditions: clear sky, overcast sky.
Class 6.0	The Ladoga Lake: areas near the Volkhov and Svir rivers mouths. C(cl) = 5.0 µg/l, C(ms) = 3.0 µg/l. N = 66, H = (300/300/300), Z = (36/47/63). Observation to nadir. Weather conditions: clear sky.
Class 7.0	The Mingechaursky reservoir in the period of "water blossom", the Sivash Gulf. C(cl) = 3.5 µg/l, C(ms) = 3.0 µg/l. N = 35, H = (300/357/500), Z = (40/53/63). Observation to nadir. Weather conditions: clear sky.
Class 8.0	The Mingechaursky reservoir in the period of "water blossom". C(cl) = 4.0 µg/l, C(ms) = 4.0 µg/l. N = 43, H = (300/300/300), Z = (40/49/61). Observation to nadir. Weather conditions: clear sky.
Class 9.0	The Mingechaursky reservoir in the period of "water blossom". C(cl) = 5.0 µg/l, C(ms) = 6.0 µg/l. N = 43, H = (300/300/300), Z = (41/48/56). Observation to nadir. Weather conditions: clear sky.
Class 10.0	The Mingechaursky reservoir in the period of "water blossom". C(cl) = 9.0 µg/l, C(ms) = 14.0 µg/l. N = 22, H = (300/300/300), Z = (42/48/56). Observation to nadir. Weather conditions: clear sky.

Table A.4. (*continued*)

Class 11.0	The Tsimlyansky reservoir (water has light green color). There is no data about C(cl) and C(ms). N = 6, H = (200/200/200), Z = (37/37/37). Weather conditions: transparent cloudiness.
Class 12.0	The Volkhov river. There is no data about C(cl) and C(ms). N = 9, H = (300/300/300), Z = (37/41/49). Observation to nadir. Weather conditions: clear sky.
Class 13.0	The Don river (water has asphalt color). There is no data about C(cl) and C(ms). N = 9, H = (100/100/100), Z = (36/37/38). Observation to nadir. Weather conditions: overcast sky.
Class 14.0	The Black Sea (green water, i.e. the standard color of sea water), there is no data about C(cl) and C(ms). N = 23, H = (150/470/500), Z = (25/31/38). Observations to nadir, from nadir to 45° at azimuth angles 90° and 135°, from nadir to 22.5° at azimuth angle 180°. Weather conditions: clear sky.
Class 14.1	The dependence of SBC upon the viewing direction for class 14.0. N = 3, H = (150/383/500), Z = (29/32/37). Observations to viewing angle 22.5° at azimuth angle 0° (the center of Sun glare). Azimuth angle 0° corresponds to flight direction "to the Sun", azimuth angle 180° – "opposite the Sun").
Class 14.2	The dependence of SBC upon the viewing direction for class 14.0. N = 2, H = (500/500/500), Z = (27/28/28). Observations to viewing angles 22.5° and 45° at azimuth angle 45° ("Sun glare").
Class 14.3	The dependence of SBC upon the viewing direction for class 14.0. N = 23, H = (150/333/566), Z = (25/30/38). Observations to viewing angles from 22.5° till 45° at azimuth angles 90° and 135°; to viewing angle 45° at azimuth angle 180°.

* There is an archive of spectra for every class (in a special binary code)

Table A.5. Spectral brightness coefficients (SBC) of the watersurface* *(continued on next page)*

λ (μm)	__	__	__	__	__	__	Class, No.										
	1	2	3	4	5	6	7	8	9	10	11	12	13	14	14.1	14.2	14.3
0.35	4.78 / 1.51	5.44 / 0.95	4.53 / 1.17	5.42 / 1.16	6.50 / 1.25	4.34 / 0.86	5.82 / 0.75	6.11 / 0.82	7.03 / 0.74	8.44 / 1.33	3.09 / 0.93	6.97 / 0.56	4.64 / 1.39	7.06 / 1.51	19.04 / 5.71	10.40 / 3.12	5.38 / 1.53
0.36	5.00 / 1.73	5.23 / 1.22	4.37 / 1.33	4.83 / 1.45	4.71 / 0.92	4.55 / 0.69	5.47 / 1.10	6.78 / 0.85	7.95 / 0.83	9.77 / 1.57	2.42 / 0.48	7.00 / 0.56	5.09 / 1.24	3.82 / 0.86	11.46 / 2.29	5.67 / 1.13	2.67 / 0.67
0.37	4.73 / 1.66	4.64 / 1.20	3.98 / 1.28	4.30 / 1.39	4.01 / 0.84	4.38 / 0.65	4.74 / 0.94	5.97 / 0.70	7.10 / 0.72	8.89 / 1.42	1.98 / 0.34	6.90 / 0.55	5.15 / 0.95	3.66 / 0.70	11.43 / 1.94	5.67 / 0.96	2.62 / 0.67
0.38	4.53 / 1.56	4.36 / 1.17	3.84 / 1.17	4.14 / 1.31	3.93 / 0.74	4.25 / 0.65	4.36 / 0.49	5.25 / 0.56	6.39 / 0.64	8.18 / 1.30	1.94 / 0.43	6.91 / 0.62	5.09 / 1.12	3.54 / 0.65	11.69 / 2.57	5.64 / 1.24	2.53 / 0.62
0.39	4.13 / 1.38	4.07 / 1.10	3.56 / 1.05	3.87 / 1.26	3.72 / 0.73	3.95 / 0.61	4.14 / 0.35	4.91 / 0.50	6.14 / 0.63	8.03 / 1.27	2.01 / 0.46	6.99 / 0.70	4.77 / 1.10	3.91 / 0.73	13.93 / 3.20	6.27 / 1.44	2.77 / 0.64
0.40	3.23 / 0.99	3.18 / 0.83	2.85 / 0.75	3.12 / 0.92	3.19 / 0.65	3.14 / 0.49	3.35 / 0.41	3.76 / 0.36	4.81 / 0.49	6.45 / 1.01	2.36 / 0.59	7.04 / 0.70	3.83 / 0.96	5.31 / 0.91	19.89 / 4.97	8.69 / 2.17	3.83 / 0.87
0.41	3.14 / 0.82	3.28 / 0.71	3.01 / 0.70	3.35 / 0.90	3.56 / 0.62	3.45 / 0.43	3.55 / 0.58	4.77 / 0.47	6.30 / 0.70	8.34 / 1.21	2.63 / 0.66	6.89 / 0.77	4.16 / 1.04	6.31 / 1.17	25.42 / 6.36	10.16 / 2.54	4.22 / 0.91
0.42	3.10 / 0.78	3.32 / 0.62	3.14 / 0.68	3.47 / 0.85	3.80 / 0.72	3.66 / 0.49	3.90 / 1.06	5.73 / 0.58	7.71 / 0.90	10.16 / 1.42	2.73 / 0.68	6.58 / 0.86	4.38 / 1.09	6.22 / 1.34	25.68 / 6.42	9.86 / 2.47	3.97 / 0.84
0.43	2.81 / 0.69	3.06 / 0.50	2.97 / 0.56	3.28 / 0.69	3.67 / 0.69	3.50 / 0.49	3.87 / 1.05	5.79 / 0.57	7.98 / 0.90	10.59 / 1.42	2.59 / 0.44	6.46 / 0.97	4.17 / 0.71	6.31 / 1.44	26.73 / 4.54	10.06 / 1.71	3.98 / 0.81
0.44	2.61 / 0.64	2.92 / 0.40	2.83 / 0.45	3.19 / 0.58	3.66 / 0.62	3.36 / 0.45	3.88 / 0.92	5.82 / 0.55	8.21 / 0.87	11.05 / 1.41	2.59 / 0.39	6.52 / 1.11	3.98 / 0.60	6.43 / 1.35	28.19 / 4.23	10.45 / 1.57	4.13 / 0.81
0.45	2.46 / 0.58	2.85 / 0.33	2.76 / 0.38	3.17 / 0.50	3.77 / 0.56	3.28 / 0.42	3.92 / 0.83	5.89 / 0.55	8.49 / 0.83	11.52 / 1.39	2.76 / 0.39	6.43 / 1.29	3.98 / 0.56	6.14 / 1.10	27.60 / 3.86	10.16 / 1.42	4.03 / 0.75
0.46	2.33 / 0.53	2.74 / 0.28	2.68 / 0.31	3.11 / 0.43	3.82 / 0.52	3.23 / 0.41	3.83 / 0.71	5.74 / 0.52	8.40 / 0.77	11.46 / 1.32	2.90 / 0.38	6.02 / 0.90	4.05 / 0.53	5.85 / 0.92	26.14 / 3.40	9.86 / 1.28	3.89 / 0.67
0.47	2.22 / 0.48	2.62 / 0.25	2.60 / 0.25	3.06 / 0.37	3.85 / 0.51	3.18 / 0.39	3.74 / 0.60	5.60 / 0.51	8.28 / 0.71	11.27 / 1.21	3.00 / 0.36	5.21 / 0.52	4.06 / 0.49	5.82 / 0.82	25.26 / 3.03	9.91 / 1.19	3.91 / 0.63
0.48	2.15 / 0.45	2.57 / 0.23	2.57 / 0.23	3.05 / 0.34	3.94 / 0.52	3.17 / 0.36	3.80 / 0.56	5.68 / 0.51	8.48 / 0.69	11.51 / 1.17	3.08 / 0.31	4.20 / 0.42	4.09 / 0.41	5.82 / 0.70	24.71 / 2.47	10.16 / 1.02	3.95 / 0.60

Table A.5. (continued)

λ (μm)	1	2	3	4	5	6	7	8	9	10	11	12	13	14	14.1	14.2	14.3
0.49	2.15	2.60	2.60	3.10	4.05	3.25	3.98	6.02	9.05	12.30	3.15	3.52	4.30	5.73	24.32	10.25	3.89
	0.44	0.23	0.23	0.34	0.54	0.36	0.60	0.54	0.72	1.20	0.28	0.53	0.39	0.61	2.19	0.92	0.57
0.50	2.14	2.62	2.63	3.14	4.10	3.35	4.17	6.39	9.70	13.21	3.29	3.27	4.52	5.54	23.86	10.11	3.71
	0.44	0.23	0.24	0.34	0.56	0.38	0.66	0.57	0.77	1.24	0.25	0.82	0.36	0.60	1.67	0.71	0.53
0.51	2.13	2.60	2.64	3.17	4.15	3.46	4.33	6.67	10.22	14.05	3.57	3.25	4.71	5.40	23.93	10.01	3.53
	0.43	0.23	0.25	0.34	0.57	0.43	0.70	0.61	0.82	1.32	0.25	0.98	0.38	0.66	1.67	0.70	0.50
0.52	2.11	2.59	2.65	3.18	4.18	3.60	4.51	6.97	10.75	14.92	3.84	3.35	4.99	5.41	24.61	10.21	3.47
	0.42	0.22	0.25	0.33	0.57	0.49	0.73	0.64	0.87	1.40	0.27	0.90	0.36	0.73	1.72	0.71	0.50
0.53	2.09	2.57	2.65	3.17	4.20	3.74	4.68	7.26	11.28	15.80	4.09	3.41	5.26	5.36	25.13	10.40	3.41
	0.41	0.21	0.24	0.31	0.59	0.53	0.75	0.66	0.95	1.50	0.30	0.58	0.37	0.74	1.76	0.73	0.49
0.54	2.07	2.56	2.65	3.18	4.23	3.90	4.78	7.42	11.61	16.37	4.34	3.41	5.48	5.21	25.55	10.40	3.28
	0.40	0.22	0.24	0.31	0.61	0.53	0.75	0.69	0.99	1.58	0.34	0.51	0.38	0.70	1.79	0.73	0.46
0.55	2.05	2.55	2.67	3.21	4.29	4.06	4.76	7.34	11.65	16.58	4.64	3.41	5.71	5.10	26.63	10.40	3.16
	0.38	0.24	0.26	0.31	0.63	0.56	0.72	0.68	1.03	1.60	0.41	0.41	0.40	0.69	1.86	0.73	0.44
0.56	2.04	2.53	2.70	3.25	4.36	4.23	4.64	7.10	11.40	16.42	4.99	3.36	5.95	4.99	28.06	10.35	3.03
	0.38	0.24	0.28	0.31	0.65	0.63	0.66	0.66	1.05	1.60	0.46	0.40	0.47	0.68	1.96	0.72	0.42
0.57	2.01	2.49	2.70	3.24	4.35	4.36	4.40	6.65	10.83	15.83	5.24	3.14	6.20	4.83	29.26	10.25	2.84
	0.37	0.23	0.30	0.32	0.67	0.72	0.60	0.63	1.03	1.60	0.48	0.31	0.50	0.66	2.05	0.72	0.40
0.58	1.98	2.44	2.68	3.23	4.35	4.46	3.99	5.96	9.85	14.70	5.27	2.87	6.48	4.60	30.47	10.11	2.59
	0.37	0.21	0.31	0.32	0.68	0.74	0.53	0.56	1.00	1.58	0.43	0.37	0.52	0.62	2.44	0.81	0.38
0.59	1.93	2.40	2.64	3.21	4.35	4.52	3.43	5.03	8.41	12.94	5.03	2.73	6.75	4.41	32.36	10.01	2.32
	0.36	0.21	0.31	0.32	0.69	0.76	0.44	0.47	0.95	1.59	0.40	0.44	0.54	0.60	2.59	0.80	0.39
0.60	1.86	2.34	2.57	3.16	4.31	4.50	2.83	4.03	6.74	10.78	4.70	2.69	4.14	4.28	34.05	10.06	2.14
	0.35	0.21	0.30	0.32	0.69	0.79	0.34	0.38	0.90	1.59	0.38	0.54	1.26	0.60	2.72	0.80	0.40
0.61	1.78	2.28	2.51	3.09	4.27	4.44	2.40	3.31	5.45	9.05	4.43	6.97	3.18	4.15	35.06	10.11	2.04
	0.32	0.21	0.29	0.31	0.69	0.80	0.27	0.32	0.85	1.61	0.35	0.56	0.79	0.59	2.80	0.81	0.39
0.62	1.71	2.22	2.45	3.04	4.29	4.43	2.17	2.94	4.77	8.10	4.25	7.00	2.63	4.06	36.26	10.35	1.96
	0.31	0.21	0.28	0.32	0.71	0.80	0.24	0.29	0.81	1.58	0.37	0.56	0.52	0.58	2.90	0.83	0.39

Class, No.

λ																	
0.63	1.91 / 0.39	10.69 / 0.96	37.76 / 3.40	4.02 / 0.58	2.62 / 0.58	6.90 / 0.55	4.15 / 0.44	7.64 / 1.55	4.45 / 0.78	2.76 / 0.28	2.05 / 0.23	4.46 / 0.84	4.30 / 0.73	3.01 / 0.33	2.41 / 0.28	2.18 / 0.22	1.67 / 0.30
0.64	1.86 / 0.38	10.89 / 1.09	39.26 / 3.93	3.97 / 0.57	3.02 / 0.69	6.91 / 0.62	4.01 / 0.58	7.29 / 1.50	4.23 / 0.76	2.64 / 0.27	1.96 / 0.22	4.50 / 0.91	4.29 / 0.75	3.00 / 0.35	2.37 / 0.29	2.17 / 0.22	1.63 / 0.30
0.65	1.82 / 0.38	10.99 / 1.10	40.76 / 4.08	3.93 / 0.55	3.46 / 0.87	6.99 / 0.70	3.70 / 0.59	6.88 / 1.46	3.98 / 0.72	2.50 / 0.25	1.88 / 0.22	4.52 / 0.95	4.27 / 0.76	2.97 / 0.37	2.33 / 0.29	2.15 / 0.22	1.59 / 0.30
0.66	1.80 / 0.38	11.18 / 1.23	42.32 / 4.65	3.90 / 0.54	3.57 / 0.89	7.04 / 0.70	3.34 / 0.39	6.47 / 1.42	3.74 / 0.68	2.39 / 0.23	1.81 / 0.22	4.46 / 0.95	4.21 / 0.74	2.91 / 0.37	2.26 / 0.28	2.10 / 0.22	1.53 / 0.29
0.67	1.78 / 0.37	11.28 / 1.47	43.62 / 5.67	3.82 / 0.54	3.31 / 0.83	6.89 / 0.77	3.13 / 0.41	6.10 / 1.39	3.54 / 0.64	2.32 / 0.22	1.77 / 0.24	4.32 / 0.91	4.16 / 0.72	2.85 / 0.38	2.17 / 0.27	2.06 / 0.23	1.48 / 0.28
0.68	1.74 / 0.35	11.28 / 1.69	44.43 / 6.67	3.71 / 0.56	2.99 / 0.54	6.58 / 0.86	3.10 / 0.47	5.77 / 1.35	3.38 / 0.62	2.26 / 0.21	1.74 / 0.25	4.22 / 0.92	4.24 / 0.75	2.83 / 0.38	2.14 / 0.27	2.05 / 0.24	1.43 / 0.28
0.69	1.72 / 0.35	11.67 / 1.98	45.87 / 7.80	3.71 / 0.52	3.09 / 0.48	6.46 / 0.97	3.08 / 0.52	6.06 / 1.47	3.53 / 0.65	2.39 / 0.23	1.83 / 0.31	4.21 / 0.96	4.41 / 0.80	2.83 / 0.41	2.11 / 0.29	2.03 / 0.29	1.36 / 0.29
0.70	1.71 / 0.37	12.06 / 2.41	45.74 / 9.15	3.69 / 0.45	3.55 / 0.50	6.52 / 1.11	3.15 / 0.63	6.35 / 1.61	3.64 / 0.69	2.52 / 0.24	1.94 / 0.39	4.16 / 0.99	4.55 / 0.83	2.83 / 0.43	2.07 / 0.30	2.04 / 0.33	1.28 / 0.29
0.71	1.63 / 0.38	12.06 / 1.81	46.68 / 7.00	3.56 / 0.46	3.99 / 0.52	6.43 / 1.29	3.12 / 0.47	5.26 / 1.34	3.03 / 0.56	2.20 / 0.21	1.77 / 0.33	3.89 / 0.94	4.55 / 0.84	2.83 / 0.40	2.03 / 0.27	2.08 / 0.29	1.26 / 0.25
0.72	1.59 / 0.41	12.74 / 1.27	56.28 / 5.63	3.68 / 0.63	4.37 / 0.52	6.02 / 0.90	2.90 / 0.41	3.78 / 0.88	2.33 / 0.37	1.82 / 0.18	1.53 / 0.27	3.32 / 0.78	4.38 / 0.81	2.71 / 0.39	1.88 / 0.24	2.04 / 0.24	1.19 / 0.22
0.73	1.65 / 0.45	13.57 / 1.36	64.88 / 6.49	3.91 / 0.62	4.82 / 0.48	5.21 / 0.52	2.68 / 0.38	3.05 / 0.58	2.09 / 0.28	1.72 / 0.17	1.48 / 0.30	2.69 / 0.58	4.16 / 0.73	2.52 / 0.42	1.68 / 0.23	1.97 / 0.23	1.09 / 0.21
0.74	1.64 / 0.42	12.89 / 1.93	59.05 / 8.86	3.74 / 0.46	5.28 / 0.48	4.20 / 0.42	2.50 / 0.37	2.72 / 0.44	2.01 / 0.24	1.69 / 0.15	1.49 / 0.32	2.23 / 0.47	4.02 / 0.67	2.42 / 0.46	1.57 / 0.25	1.96 / 0.24	1.03 / 0.21
0.75	1.58 / 0.36	12.11 / 3.03	56.121 / 4.03	3.52 / 0.46	5.66 / 0.40	3.52 / 0.53	2.28 / 0.57	2.44 / 0.37	1.85 / 0.21	1.56 / 0.14	1.41 / 0.30	2.07 / 0.46	3.90 / 0.65	2.34 / 0.48	1.51 / 0.25	1.92 / 0.24	0.99 / 0.21
0.76	1.64 / 0.42	13.13 / 3.94	68.492 / 0.55	3.71 / 0.60	6.10 / 0.46	3.27 / 0.82	2.21 / 0.66	2.32 / 0.35	1.76 / 0.21	1.49 / 0.14	1.33 / 0.29	2.02 / 0.51	3.76 / 0.68	2.27 / 0.49	1.48 / 0.24	1.88 / 0.24	0.96 / 0.21
0.77	1.71 / 0.46	14.01 / 3.78	72.721 / 9.63	3.82 / 0.54	6.65 / 0.57	3.25 / 0.98	2.40 / 0.65	2.47 / 0.41	1.86 / 0.24	1.60 / 0.14	1.41 / 0.32	2.07 / 0.52	3.77 / 0.64	2.27 / 0.51	1.47 / 0.26	1.89 / 0.25	0.96 / 0.21
0.78	1.66 / 0.42	13.57 / 2.31	64.191 / 1.54	3.57 / 0.42	7.35 / 0.60	3.35 / 0.90	2.59 / 0.44	2.55 / 0.44	1.90 / 0.25	1.64 / 0.15	1.45 / 0.32	2.10 / 0.48	3.86 / 0.65	2.30 / 0.51	1.47 / 0.25	1.89 / 0.26	0.96 / 0.20

Table A.5. (continued)

Class, No.

λ (μm)	1	2	3	4	5	6	7	8	9	10	11	12	13	14	14.1	14.2	14.3
0.63	1.67	2.18	2.41	3.01	4.30	4.46	2.05	2.76	4.45	7.64	4.15	6.90	2.62	4.02	37.76	10.69	1.91
	0.30	0.22	0.28	0.33	0.73	0.84	0.23	0.28	0.78	1.55	0.44	0.55	0.58	0.58	3.40	0.96	0.39
0.64	1.63	2.17	2.37	3.00	4.29	4.50	1.96	2.64	4.23	7.29	4.01	6.91	3.02	3.97	39.26	10.89	1.86
	0.30	0.22	0.29	0.35	0.75	0.91	0.22	0.27	0.76	1.50	0.58	0.62	0.69	0.57	3.93	1.09	0.38
0.65	1.59	2.15	2.33	2.97	4.27	4.52	1.88	2.50	3.98	6.88	3.70	6.99	3.46	3.93	40.76	10.99	1.82
	0.30	0.22	0.29	0.37	0.76	0.95	0.22	0.25	0.72	1.46	0.59	0.70	0.87	0.55	4.08	1.10	0.38
0.66	1.53	2.10	2.26	2.91	4.21	4.46	1.81	2.39	3.74	6.47	3.34	7.04	3.57	3.90	42.32	11.18	1.80
	0.29	0.22	0.28	0.37	0.74	0.95	0.22	0.23	0.68	1.42	0.39	0.70	0.89	0.54	4.65	1.23	0.38
0.67	1.48	2.06	2.17	2.85	4.16	4.32	1.77	2.32	3.54	6.10	3.13	6.89	3.31	3.82	43.62	11.28	1.78
	0.28	0.23	0.27	0.38	0.72	0.91	0.24	0.22	0.64	1.39	0.41	0.77	0.83	0.54	5.67	1.47	0.37
0.68	1.43	2.05	2.14	2.83	4.24	4.22	1.74	2.26	3.38	5.77	3.10	6.58	2.99	3.71	44.43	11.28	1.74
	0.28	0.24	0.27	0.38	0.75	0.92	0.25	0.21	0.62	1.35	0.47	0.86	0.54	0.56	6.67	1.69	0.35
0.69	1.36	2.03	2.11	2.83	4.41	4.21	1.83	2.39	3.53	6.06	3.08	6.46	3.09	3.71	45.87	11.67	1.72
	0.29	0.29	0.29	0.41	0.80	0.96	0.31	0.23	0.65	1.47	0.52	0.97	0.48	0.52	7.80	1.98	0.35
0.70	1.28	2.04	2.07	2.83	4.55	4.16	1.94	2.52	3.64	6.35	3.15	6.52	3.55	3.69	45.74	12.06	1.71
	0.29	0.33	0.30	0.43	0.83	0.99	0.39	0.24	0.69	1.61	0.63	1.11	0.50	0.45	9.15	2.41	0.37
0.71	1.26	2.08	2.03	2.83	4.55	3.89	1.77	2.20	3.03	5.26	3.12	6.43	3.99	3.56	46.68	12.06	1.63
	0.25	0.29	0.27	0.40	0.84	0.94	0.33	0.21	0.56	1.34	0.47	1.29	0.52	0.46	7.00	1.81	0.38
0.72	1.19	2.04	1.88	2.71	4.38	3.32	1.53	1.82	2.33	3.78	2.90	6.02	4.37	3.68	56.28	12.74	1.59
	0.22	0.24	0.24	0.39	0.81	0.78	0.27	0.18	0.37	0.88	0.41	0.90	0.52	0.63	5.63	1.27	0.41
0.73	1.09	1.97	1.68	2.52	4.16	2.69	1.48	1.72	2.09	3.05	2.68	5.21	4.82	3.91	64.88	13.57	1.65
	0.21	0.23	0.23	0.42	0.73	0.58	0.30	0.17	0.28	0.58	0.38	0.52	0.48	0.62	6.49	1.36	0.45
0.74	1.03	1.96	1.57	2.42	4.02	2.28	1.49	1.69	2.01	2.72	2.50	4.20	5.28	3.74	59.05	12.89	1.64
	0.21	0.24	0.25	0.46	0.67	0.47	0.32	0.15	0.24	0.44	0.37	0.42	0.48	0.46	8.86	1.93	0.42
0.75	0.99	1.92	1.51	2.34	3.90	2.07	1.41	1.56	1.85	2.44	2.28	3.52	5.66	3.52	56.121	12.11	1.58
	0.21	0.24	0.25	0.48	0.65	0.46	0.30	0.14	0.21	0.37	0.57	0.53	0.40	0.46	4.03	3.03	0.36
0.76	0.96	1.88	1.48	2.27	3.76	2.02	1.33	1.49	1.76	2.32	2.21	3.27	6.10	3.71	68.492	13.13	1.64
	0.21	0.24	0.24	0.49	0.68	0.51	0.29	0.14	0.21	0.35	0.66	0.82	0.46	0.60	0.55	3.94	0.42

	1	2	3	4	5	6	7	8	9	10	11	12	13	14	15	16	17
0.77	0.96	1.89	1.47	2.27	3.77	2.07	1.41	1.60	1.86	2.47	2.40	3.25	6.65	3.82	72.721	14.01	1.71
	0.21	0.25	0.26	0.51	0.64	0.52	0.32	0.14	0.24	0.41	0.65	0.98	0.57	0.54	9.63	3.78	0.46
0.78	0.96	1.89	1.47	2.30	3.86	2.10	1.45	1.64	1.90	2.55	2.59	3.35	7.35	3.57	64.191	13.57	1.66
	0.20	0.26	0.25	0.51	0.65	0.48	0.32	0.15	0.25	0.44	0.44	0.90	0.60	0.42	1.54	2.31	0.42
0.79	0.94	1.86	1.45	2.30	3.87	2.12	1.39	1.57	1.83	2.52	2.49	3.41	8.11	3.25	56.64	12.99	1.57
	0.19	0.25	0.24	0.50	0.68	0.49	0.30	0.15	0.24	0.42	0.37	0.58	0.65	0.38	8.50	2.05	0.37
0.80	0.92	1.84	1.44	2.28	3.88	2.13	1.31	1.50	1.77	2.46	2.28	3.41	8.82	3.00	51.89	12.89	1.50
	0.19	0.24	0.24	0.49	0.70	0.49	0.28	0.15	0.22	0.42	0.32	0.51	0.71	0.34	6.23	2.25	0.32
0.81	0.89	1.83	1.42	2.25	3.88	2.06	1.23	1.43	1.69	2.31	2.13	3.41	9.67	2.96	53.71	13.38	1.46
	0.19	0.25	0.23	0.48	0.73	0.44	0.26	0.15	0.19	0.39	0.30	0.41	0.78	0.38	6.45	1.86	0.29
0.82	0.87	1.82	1.37	2.20	3.82	1.88	1.21	1.42	1.66	2.21	2.07	3.36	10.33	3.21	61.43	14.75	1.50
	0.19	0.25	0.23	0.47	0.73	0.38	0.28	0.16	0.17	0.33	0.29	0.40	0.82	0.47	6.14	1.47	0.33
0.83	0.85	1.83	1.32	2.14	3.71	1.69	1.27	1.46	1.69	2.17	2.04	3.14	10.73	3.37	65.98	15.58	1.57
	0.20	0.24	0.24	0.50	0.67	0.33	0.33	0.17	0.17	0.28	0.29	0.31	0.90	0.45	8.58	2.02	0.36
0.84	0.83	1.81	1.28	2.10	3.63	1.56	1.29	1.47	1.66	2.08	2.00	2.87	10.77	3.26	67.291	15.28	1.56
	0.20	0.24	0.26	0.53	0.64	0.33	0.35	0.18	0.20	0.27	0.30	0.37	0.92	0.38	0.09	2.59	0.35
0.85	0.81	1.76	1.24	2.06	3.56	1.48	1.26	1.42	1.58	1.95	1.94	2.73	10.47	3.12	69.731	14.84	1.56
	0.20	0.24	0.27	0.55	0.64	0.35	0.36	0.17	0.21	0.30	0.39	0.44	0.91	0.37	3.95	3.22	0.35

* Upper value is the arithmetic mean SBC (percent), lower value is the standard deviation over the class

Table A.6. Description of the classes of the spectral brightness coefficients (SBC) of the ground surface*

Class 15.0	Snow surface. N = 41, H = (200/354/500), Z = (60/62/65). Observation to nadir, and from nadir to 45° at azimuth angles 0°, 45°, 90°, 135°, 180°. Variation of weather conditions: clear sky, overcast sky.
Class 16.0	Sand surface (dunes of the Kara-Kum desert). N = 63, H = (500/500/500), Z = (52/54/62). Observation to nadir. Weather conditions: clear sky.
Class 17.0	Tillage (fields of the black soil). N = 134, H = (100/354/500), Z = (34/50/72). Observation to nadir. Variation of weather conditions: clear sky, transparent cloudiness.
Class 18.0	Field with weak green young growth. N = 6, H = (500/500/500), Z = (31/32/36). Observation to nadir. Weather conditions: transparent cloudiness.
Class 19.0	Field with green young growth. N = 9, H = (100/144/200), Z = (41/46/50). Observation to nadir Weather conditions: clear sky.
Class 20.0	Fields with continuous green grass cover (without blossom). N = 26, H = (100/438/500), Z = (28/45/58). Observation to nadir. Variation of weather conditions: clear sky, transparent cloudiness.
Class 21.0	Moss marsh of brownish-green color. N = 89, H = (500/500/500), Z = (37/41/54). Observation to nadir at azimuthal angles from 0° till 90°, viewing angles from 15° till 45°. Weather conditions: clear sky.
Class 21.1	Dependence of SBC upon the viewing direction for class 21: at azimuth angle 135° viewing angles from 15° to 45°; at azimuth angle 180° viewing angles from 15° to 30°. N = 5, H = (500/500/500), Z = (37/38/39).
Class 21.2	Dependence of SBC upon the viewing direction for class 21 (direction is opposite to the Sun): at azimuth angle 180° viewing angles from 30° to 45°. N = 6, H = (500/500/500), Z = (37/38/39).
Class 22.0	Fields with sunflower (blossom period, yellowish-green color). N = 46, H = (200/200/200), Z = (30/39/57). Observation at azimuth angle 0°. The viewing angles from nadir to 45°. Weather conditions: clear sky.
Class 23.0	Fields with maize (period of the corn ripening, yellowish-green color). N = 6, H = (200/200/200), Z = (37/40/47). Observation at azimuth angle 0° from nadir to 45°. Weather conditions: clear sky.
Class 24.0	Ripe grain crop (wheat, barley). N = 57, H = (200/200/200), Z = (25/38/50). Observation: at azimuth angle 0° viewing angles from nadir to 45°. Weather conditions: clear sky.
Class 25.0	Stubble (fields after harvesting of grain crop). N = 14, H = (200/200/200), Z = (31/40/44). Observation: at azimuth angle 0° the viewing angles from nadir to 45°. Weather conditions: clear sky.
Class 26.0	Asphalt road (gray color, dry). N = 9, H = (50/50/50), Z = (34/44/52). Observation to nadir. Weather conditions: clear sky, transparent cloudiness.
Class 27.0	Concrete road (clean, light gray color). N = 3, H = (50/50/50), Z = (31/31/32). Observation to nadir. Weather conditions: clear sky.

* There is an archive of spectra for every class (in a special binary code)

Table A.7. Spectral brightness coefficients (SBC) of the ground surface* *(continued on next page)*

								Class, No.							
λ (μm)	15	16	17	18	19	20	21	21.1	21.2	22	23	24	25	26	27
0.35	60.75	11.49	4.49	4.36	1.89	3.21	9.24	12.66	15.04	1.81	2.21	3.15	2.51	6.58	6.16
	10.64	1.41	1.40	1.31	0.57	0.85	1.35	3.80	4.51	0.53	0.91	0.74	0.40	1.97	1.85
0.36	48.46	7.38	3.44	3.63	1.45	2.54	6.33	8.70	10.34	1.53	1.43	1.96	2.03	5.58	5.75
	8.10	0.61	0.79	0.73	0.35	0.69	0.84	1.74	2.07	0.35	0.29	0.36	0.26	1.12	1.15
0.37	50.88	6.83	3.10	3.38	1.20	2.23	4.85	6.71	7.91	1.34	1.16	1.52	1.83	5.12	5.35
	8.29	0.55	0.73	0.58	0.22	0.60	0.63	1.14	1.34	0.30	0.20	0.25	0.20	1.13	0.91
0.38	58.94	8.45	3.52	3.71	1.09	2.38	4.87	6.77	7.98	1.35	1.19	1.49	1.95	5.87	5.84
	8.59	0.84	0.91	0.82	0.24	0.69	0.64	1.49	1.75	0.29	0.26	0.23	0.23	1.61	1.28
0.39	61.64	9.91	3.73	3.99	1.17	2.47	4.42	6.22	7.39	1.52	1.38	1.75	2.29	6.67	7.14
	10.55	0.91	0.97	0.92	0.27	0.69	0.57	1.43	1.70	0.32	0.32	0.30	0.19	1.55	1.64
0.40	58.95	10.48	3.79	4.95	1.43	2.55	2.65	3.79	4.53	1.77	1.64	2.20	2.73	7.17	9.15
	11.02	1.01	0.93	1.24	0.36	0.65	0.35	0.95	1.13	0.35	0.41	0.40	0.15	1.79	2.29
0.41	47.39	10.87	3.18	4.60	2.15	2.21	5.11	7.88	9.49	1.17	1.05	1.34	2.00	6.88	11.04
	6.88	2.27	0.64	1.15	0.54	0.43	0.51	1.97	2.37	0.28	0.26	0.45	0.24	1.72	2.76
0.42	46.58	10.59	3.05	4.13	2.43	2.22	7.62	11.84	14.50	0.72	0.46	0.73	1.40	7.09	12.01
	6.27	2.59	0.56	1.03	0.61	0.46	0.82	2.96	3.63	0.39	0.12	0.50	0.33	1.77	3.00
0.43	50.24	10.30	3.22	3.97	2.36	2.46	8.48	13.08	16.55	0.82	0.48	1.02	1.63	7.44	12.21
	6.53	1.67	0.54	0.68	0.40	0.54	0.89	2.22	2.81	0.47	0.13	0.69	0.29	2.02	2.08
0.44	56.75	11.23	3.63	4.27	2.39	2.85	7.90	12.09	15.77	1.33	0.97	2.02	2.35	8.27	11.72
	6.64	1.17	0.57	0.64	0.36	0.64	0.85	1.81	2.37	0.49	0.15	0.94	0.27	2.26	1.76
0.45	62.69	12.77	4.09	4.92	2.53	3.20	7.04	10.85	14.44	2.00	1.61	3.23	3.25	9.24	11.85
	6.49	1.13	0.61	0.69	0.39	0.70	0.77	1.52	2.02	0.46	0.23	1.15	0.29	2.03	1.66
0.46	66.49	13.92	4.45	5.53	2.56	3.45	6.54	10.15	13.61	2.65	2.20	4.30	4.07	9.98	12.63
	6.38	1.19	0.63	0.72	0.38	0.75	0.73	1.32	1.77	0.41	0.29	1.33	0.31	1.88	1.64
0.47	69.49	14.92	4.74	5.89	2.59	3.64	6.35	9.93	13.15	3.20	2.69	5.23	4.73	10.69	13.70
	6.42	1.25	0.66	0.71	0.31	0.76	0.69	1.19	1.58	0.39	0.32	1.47	0.35	1.95	1.64
0.48	72.48	16.10	5.04	6.23	2.72	3.82	6.48	10.13	13.17	3.68	3.11	6.14	5.33	11.49	14.81
	6.76	1.31	0.69	0.62	0.27	0.78	0.65	1.01	1.32	0.41	0.31	1.64	0.43	2.10	1.48

Table A.7. (continued)

λ (μm)	15	16	17	18	19	20	21	21.1	21.2	22	23	24	25	26	27
0.49	74.45	17.29	5.31	6.69	2.90	3.97	6.95	10.83	13.79	4.13	3.46	7.00	5.90	12.15	15.66
	7.05	1.41	0.71	0.60	0.26	0.81	0.64	0.97	1.24	0.48	0.31	1.83	0.47	2.15	1.41
0.50	75.51	18.41	5.58	7.18	3.08	4.19	7.80	12.14	15.09	4.71	3.94	7.86	6.55	12.62	16.54
	7.15	1.61	0.73	0.59	0.25	0.85	0.67	0.88	1.06	0.61	0.28	2.02	0.53	2.22	1.16
0.51	76.66	19.86	5.92	7.93	3.42	4.84	9.20	14.36	17.56	5.78	4.72	8.83	7.35	13.21	17.71
	7.38	1.72	0.77	0.83	0.27	0.96	0.80	1.10	1.37	0.81	0.33	2.22	0.62	2.34	1.24
0.52	77.69	21.60	6.26	8.85	4.06	6.20	10.49	16.33	19.97	7.38	5.67	9.83	8.26	13.70	18.91
	7.61	1.71	0.82	0.62	0.32	1.16	0.96	1.31	1.62	1.12	0.42	2.43	0.65	2.29	1.32
0.53	78.28	23.18	6.57	9.34	4.91	8.11	11.18	17.45	21.37	9.48	6.99	11.03	9.46	14.10	20.57
	7.59	1.73	0.86	0.65	0.36	1.45	1.02	1.45	1.62	1.54	0.57	2.65	0.63	2.15	1.44
0.54	78.80	24.69	6.91	9.34	5.59	9.86	11.38	17.94	21.94	11.55	8.59	12.28	10.62	14.74	22.10
	7.60	1.92	0.93	0.65	0.39	1.76	1.02	1.46	1.54	1.93	0.60	2.85	0.67	2.19	1.55
0.55	79.78	26.50	7.34	9.26	5.92	10.96	11.31	17.90	22.05	13.08	9.99	13.46	11.53	15.57	23.37
	7.97	2.12	1.01	0.65	0.41	2.02	0.99	1.41	1.54	2.15	0.70	3.05	0.70	2.30	1.64
0.56	80.49	28.39	7.81	9.00	5.98	11.25	11.26	17.84	22.22	13.97	10.76	14.65	12.35	16.30	24.32
	8.41	2.04	1.11	0.63	0.42	2.13	0.94	1.37	1.56	2.29	0.75	3.23	0.72	2.24	1.70
0.57	79.94	29.85	8.19	8.98	5.75	10.51	11.15	17.87	22.33	13.90	10.68	15.58	12.90	16.67	24.71
	8.43	2.08	1.20	0.88	0.40	2.04	0.96	1.50	1.56	2.40	0.94	3.38	0.82	2.21	1.73
0.58	79.54	31.25	8.54	9.34	5.47	9.22	11.26	18.39	22.97	13.48	10.25	16.63	13.46	17.00	26.11
	8.26	2.36	1.26	0.75	0.44	1.81	1.01	1.56	1.84	2.53	1.28	3.56	1.02	2.13	2.09
0.59	79.95	32.89	8.91	10.04	5.41	8.18	11.44	19.00	23.76	13.21	9.85	17.79	13.92	17.49	28.91
	8.21	2.45	1.36	0.80	0.45	1.58	1.10	1.52	1.90	2.64	1.13	3.79	1.11	2.04	2.98
0.60	80.03	34.16	9.21	10.79	5.51	7.58	11.39	19.28	24.10	13.19	9.57	19.04	14.38	18.14	30.73
	8.29	2.81	1.45	0.86	0.45	1.38	1.11	1.54	1.93	2.73	0.77	3.97	1.20	2.14	2.62
0.61	80.43	34.97	9.49	11.33	5.51	7.13	11.29	19.51	24.40	13.32	9.62	20.49	15.19	18.75	30.50
	8.70	3.13	1.51	0.91	0.44	1.23	1.04	1.56	1.95	2.85	0.78	4.13	1.32	2.14	2.44
0.62	81.20	35.72	9.80	11.62	5.41	6.68	11.36	19.77	24.67	13.27	9.72	21.74	16.02	19.18	30.73
	9.17	3.05	1.60	0.93	0.43	1.08	1.01	1.58	2.10	2.90	0.93	4.33	1.31	2.02	2.46

Class, No.

λ	1	2	3	4	5	6	7	8	9	10	11	12	13	14	15
0.63	80.96 / 8.93	36.43 / 3.01	10.05 / 1.67	11.70 / 1.05	5.31 / 0.48	6.24 / 0.98	11.16 / 1.07	19.57 / 1.76	23.71 / 2.28	13.04 / 2.95	9.57 / 1.05	22.85 / 4.53	16.64 / 1.31	19.51 / 1.96	32.88 / 2.96
0.64	80.60 / 8.59	37.06 / 3.01	10.25 / 1.74	11.87 / 1.19	5.14 / 0.51	5.91 / 0.97	10.72 / 1.15	18.93 / 1.89	22.22 / 2.74	12.89 / 3.04	9.46 / 1.26	23.94 / 4.65	17.34 / 1.52	19.80 / 2.14	34.77 / 3.48
0.65	80.69 / 8.57	37.79 / 3.18	10.46 / 1.81	11.88 / 1.36	4.90 / 0.49	5.45 / 0.95	10.47 / 1.22	18.60 / 1.86	21.57 / 3.08	12.50 / 2.92	9.15 / 1.54	24.59 / 4.73	17.71 / 1.76	20.04 / 2.27	35.06 / 3.51
0.66	80.45 / 8.48	38.54 / 3.38	10.62 / 1.84	11.05 / 1.41	4.78 / 0.53	4.84 / 0.92	10.29 / 1.32	18.44 / 2.03	21.14 / 3.23	11.68 / 2.69	8.64 / 1.53	25.00 / 4.83	17.68 / 1.64	20.24 / 2.23	35.03 / 3.85
0.67	81.08 / 9.19	39.31 / 3.28	10.86 / 1.91	9.88 / 1.28	4.75 / 0.62	4.35 / 0.86	10.41 / 1.42	18.75 / 2.44	21.00 / 3.20	10.92 / 2.55	8.32 / 1.39	25.81 / 4.85	18.28 / 1.59	20.75 / 2.70	36.59 / 4.76
0.68	82.57 / 10.96	40.04 / 3.67	11.31 / 2.05	9.60 / 1.44	4.71 / 0.71	4.22 / 0.85	12.65 / 1.60	22.22 / 3.33	25.11 / 3.77	10.58 / 2.56	8.28 / 1.35	26.60 / 4.88	19.18 / 1.91	21.42 / 3.21	38.80 / 5.82
0.69	81.95 / 11.27	40.62 / 3.74	11.66 / 2.18	10.12 / 1.72	5.26 / 0.89	5.34 / 1.25	19.18 / 2.00	30.92 / 5.26	36.00 / 6.12	11.36 / 2.86	8.45 / 1.44	26.99 / 4.83	19.57 / 1.94	21.67 / 3.68	38.70 / 6.58
0.70	80.58 / 9.86	41.98 / 3.27	12.03 / 2.25	10.68 / 2.14	7.45 / 1.49	9.99 / 2.13	30.41 / 2.92	43.27 / 8.65	50.67 / 10.13	15.70 / 4.03	10.90 / 2.18	28.48 / 4.53	20.93 / 1.43	22.08 / 4.42	36.17 / 7.23
0.71	81.30 / 9.87	43.55 / 3.64	12.54 / 2.37	11.49 / 1.72	11.09 / 1.66	19.28 / 3.85	45.32 / 4.76	58.91 / 8.84	68.33 / 10.25	23.43 / 5.56	16.96 / 4.28	30.45 / 4.93	22.85 / 1.05	23.31 / 5.46	32.26 / 4.84
0.72	81.00 / 10.02	43.62 / 3.80	12.81 / 2.42	12.71 / 1.27	15.58 / 1.56	31.70 / 6.32	60.20 / 6.63	75.71 / 7.57	87.22 / 8.72	30.45 / 6.26	24.27 / 5.22	30.19 / 5.01	22.87 / 1.98	23.97 / 6.66	28.29 / 2.83
0.73	80.77 / 9.77	43.93 / 4.33	13.09 / 2.45	13.87 / 1.58	20.66 / 2.19	46.15 / 8.46	65.47 / 7.07	81.59 / 8.28	93.64 / 9.36	37.76 / 6.90	34.83 / 3.94	30.65 / 4.83	25.09 / 3.76	24.23 / 7.45	29.75 / 2.98
0.74	82.16 / 10.23	45.32 / 4.65	13.74 / 2.59	15.09 / 2.26	25.04 / 3.76	60.30 / 10.16	59.86 / 6.09	74.03 / 11.11	84.77 / 12.71	50.54 / 8.65	50.93 / 7.64	35.79 / 5.62	32.16 / 5.68	25.29 / 7.46	34.99 / 5.25
0.75	84.01 / 11.15	44.36 / 4.42	13.93 / 2.85	16.49 / 4.12	26.83 / 6.71	66.98 / 10.64	64.26 / 6.89	78.77 / 19.69	90.61 / 22.65	60.08 / 10.68	61.23 / 15.31	39.01 / 6.12	34.73 / 5.59	26.26 / 6.56	33.79 / 8.45
0.76	87.64 / 11.34	42.64 / 3.96	13.80 / 3.13	17.51 / 5.25	27.66 / 8.30	67.24 / 10.27	80.84 / 9.06	99.12 / 29.74	114.43 / 4.32	56.69 / 11.01	59.21 / 17.76	36.29 / 5.74	30.43 / 4.66	26.24 / 7.87	28.91 / 8.67
0.77	88.75 / 10.71	44.94 / 4.26	14.47 / 3.11	18.68 / 5.04	29.84 / 8.06	71.01 / 10.89	83.22 / 8.15	102.92 / 7.79	116.93 / 1.57	54.17 / 9.72	58.63 / 15.83	36.43 / 5.61	30.22 / 4.86	26.38 / 7.12	30.31 / 8.18
0.78	83.74 / 10.59	47.61 / 4.40	15.43 / 3.00	20.00 / 3.40	30.92 / 5.26	75.93 / 12.13	73.79 / 6.74	92.06 / 15.65	103.51 / 7.60	60.12 / 10.54	67.25 / 11.43	41.09 / 6.67	35.65 / 6.59	27.01 / 5.16	36.30 / 6.17

Table A.7. (*continued*)

λ (µm)		15	16	17	18	19	20	21	21.1	21.2	22	23	24	25	26	27
									Class, No.							
0.79		79.28	47.92	15.84	20.12	30.87	76.43	67.89	84.90	96.32	66.24	75.29	45.35	40.77	27.18	40.46
		10.97	4.45	2.96	3.02	4.63	12.80	6.82	12.73	14.45	11.26	11.29	7.79	8.72	4.44	6.07
0.80		77.91	47.93	16.13	19.97	31.72	76.01	64.70	81.12	92.50	70.89	78.58	49.11	43.48	27.19	40.56
		11.22	4.76	2.95	2.40	3.81	12.32	6.66	9.73	11.10	12.04	10.56	8.21	9.21	3.84	4.87
0.81		76.27	47.45	16.26	20.46	32.39	75.70	68.25	84.85	97.27	70.71	76.73	49.23	41.03	27.16	36.72
		11.05	4.74	2.97	2.46	3.89	11.58	7.00	10.18	11.67	12.86	10.11	8.00	7.68	3.57	4.41
0.82		75.01	46.56	16.25	21.16	32.57	74.64	77.65	95.33	110.51	64.82	69.84	45.42	35.90	26.69	32.65
		9.88	4.54	3.01	2.12	3.87	11.55	8.04	11.04	1.04	12.20	9.29	7.39	6.32	4.28	3.26
0.83		76.22	46.52	16.63	22.66	33.30	75.11	81.29	99.07	115.31	62.47	66.78	44.86	34.97	26.70	32.65
		9.47	4.13	3.04	2.95	4.33	11.82	8.14	12.88	4.99	10.72	8.68	7.38	6.46	5.05	4.24
0.84		77.94	47.14	17.35	23.88	34.52	77.30	78.50	96.57	111.91	66.17	71.96	48.34	38.13	27.44	34.93
		9.71	3.78	3.07	3.58	5.18	11.90	7.59	14.49	6.78	10.99	10.79	8.00	7.14	5.40	5.24
0.85		78.06	47.19	17.88	24.15	35.38	79.00	77.13	95.71	110.92	70.01	79.51	51.26	41.55	28.17	36.49
		9.54	4.09	3.10	4.83	7.08	12.05	7.50	19.14	2.18	11.60	15.90	8.44	7.52	5.63	7.30

* Upper value is the arithmetic mean SBC (percent), lower value is the standard deviation over the class

Table A.8. Volume coefficients of the aerosol scattering (km^{-1}) retrieved from the data of airborne radiative sounding 16 Oct. 1983 (The Kara-Kum desert)* (*continued on next page*)

λ (nm)	950	900	850	800	700	600	500
325	0.0278	0.0206	0.0120	0.1100	0.0470	0.0480	0.0021
	0.0069	0.0063	0.0042	0.0063	0.0038	0.0037	0.0021
345	0.0280	0.0209	0.0120	0.1140	0.0492	0.0485	0.0027
	0.0068	0.0325	0.0042	0.0064	0.0038	0.0037	0.0021
365	0.0269	0.0200	0.0118	0.1141	0.0477	0.0502	0.0033
	0.0066	0.0060	0.0040	0.0062	0.0038	0.0037	0.0020
385	0.0259	0.0192	0.0115	0.1163	0.0484	0.0518	0.0055
	0.0066	0.0058	0.0040	0.0060	0.0037	0.0038	0.0019
405	0.0255	0.0189	0.0114	0.1189	0.0493	0.0531	0.0074
	0.0066	0.0058	0.0039	0.0060	0.0037	0.0039	0.0019
425	0.0262	0.0194	0.0115	0.1220	0.0498	0.0542	0.0073
	0.0063	0.0058	0.0039	0.0061	0.0038	0.0038	0.0019
445	0.0270	0.0200	0.0118	0.1264	0.0524	0.0558	0.0069
	0.0062	0.0058	0.0039	0.0061	0.0038	0.0038	0.0019
465	0.0279	0.0207	0.0120	0.1241	0.0540	0.0553	0.0059
	0.0062	0.0059	0.0038	0.0063	0.0039	0.0038	0.00195
485	0.0287	0.0213	0.0122	0.1277	0.0567	0.0556	0.0046
	0.0063	0.0060	0.0038	0.0064	0.0039	0.0038	0.00199
505	0.0294	0.0221	0.0124	0.1341	0.0575	0.0527	0.0041
	0.0062	0.0060	0.0039	0.0065	0.0040	0.0039	0.00203
525	0.0296	0.0222	0.0125	0.1390	0.0582	0.0539	0.0046
	0.0062	0.0061	0.0038	0.0066	0.0040	0.0039	0.00202
545	0.0299	0.0224	0.0126	0.1388	0.0567	0.0539	0.0047
	0.0061	0.0061	0.0038	0.0067	0.0040	0.0039	0.00202
565	0.0301	0.0225	0.0126	0.1412	0.0582	0.0534	0.0049
	0.0061	0.0061	0.0038	0.0067	0.0041	0.0040	0.0020
585	0.0302	0.0227	0.0127	0.1420	0.0591	0.0524	0.0049
	0.0061	0.0060	0.0038	0.0067	0.0041	0.0040	0.0021
605	0.0303	0.0228	0.0128	0.1504	0.0585	0.0505	0.0053
	0.0060	0.0060	0.0038	0.0068	0.0040	0.0040	0.0021
625	0.0306	0.0230	0.0129	0.1501	0.0581	0.0523	0.0056
	0.0061	0.0061	0.0039	0.0068	0.0041	0.0040	0.0021
645	0.0308	0.0231	0.0130	0.1540	0.0585	0.0533	0.0057
	0.0062	0.0061	0.0040	0.0068	0.0041	0.0041	0.0021
665	0.0309	0.0235	0.0131	0.1532	0.0586	0.0498	0.0055
	0.0063	0.0062	0.0040	0.0069	0.0041	0.0041	0.0021
685	0.0311	0.0236	0.0132	0.1568	0.0582	0.0485	0.00565
	0.0065	0.0063	0.0041	0.0070	0.0042	0.0041	0.0021
725	0.0312	0.0237	0.0132	0.1595	0.0573	0.0490	0.0059
	0.0066	0.0062	0.0041	0.0073	0.0042	0.0042	0.0021
765	0.0311	0.0236	0.0131	0.1581	0.0568	0.0459	0.0057
	0.0068	0.0062	0.0041	0.0071	0.0042	0.0042	0.0021
805	0.0312	0.0237	0.0131	0.1580	0.0546	0.0444	0.0059
	0.0069	0.0063	0.0042	0.0073	0.0041	0.0043	0.0021
845	0.0313	0.0239	0.0132	0.1603	0.0537	0.0426	0.0058
	0.0073	0.0063	0.0044	0.0074	0.0042	0.0043	0.0021
885	0.0316	0.0241	0.0133	0.1547	0.0503	0.0404	0.0058
	0.0080	0.0065	0.0045	0.0074	0.0042	0.0045	0.0022

Table A.8. (*continued*)

λ (nm)	950	900	850	P (mbar) 800	700	600	500
925	0.0318	0.0243	0.0135	0.1530	0.0477	0.0387	0.0057
	0.0086	0.0066	0.0049	0.0074	0.0043	0.0047	0.0022
945	0.0403	0.0245	0.0136	0.1502	0.0470	0.0381	0.0058
	0.0089	0.0067	0.0051	0.0076	0.0043	0.0048	0.0022
955	0.0374	0.0246	0.0136	0.1494	0.0464	0.0352	0.0054
	0.0093	0.0069	0.0053	0.0078	0.0043	0.0049	0.0022
985	0.0321	0.0247	0.0137	0.1508	0.0448	0.0345	0.0052
	0.0099	0.0070	0.0057	0.0077	0.0044	0.0050	0.0023

* Upper value is the arithmetic mean scattering coefficient, lower value is the standard deviation (km^{-1})

Table A.9. Volume coefficients of the aerosol absorption (km^{-1}) retrieved from the airborne radiative sounding 16 Oct. 1983 above the Kara-Kum desert* (*continued on next page*)

λ (nm)	950	900	850	800	700	600	500
				P (mbar)			
325	0.0053	0.0044	0.0046	0.0050	0.0012	0.0009	0.0013
	0.0160	0.0108	0.0088	0.0075	0.0037	0.0040	0.0016
345	0.0051	0.0043	0.0046	0.0040	0.0012	0.0010	0.0016
	0.0161	0.0107	0.0088	0.0076	0.0037	0.0041	0.0016
365	0.0064	0.0053	0.0049	0.0066	0.0014	0.0010	0.0004
	0.0160	0.0102	0.0086	0.0074	0.0037	0.0038	0.0016
385	0.0076	0.0062	0.0052	0.0082	0.0015	0.0011	0.0005
	0.0153	0.0968	0.0083	0.0070	0.0035	0.0036	0.0016
405	0.0081	0.0066	0.0053	0.0119	0.0016	0.0011	0.0005
	0.0152	0.0946	0.0081	0.0069	0.0035	0.0035	0.0016
425	0.0075	0.0061	0.0052	0.0095	0.0015	0.0010	0.00045
	0.0156	0.0974	0.0083	0.0071	0.0036	0.0036	0.0016
445	0.0068	0.0056	0.0050	0.0074	0.0014	0.0010	0.0004
	0.0157	0.0100	0.0085	0.0072	0.0036	0.0037	0.0016
465	0.0061	0.0050	0.0048	0.0052	0.0013	0.0010	0.0004
	0.0161	0.0103	0.0087	0.0074	0.0037	0.0038	0.0016
485	0.0053	0.0044	0.0046	0.0031	0.0012	0.0009	0.0013
	0.0162	0.0106	0.0089	0.0075	0.0037	0.0039	0.0016
505	0.0046	0.0038	0.0044	0.0029	0.0011	0.0009	0.0023
	0.0158	0.0107	0.0089	0.0075	0.0036	0.0039	0.0016
525	0.0045	0.0037	0.0043	0.0029	0.0011	0.0008	0.0028
	0.0159	0.0107	0.0089	0.0075	0.0036	0.0039	0.0016
545	0.0043	0.0036	0.0043	0.0028	0.0011	0.0008	0.0029
	0.0157	0.0106	0.0089	0.0075	0.0035	0.0038	0.0016
565	0.0042	0.0035	0.0042	0.0026	0.0010	0.0008	0.0030
	0.0155	0.0105	0.0089	0.0075	0.0035	0.0038	0.0015
585	0.0041	0.0034	0.0042	0.0027	0.0010	0.0008	0.0030
	0.0153	0.0105	0.0088	0.0074	0.0034	0.0037	0.0015
625	0.0038	0.0041	0.0040	0.0026	0.0009	0.0008	0.0034
	0.0147	0.0103	0.0087	0.0074	0.0033	0.0035	0.0015
645	0.0037	0.0031	0.0039	0.0025	0.0009	0.0008	0.0034
	0.0144	0.0102	0.0085	0.0073	0.0032	0.0034	0.0014
665	0.0035	0.0040	0.0039	0.0025	0.0009	0.0007	0.0033
	0.0140	0.0100	0.0084	0.0072	0.0031	0.0033	0.0014
685	0.0034	0.0034	0.0038	0.0024	0.0008	0.0007	0.0033
	0.0135	0.0982	0.0083	0.0071	0.0030	0.0032	0.0014
725	0.0034	0.0040	0.0038	0.0024	0.0008	0.0007	0.0035
	0.0136	0.0980	0.0083	0.0071	0.0030	0.0032	0.0014
765	0.0034	0.0042	0.0038	0.0024	0.0009	0.0007	0.0035
	0.0137	0.0987	0.0084	0.0072	0.0030	0.0032	0.0014
805	0.0034	0.0029	0.0038	0.0024	0.0009	0.0007	0.0036
	0.0137	0.0985	0.0084	0.0072	0.0030	0.0032	0.0014
845	0.0033	0.0030	0.0037	0.00237	0.00082	0.0007	0.0035
	0.0134	0.0967	0.0083	0.00712	0.00296	0.0031	0.0013
885	0.0031	0.0026	0.0036	0.0023	0.0008	0.0007	0.0035
	0.0127	0.0931	0.0081	0.0070	0.0028	0.0029	0.0013
925	0.0029	0.0024	0.0034	0.0021	0.0007	0.0006	0.0034
	0.0119	0.0885	0.0077	0.0067	0.0026	0.0026	0.0012

Table A.9. (*continued*)

λ (nm)	P (mbar)						
	950	900	850	800	700	600	500
945	0.0028	0.0037	0.0033	0.0021	0.0007	0.0006	0.0035
	0.0115	0.0866	0.0075	0.0066	0.0025	0.0025	0.0012
955	0.0027	0.0023	0.0032	0.0020	0.0007	0.0006	0.0032
	0.0114	0.0851	0.0074	0.0066	0.0025	0.0025	0.0011
985	0.0026	0.0022	0.0031	0.0019	0.0006	0.0006	0.0032
	0.0108	0.0817	0.0071	0.0064	0.0024	0.0023	0.0010

* The upper value is the arithmetic mean absorption coefficient, the lower value is the standard deviation (km^{-1})

Table A.10. Volume coefficients of aerosol scattering (km^{-1}) retrieved from airborne radiative sounding 29 Apr. 1985 above the Ladoga Lake* (*continued on next page*)

λ (nm)	950	900	850	800	700	600	500
				P (mbar)			
325	0.563	0.435	0.309	0.244	0.0982	0.0689	0.0082
	0.027	0.013	0.012	0.009	0.0051	0.0032	0.0016
345	0.537	0.419	0.309	0.239	0.0989	0.0726	0.0074
	0.026	0.012	0.012	0.009	0.0051	0.0033	0.0016
365	0.501	0.404	0.308	0.243	0.0925	0.0718	0.0064
	0.026	0.012	0.012	0.009	0.0050	0.0033	0.0016
385	0.455	0.408	0.307	0.248	0.0102	0.0705	0.0076
	0.024	0.013	0.012	0.009	0.0050	0.0032	0.0016
405	0.465	0.388	0.292	0.241	0.0955	0.0720	0.0079
	0.024	0.013	0.012	0.009	0.0050	0.0032	0.0015
425	0.450	0.383	0.295	0.235	0.0101	0.0716	0.0077
	0.024	0.013	0.012	0.009	0.0048	0.0032	0.0015
445	0.449	0.373	0.294	0.237	0.0950	0.0727	0.0080
	0.023	0.011	0.011	0.009	0.0049	0.0032	0.0014
465	0.428	0.371	0.280	0.235	0.0949	0.0736	0.0075
	0.023	0.011	0.011	0.009	0.0047	0.0032	0.0014
485	0.418	0.361	0.277	0.237	0.0939	0.0685	0.0068
	0.023	0.011	0.011	0.009	0.0047	0.0031	0.0014
505	0.418	0.361	0.280	0.236	0.0104	0.0725	0.0070
	0.023	0.011	0.011	0.009	0.0048	0.0032	0.0014
525	0.385	0.349	0.272	0.246	0.0961	0.0669	0.0071
	0.022	0.011	0.011	0.009	0.0047	0.0031	0.0014
545	0.378	0.342	0.275	0.236	0.0943	0.0666	0.0071
	0.022	0.011	0.011	0.009	0.0046	0.0031	0.0014
565	0.370	0.341	0.275	0.233	0.0924	0.0662	0.0074
	0.022	0.011	0.011	0.009	0.0047	0.0031	0.0014
385	0.455	0.408	0.307	0.248	0.0102	0.0705	0.0076
	0.024	0.013	0.012	0.009	0.0050	0.0032	0.0016
405	0.465	0.388	0.292	0.241	0.0955	0.0720	0.0079
	0.024	0.013	0.012	0.009	0.0050	0.0032	0.0015
425	0.450	0.383	0.295	0.235	0.0101	0.0716	0.0077
	0.024	0.013	0.012	0.009	0.0048	0.0032	0.0015
445	0.449	0.373	0.294	0.237	0.0950	0.0727	0.0080
	0.023	0.011	0.011	0.009	0.0049	0.0032	0.0014
465	0.428	0.371	0.280	0.235	0.0949	0.0736	0.0075
	0.023	0.011	0.011	0.009	0.0047	0.0032	0.0014
485	0.418	0.361	0.277	0.237	0.0939	0.0685	0.0068
	0.023	0.011	0.011	0.009	0.0047	0.0031	0.0014
505	0.418	0.361	0.280	0.236	0.0104	0.0725	0.0070
	0.023	0.011	0.011	0.009	0.0048	0.0032	0.0014
525	0.385	0.349	0.272	0.246	0.0961	0.0669	0.0071
	0.022	0.011	0.011	0.009	0.0047	0.0031	0.0014
545	0.378	0.342	0.275	0.236	0.0943	0.0666	0.0071
	0.022	0.011	0.011	0.009	0.0046	0.0031	0.0014
565	0.370	0.341	0.275	0.233	0.0924	0.0662	0.0074
	0.022	0.011	0.011	0.009	0.0047	0.0031	0.0014
585	0.339	0.332	0.265	0.229	0.0958	0.0706	0.0067
	0.021	0.010	0.011	0.009	0.0046	0.0032	0.0014

Table A.10. (*continued*)

λ (nm)	P (mbar)						
	950	900	850	800	700	600	500
605	0.322	0.322	0.262	0.231	0.0941	0.0669	0.0072
	0.021	0.010	0.011	0.009	0.0046	0.0031	0.0014
625	0.317	0.321	0.258	0.227	0.0947	0.0683	0.0071
	0.021	0.010	0.011	0.009	0.0046	0.0031	0.0014
645	0.299	0.314	0.251	0.233	0.0953	0.0715	0.0063
	0.020	0.010	0.012	0.009	0.0046	0.0032	0.0014
665	0.327	0.312	0.252	0.233	0.0941	0.0641	0.0060
	0.020	0.010	0.011	0.009	0.0046	0.0031	0.0013
685	0.312	0.311	0.248	0.231	0.0939	0.0723	0.0068
	0.020	0.010	0.010	0.009	0.0046	0.0031	0.0013
725	0.299	0.302	0.241	0.233	0.0957	0.0717	0.0072
	0.019	0.010	0.010	0.009	0.0046	0.0031	0.0013
765	0.279	0.292	0.239	0.232	0.0952	0.0723	0.0066
	0.019	0.010	0.010	0.008	0.0045	0.0032	0.0013
805	0.262	0.288	0.233	0.228	0.0959	0.0714	0.0069
	0.019	0.010	0.010	0.008	0.0045	0.0032	0.0013
845	0.243	0.283	0.228	0.229	0.0961	0.0723	0.0064
	0.019	0.010	0.010	0.008	0.0045	0.0031	0.0013
885	0.238	0.279	0.224	0.226	0.0914	0.0750	0.0072
	0.019	0.010	0.010	0.008	0.0044	0.0032	0.0013
925	0.226	0.264	0.225	0.233	0.0891	0.0733	0.0061
	0.018	0.009	0.010	0.008	0.0043	0.0031	0.0013
945	0.173	0.257	0.222	0.229	0.0942	0.0724	0.0058
	0.018	0.009	0.010	0.008	0.0045	0.0031	0.0013
955	0.168	0.257	0.221	0.225	0.0922	0.0749	0.0068
	0.018	0.009	0.010	0.008	0.0045	0.0032	0.0013
985	0.135	0.254	0.218	0.227	0.0893	0.0728	0.0067
	0.018	0.009	0.010	0.008	0.0044	0.0031	0.0013

* Upper value – arithmetic mean scattering coefficient, lower value – standard deviation (km^{-1})

Table A.11. Volume coefficients of the aerosol absorption (km^{-1}) retrieved from airborne radiative sounding 29 Apr. 1985 above the Ladoga Lake* (*continued on next page*)

λ (nm)	P (mbar)						
	950	900	850	800	700	600	500
325	0.0221	0.0076	0.0049	0.0018	0.0021	0.0027	0.0049
	0.0282	0.0106	0.0068	0.0052	0.0044	0.0027	0.0021
345	0.0228	0.0078	0.0073	0.0018	0.0021	0.0013	0.0044
	0.0280	0.0106	0.0068	0.0052	0.0044	0.0026	0.0021
365	0.0225	0.0077	0.0065	0.0018	0.0021	0.0019	0.0038
	0.0278	0.0106	0.0068	0.0052	0.0043	0.0026	0.0021
385	0.0227	0.0077	0.0050	0.0018	0.0021	0.0025	0.0046
	0.0276	0.0106	0.0068	0.0052	0.0043	0.0026	0.0021
405	0.0226	0.0078	0.0049	0.0018	0.0021	0.0021	0.0048
	0.0281	0.0105	0.0068	0.0052	0.0043	0.0026	0.0020
425	0.0224	0.0076	0.0049	0.0018	0.0022	0.0025	0.0047
	0.0280	0.0105	0.0069	0.0052	0.0042	0.0026	0.0020
445	0.0225	0.0077	0.0049	0.0018	0.0020	0.0023	0.0048
	0.0280	0.0105	0.0068	0.0052	0.0042	0.0026	0.0020
465	0.0223	0.0076	0.0049	0.0018	0.0020	0.0018	0.0045
	0.0283	0.0105	0.0068	0.0052	0.0042	0.0026	0.0020
485	0.0225	0.0076	0.0049	0.0017	0.0020	0.0017	0.0041
	0.0282	0.0104	0.0069	0.0052	0.0041	0.0027	0.0020
505	0.0222	0.0075	0.0049	0.0017	0.0020	0.0016	0.0042
	0.0284	0.0105	0.0069	0.0052	0.0041	0.0026	0.0020
525	0.0221	0.0075	0.0049	0.0017	0.0020	0.0019	0.0043
	0.0285	0.0104	0.0069	0.0052	0.0041	0.0026	0.0020
545	0.0222	0.0075	0.0049	0.0017	0.0020	0.0027	0.0043
	0.0284	0.0104	0.0070	0.0052	0.0040	0.0026	0.0020
565	0.0220	0.0075	0.0048	0.0017	0.0020	0.0021	0.0044
	0.0286	0.0104	0.0070	0.0052	0.0040	0.0026	0.0020
585	0.0222	0.0075	0.0048	0.0017	0.0020	0.0037	0.0040
	0.0282	0.0103	0.0071	0.0052	0.0040	0.0027	0.0019
605	0.0221	0.0074	0.0048	0.0017	0.0020	0.0033	0.0044
	0.0285	0.0103	0.0071	0.0052	0.0039	0.0026	0.0019
625	0.0218	0.0074	0.0048	0.0017	0.0020	0.0016	0.0043
	0.0285	0.0102	0.0071	0.0052	0.0039	0.0026	0.0019
645	0.0220	0.0074	0.0048	0.0017	0.0020	0.0036	0.0038
	0.0282	0.0100	0.0071	0.0051	0.0038	0.0027	0.0019
665	0.0220	0.0074	0.0048	0.0017	0.0019	0.0036	0.0036
	0.0282	0.0998	0.0071	0.0051	0.0038	0.0026	0.0019
685	0.0217	0.0073	0.0048	0.0017	0.0019	0.0033	0.0041
	0.0284	0.0991	0.0072	0.0051	0.0038	0.0026	0.0019
725	0.0219	0.0073	0.0048	0.0017	0.0019	0.0037	0.0043
	0.0284	0.0976	0.0072	0.0051	0.0037	0.0026	0.0018
765	0.0216	0.0072	0.0048	0.0017	0.0019	0.0032	0.0040
	0.0283	0.0959	0.0072	0.0050	0.0036	0.0026	0.0018
805	0.0217	0.0072	0.0047	0.0021	0.0019	0.0038	0.0042
	0.0280	0.0943	0.0072	0.0050	0.0035	0.0026	0.0018
845	0.0217	0.0072	0.0047	0.0016	0.0019	0.0036	0.0038
	0.0279	0.0927	0.0071	0.0049	0.0034	0.0026	0.0017
885	0.0214	0.0071	0.0047	0.0016	0.0019	0.0027	0.0044
	0.0281	0.0917	0.0072	0.0049	0.0034	0.0026	0.0017

Table A.11. (*continued*)

λ (nm)	P (mbar)						
	950	900	850	800	700	600	500
925	0.0214	0.0071	0.0047	0.0016	0.0019	0.0037	0.0037
	0.0278	0.0909	0.0072	0.0049	0.0033	0.0026	0.0017
945	0.0215	0.0071	0.0047	0.0018	0.0020	0.0038	0.0035
	0.0279	0.0898	0.0072	0.0049	0.0033	0.0026	0.0017
955	0.0216	0.0071	0.0047	0.0016	0.0027	0.0034	0.0041
	0.0276	0.0894	0.0073	0.0048	0.0032	0.0026	0.0017
985	0.0215	0.0071	0.0047	0.0016	0.0023	0.0037	0.0041
	0.0277	0.0883	0.0072	0.0048	0.0032	0.0026	0.0016

* The upper value is the arithmetic mean scattering coefficient, the lower value is the standard deviation

Table A.12. Volume coefficients of absorption κ and scattering α (km^{-1}) retrieved from the airborne radiative observation in the overcast sky (*continued on next page*)

λ (nm)	The Black Sea 10.04.1971 κ	α	The Azov Sea 05.10.1972 κ	α	Rustavi city 05.12.1972 κ	α	GATE 12.07.1974 κ	α	GATE 4.08.1974 κ	α	The Ladoga Lake 20.04.1985 κ	α
400	0.0260	67.5	0.2903	60.7	0.1000	7.80	0.3069	31.9	0.2340	32.8	0.0145	56.8
410	0.0253	65.7	0.2873	60.3	0.0945	5.54	0.3065	31.7	0.2400	31.0	0.0068	56.0
420	0.0257	64.8	0.1708	59.6	0.0931	5.86	0.3053	31.5	0.2387	32.4	0.0000	55.6
430	0.0372	65.7	0.1334	57.4	0.0903	5.86	0.3031	31.1	0.2357	33.0	0.0012	54.4
440	0.0123	66.5	0.1089	58.0	0.0903	5.93	0.3004	32.1	0.2240	35.6	0.0031	53.6
450	0.0371	65.0	0.1132	55.0	0.0903	6.17	0.2954	33.2	0.2110	37.6	0.0086	51.7
460	0.0395	65.2	0.0470	53.5	0.0917	6.45	0.2966	31.1	0.2301	36.8	0.0130	50.4
470	0.0486	64.7	0.0354	54.7	0.0931	5.87	0.2946	31.9	0.2281	35.6	0.0132	47.2
480	0.0426	62.0	0.0299	54.8	0.0910	5.80	0.2903	31.5	0.2013	36.8	0.0135	43.6
490	0.0432	61.5	0.0259	53.1	0.0910	5.79	0.2895	31.4	0.1994	37.8	0.0123	42.0
500	0.0274	60.4	0.0817	53.5	0.0903	5.74	0.2876	30.7	0.1976	39.3	0.0110	40.2
510	0.0183	59.7	0.1416	52.0	0.0903	5.88	0.2894	30.4	0.2045	38.6	0.0156	39.3
520	0.0510	60.1	0.1860	52.4	0.0945	6.06	0.2913	30.0	0.2355	38.3	0.0224	38.6
530	0.0366	56.0	0.2043	50.9	0.0890	5.83	0.2938	29.8	0.2519	37.9	0.0198	38.5
540	0.0335	55.3	0.1177	48.5	0.0876	5.98	0.2980	29.7	0.2528	37.0	0.0309	37.2
550	0.0478	54.2	0.1177	47.5	0.0876	5.97	0.2945	28.5	0.2544	36.8	0.0367	37.3
560	0.0426	54.2	0.1177	45.7	0.0855	6.10	0.2965	27.7	0.2585	36.8	0.0405	36.6
570	0.0500	53.2	0.1330	46.1	0.0862	6.07	0.2983	27.0	0.2580	35.5	0.0435	36.1
580	0.0481	52.2	0.1330	43.9	0.0862	5.82	0.2952	27.4	0.2558	34.8	0.0468	35.4
590	0.0585	50.8	0.1369	44.8	0.0876	5.83	0.2962	26.8	0.2560	34.6	0.0477	34.8
600	0.0558	50.0	0.1404	44.0	0.0848	6.16	0.2904	27.3	0.2435	34.4	0.0497	34.1
610	0.0634	50.1	0.1404	43.2	0.0864	5.97	0.2897	28.4	0.2406	34.1	0.0488	33.9
620	0.0478	48.9	0.1437	42.8	0.0855	5.92	0.2914	28.4	0.2477	35.7	0.0491	33.6
630	0.0622	48.5	0.1465	42.6	0.0869	5.91	0.2936	28.2	0.2510	36.3	0.0532	32.3
640	0.0375	48.3	0.1368	42.4	0.0765	5.72	0.2924	28.3	0.2479	35.4	0.0559	31.1
650	0.0422	47.6	0.1261	42.0	0.0871	6.05	0.2918	28.6	0.2381	33.7	0.0603	30.6
660	0.0476	47.5	0.1052	41.3	0.0945	6.12	0.2885	29.0	0.2327	35.8	0.0644	30.8
670	0.0419	47.0	0.0950	41.8	0.0863	5.84	0.2960	28.4	0.2438	36.2	0.0688	30.8
680	0.0514	47.5	0.1411	42.0	0.0848	5.79	0.3120	25.6	0.2302	31.9	0.0724	32.4
690	0.0959	44.9	0.1205	41.3	0.0821	5.40	0.3012	30.0	0.2677	35.9	0.0746	31.5
700	0.0607	45.0	0.150	41.0	0.1099	5.20	0.2940	20.3	0.2420	25.6	0.0786	28.2
710	0.0552	44.3	0.1614	41.1	0.0876	5.73	0.2903	20.5	0.2563	24.9	0.0814	28.8
720	0.1170	44.5	0.1603	40.7	0.0882	5.75	0.3128	21.3	0.2688	24.7	0.0909	29.6
730	0.1105	40.7	0.1463	40.9	0.1523	5.98	0.3077	23.0	0.2490	29.3	0.0552	26.4
740	0.0628	41.3	0.1104	40.2	0.0876	5.77	0.2977	23.3	0.2432	24.2	0.0451	25.4
750	0.0672	42.5	0.1182	40.6	0.0869	5.74	0.3330	24.0	0.2845	23.5	0.0467	25.7
760	0.2165	42.0	0.1463	40.7	0.0848	5.71	0.3723	25.8	0.3075	34.5	0.0667	25.4
770	0.0691	41.8	0.0796	40.1	0.0828	6.82	0.3102	24.3	0.2889	33.2	0.0559	25.1
780	0.0629	40.6	0.0280	39.5	0.0876	6.47	0.2877	24.0	0.2547	32.6	0.0405	24.6
790	0.0660	40.7	0.0206	38.9	0.0848	6.18	0.2900	23.2	0.2300	31.7	0.0542	23.8
800	0.0589	40.2	0.0542	39.0	0.0910	6.30	0.2933	20.5	0.2474	30.8	0.0637	23.4

Table A.12. (*continued*)

λ (nm)	The Black Sea 10.04.1971		The Azov Sea 05.10.1972		Rustavi city 05.12.1972		GATE 12.07.1974		GATE 4.08.1974		The Ladoga Lake 20.04.1985	
nm	κ	α	κ	α	κ	α	κ	α	κ	α	κ	α
810	0.0773	40.2	0.1502	39.1	0.0862	7.05	0.2959	18.0	0.2500	30.0	0.0788	23.2
820	0.1050	39.7	0.1685	38.4	0.0841	8.20	0.2944	19.7	0.2378	29.3	0.0940	23.5
830	0.0736	38.9	0.1585	38.8	0.0841	8.12	0.3012	18.6	0.2317	28.8	0.0647	23.4
840	0.0519	39.2	0.1284	38.5	0.0856	8.15	0.3094	19.7	0.2000	28.0	0.0583	23.8
850	0.0661	39.2	0.1073	38.3	–	–	0.3142	18.3	0.2213	26.8	0.0521	23.0
860	0.0489	39.3	0.0358	38.3	–	–	0.3088	19.2	0.2162	27.3	0.0453	23.2
870	0.0593	38.7	0.2950	38.1	–	–	0.3152	18.4	0.2027	26.5	0.0437	23.5
880	0.0690	38.9	0.0850	38.0	–	–	0.3020	20.2	0.1920	27.4	0.0418	23.4
890	0.0796	38.1	0.0397	37.9	–	–	–	–	–	–	0.0744	22.8
900	0.0992	38.0	0.2187	37.7	–	–	–	–	–	–	0.0859	21.4
910	0.2527	36.7	0.2371	37.6	–	–	–	–	–	–	0.0888	20.5
920	0.1616	37.5	0.2559	37.5	–	–	–	–	–	–	0.0863	20.2
930	0.2801	34.3	0.2672	37.3	–	–	–	–	–	–	0.0932	21.7
950	0.2020	35.0	0.3574	37.1	–	–	–	–	–	–	0.0953	20.4

Table A.13. The single scattering albedo and optical thickness of the stratus clouds from the ground spectral irradiance observation

Date	13 Aug. 1979		08 Oct. 1979		12 Apr. 1996
λ (nm)	ω_0	τ_0	ω_0	τ_0	ω_0
350	0.9989	25.5	1.0000	26.2	–
400	0.9981	22.2	1.0000	20.6	0.9956
450	0.9987	21.0	0.9963	20.0	0.9919
500	0.9994	20.2	0.9985	19.3	1.0000
550	0.9990	19.7	0.9974	18.5	0.9975
600	0.9985	17.6	0.9987	17.4	0.9978
650	0.9907	17.3	0.9957	17.7	0.9972
700	0.9930	17.8	0.9968	17.0	0.9977
750	0.9894	16.5	0.9921	17.6	0.9957
800	0.9919	17.1	0.9919	17.4	–
900	0.9844	14.9	0.9868	18.0	–

Table A.14. The single scattering albedo and optical thickness at different sublayers of the stratus cloud from airborne spectral irradiance observation 24 Sept. 1972 above the Ladoga Lake

z (km)	4.1–3.0		3.0–1.6		1.6–0.6		0.6–0.05	
λ (nm)	ω_0	τ_1	ω_0	τ_2	ω_0	τ_3	ω_0	τ_4
450	0.9972	21	0.9912	38	0.9947	28	0.9961	3.9
460	0.9968	18	0.9953	39	0.9945	27	0.9950	3.3
470	0.9956	20	0.9979	46	0.9943	28	0.9953	4.2
480	0.9959	23	1.0000	43	0.9939	27	0.9930	3.7
490	0.9966	19	0.9972	41	0.9827	27	0.9914	3.3
500	0.9973	24	1.0000	44	0.9932	31	0.9903	3.0
510	0.9974	22	1.0000	43	0.9913	27	0.9821	3.6
520	0.9971	21	0.9952	41	0.9951	34	0.9951	4.7
530	0.9971	21	0.9950	40	0.9950	33	0.9950	4.6
540	0.9972	18	0.9967	42	0.9931	31	0.9945	4.2
550	0.9970	21	0.9948	38	0.9949	32	0.9927	3.7
560	0.9964	22	0.9952	40	0.9957	33	0.9949	4.1
570	0.9951	17	0.9968	42	0.9936	30	0.9931	3.7
580	0.9955	19	0.9971	43	0.9953	29	0.9923	3.7
590	0.9962	18	0.9969	43	0.9934	27	0.9963	4.6
600	0.9954	18	0.9969	43	0.9934	27	0.9963	4.6
610	0.9954	20	0.9975	44	0.9937	25	0.9938	4.0
620	0.9963	20	0.9973	46	0.9943	20	1.0000	5.1
630	0.9947	24	1.0000	44	0.9944	26	0.9993	5.3
640	0.9954	18	0.9971	48	0.9930	28	0.9972	5.4
650	0.9964	25	0.9970	44	0.9925	29	0.9959	5.2
660	0.9952	19	0.9973	44	0.9938	31	0.9929	4.1
670	0.9964	22	0.9984	51	0.9921	29	0.9921	4.2
680	0.9957	17	0.9968	44	0.9924	27	0.9962	5.1
690	0.9952	17	0.9970	41	0.9912	26	0.9934	4.7
700	0.9964	22	0.9891	38	0.9898	23	0.9960	5.3
710	0.9961	22	0.9882	39	0.9912	24	0.9977	6.4
720	0.9815	17	0.9916	40	0.9882	23	0.9949	5.7
730	0.9933	19	0.9913	41	0.9913	23	0.9989	5.9
740	0.9962	17	0.9949	37	0.9907	25	0.9962	6.1
750	0.9966	18	0.9956	44	0.9919	26	0.9989	7.3
760	0.9808	20	0.9740	47	0.9600	24	0.9972	8.2
770	0.9964	20	0.9974	45	0.9928	25	0.9993	7.5
780	0.9968	20	0.9964	45	0.9866	23	0.9984	8.1
790	0.9988	21	0.9953	42	0.9899	26	0.9972	7.3
800	0.9974	21	0.9996	47	0.9863	24	0.9965	6.3
810	0.9962	17	0.9980	45	0.9841	30	0.9948	5.7
820	0.9958	18	0.9969	46	0.9875	27	0.9944	4.4
830	0.9959	20	0.9995	45	0.9882	27	0.9950	5.6
840	0.9978	18	0.9956	42	0.9936	31	0.9940	4.3
850	0.9954	17	0.9969	45	0.9918	34	0.9931	2.1

Table A.15. The single scattering albedo and optical thickness at different levels of the stratus cloud from airborne spectral irradiance observation 20 Apr. 1985 above the Ladoga Lake

z (km)	1.4–1.3		1.3–1.2		1.2–1.1		1.1–0.9		0.9–0.8	
λ (nm)	ω_0	τ_1	ω_0	τ_2	ω_0	τ_3	ω_0	τ_4	ω_0	τ_5
350	0.9831	2	0.9968	4.9	0.9949	5.3	0.9974	4.6	–	1.8
360	0.9891	1.5	0.9996	3.9	0.9942	4.7	0.9964	3.8	–	3.8
370	0.9876	1.5	0.9991	4.2	0.9944	5.0	0.9971	3.6	–	1.9
380	0.9971	1.7	0.9966	3.4	0.9899	3.9	0.9937	3.1	–	2.2
390	0.9925	1.4	0.9949	3.5	0.9881	4.0	0.9936	3.6	–	1.3
400	0.9961	1.4	0.9962	3.5	0.9906	4.0	0.9934	3.3	–	2.4
410	0.9977	1.7	0.9976	3.8	0.9920	3.9	0.9944	3.3	–	2.3
430	0.9920	2.1	0.9937	5.0	0.9957	2.4	0.9973	1.8	–	2.3
450	0.9889	1.8	0.9918	4.4	0.9956	2.1	0.9967	1.7	–	2.0
470	0.9876	1.8	0.9916	4.4	0.9960	2.0	0.9974	1.5	–	2.0
490	0.9867	1.8	0.9914	4.6	0.9959	2.0	0.9962	1.7	–	2.1
500	0.9876	1.9	0.9918	4.9	0.9964	2.1	0.9971	1.6	–	2.3
510	0.9873	1.9	0.9917	4.8	0.9960	2.0	0.9966	1.6	–	2.2
530	0.9859	1.9	0.9910	4.7	0.9963	1.8	0.9973	1.4	–	2.1
550	0.9857	1.8	0.9908	4.6	0.9961	1.7	0.9965	1.4	–	2.1
570	0.9858	1.8	0.9906	4.6	0.9961	1.7	0.9954	1.5	–	2.1
590	0.9854	1.8	0.9904	4.5	0.9960	1.6	0.9939	1.6	–	2.1
600	0.9842	1.8	0.9898	4.4	0.9960	1.6	0.9945	1.5	–	2.1
610	0.9850	1.8	0.9904	4.6	0.9961	1.5	0.9943	1.5	–	2.1
630	0.9858	1.9	0.9909	4.7	0.9962	1.5	0.9939	1.6	–	2.2
650	0.9852	1.8	0.9903	4.4	0.9956	1.4	0.9928	1.5	–	2.1
670	0.9839	1.7	0.9903	4.3	0.9950	1.4	0.9919	1.5	–	2.1
690	0.9825	1.6	0.9895	4.0	0.9944	1.3	0.9908	1.4	–	1.9
700	0.9804	1.5	0.9903	3.7	0.9920	1.4	0.9865	1.6	–	1.8
710	0.9829	1.5	0.9912	3.6	0.9920	1.3	0.9849	1.9	–	1.9
730	0.9866	1.6	0.9918	3.8	0.9913	1.3	0.9819	2.1	–	1.9
750	0.9874	1.6	0.9935	3.7	0.9933	1.3	0.9849	2.1	–	2.0
770	0.9822	1.5	0.9913	3.9	0.9895	1.6	0.9830	2.2	–	2.0
790	0.9870	1.5	0.9926	3.8	0.9884	1.6	0.9815	2.3	–	2.0
800	0.9879	1.5	0.9934	3.7	0.9898	1.5	0.9808	2.3	–	2.0
810	0.9889	1.6	0.9943	3.7	0.9899	1.3	0.9795	2.3	–	2.0
830	0.9897	1.7	0.9952	3.7	0.9903	1.4	0.9777	2.5	–	2.0
850	0.9871	1.7	0.9963	3.4	0.9944	1.1	0.9790	2.4	–	2.0
870	0.9902	1.7	0.9982	3.1	0.9927	1.1	0.9774	2.3	–	1.9
890	0.9894	1.6	0.9993	2.9	0.9923	1.0	0.9720	2.3	–	1.8
900	0.9875	1.6	0.9968	2.7	0.9890	1.0	–	2.6	–	1.8
910	0.9899	1.5	0.9979	3.0	0.9836	1.1	–	2.4	–	1.7
930	0.9874	1.5	0.9962	3.0	0.9838	1.0	–	2.4	–	1.7
950	0.9826	1.4	0.9903	3.4	0.9737	1.3	–	2.7	–	1.8
960	0.9873	1.4	0.9917	3.3	–	1.3	–	2.6	–	1.8

Appendix B: Formulas Derivation

Derivation of formulas for the determination of the cloud optical parameters from the data of observations of solar irradiances at its boundaries. The initial formulas for the irradiances reflected and transmitted by the cloud layer:

$$F^{\uparrow}(0, \mu_0) \equiv F_0^{\uparrow} = a(\mu_0) - \frac{nK(\mu_0)m\bar{l}\exp(-2k\tau_0)}{1 - ll\exp(-2k\tau_0)} , \qquad (B.1)$$

$$F^{\downarrow}(\tau, \mu_0) = \frac{\bar{n}K(\mu_0)m\exp(-k\tau_0)}{1 - \bar{l}l\exp(-2k\tau_0)} . \qquad (B.2)$$

From (B.1) the following is easy to derive:

$$\exp(2k\tau_0) = \bar{l}\frac{mnK(\mu_0) + l[a(\mu_0) - F_0^{\uparrow}]}{a(\mu_0) - F_0^{\uparrow}} . \qquad (B.3)$$

According to the definition, values $\bar{n}$, assuming the ground albedo influence, are expressed with the following:

$$\bar{l} = l - \frac{Amn^2}{1 - Aa^{\infty}} \quad \text{and} \quad \bar{n} = \frac{n}{1 - Aa^{\infty}} .$$

Substituting (B.3) to (B.2) the intermediate relation is obtained:

$$F^{\downarrow} = (a(\mu_0) - F_0^{\uparrow})\frac{\exp(-k\tau)}{\bar{l}}\frac{1}{1 - Aa^{\infty}} . \qquad (B.4)$$

Raise to the square both parts of (B.4), substituting the expression for $\exp(-2k\tau)$, and deleting the common multilayer, obtain the following:

$$F^{\downarrow 2}[1 - Aa^{\infty}][l(1 - Aa^{\infty}) - Amn^2]$$
$$= mnK(\mu_0)[a(\mu_0) - F_0^{\uparrow}] + l[a(\mu_0) - F_0^{\uparrow}]^2 \qquad (B.5)$$

Substitute to (B.5) the following expansions over powers of small parameter s for values:

$$m = 8s^2 \left[1 + s \left(6 - 7.5g + \frac{3.6}{1+g} \right) \right] + O(s^4) ,$$

$$l = 1 - 6q's + 18q'^2 s^2 + O(s^3) ,$$

$$a^\infty = 1 - 4s + 12q''s^2 + O(s^3) , \tag{B.6}$$

$$n = 1 - 3q's + n_2 s^2 + O(s^3) ,$$

and for functions:

$$K(\mu) = K_0(\mu)[1 - 3q's + n_2\omega(\mu)s^2] + O(s^3) ,$$

$$a(\mu) = 1 - 4K_0(\mu)s + a_2(\mu)s^2 + O(s^3) , \tag{B.7}$$

where:

$$a_2(\mu) = 3K_0(\mu) \left(\frac{3}{1+g}(1.271\mu - 0.9) + 4q' \right) .$$

Obtain the expression:

$$[a(\mu_0) - F^\uparrow]8sK_0(\mu_0)(1 - 3q's + n_2 w(\mu_0)s^2)(1 - 3q's + n_2 s^2)$$
$$+ (1 - 6q's + 18q'^2 s^2)[a(\mu_0) - F^\uparrow]^2$$
$$= F_n^{\downarrow 2}(1 - 6q's + 18q'^2 s^2)(1 - A + 4As - 12q'As^2)^2 \tag{B.8}$$
$$- 8As(1 - 6q's + 9\delta^2 s^2 + 2n_2 s^2)(1 - A + 4As - 12q'As^2) .$$

Accomplishing the multiplication of polynomials and keeping items with the power of s not exceeding 2 the following is obtained:

$$[1 - F^\uparrow - 4K_0(\mu_0)s + a_2(\mu_0)s^2]8sK_0(\mu_0)(1 - 6q's + 9q'^2 s^2 + n_2(1 + w(\mu_0))s^2)$$
$$+ (1 - 6q's + 18q'^2 s^2)[(1 - F^\uparrow)^2 + 16K_0^2(\mu_0)s^2 + 2a_2(\mu_0)s^2] \tag{B.9}$$
$$= F^{\downarrow 2}(1 - 6q's + 18q'^2 s^2)((1 - A)^2 - 16A^2 s^2 - 24q'As^2) .$$

Divide both parts by polynomial $1 - 3\delta s + 19q'^2 s^2$, keeping the items with the first and the second power of s, collecting the likewise terms and obtain the linear equation respected to value s^2:

$$1 - 3\delta s + 18q'^2 s^2(1 - F_0^\uparrow)^2 + 2s^2(1 - F_0^\uparrow)a_2(\mu_0) - 16s^2 K_0^2(\mu_0)$$
$$= F^{\downarrow 2}(1 - A)^2 - 24q'F^{\downarrow 2}A(1 - A)s^2 - 16F^{\downarrow 2}A^2 s^2 . \tag{B.10}$$

Assuming that $1 - F_0^\uparrow = F_0$ and $F_1^\uparrow(1 - A) = F_1$ are the net fluxes at the top (subscript 0) and bottom (subscript 1) of the cloud layer correspondingly, and assuming that $F_1^\downarrow A = F_1^\uparrow$, obtain for value s^2 the following:

$$s^2 = \frac{F_0^2 - F_1^2}{16(K_0^2(\mu_0) - F_1^{\uparrow 2}) - 2F_0 a_2(\mu_0) - 24q'F_1^\uparrow F_1} . \tag{B.11}$$

The optical thickness of the cloud layer is derived from (B.3), and with a subject to expansions (B.6) and (B.7), the result is obtained as:

$$\tau_0' = 3\tau_0(1 - g) = \frac{1}{2s} \ln \left[\frac{\bar{l}(mnK(\mu_0) - l)}{a(\mu_0) - F_0^{\uparrow}} \right].$$ (B.12)

Index